Secrets of Signals Intelligence during the Cold War and Beyond

Editors
MATTHEW M. AID
Kroll Associates, Washington DC

CEES WIEBES
*Netherlands Institute for War Documentation and
University of Amsterdam*

FRANK CASS
LONDON • NEW YORK

First published 2001 in Great Britain by
FRANK CASS PUBLISHERS

Reprinted 2004
By Frank Cass
2 Park Square
Milton Park
Abingdon
Oxon
OX14 4RN

Transferred to Digital Printing 2005

Frank Cass is an imprint of the Taylor & Francis Group

Copyright © 2001 Frank Cass & Co. Ltd.

British Library Cataloguing in Publication Data

Secrets of signals intelligence during the Cold War and
beyond. – (Studies in intelligence)
1. Electronic surveillance – History – 20th century –
Congresses 2. Cold War – Electronic intelligence –
Congresses
1. Aid, Matthew M. II. Wiebes, Cees
327.1'2'09045

ISBN 0 7146 5176 1 (cloth)
ISBN 0 7146 8182 2 (paper)
ISSN 1368-9916

Library of Congress Cataloging-in-Publication Data

Secrets of signals intelligence during the Cold War and beyond / edited
by Matthew M. Aid & Cees Wiebes.
    p. cm. – (Cass series – studies in intelligence, ISSN 1368-9916)
Includes bibliographical references and index.
ISBN 0-7146-5176-1 – ISBN 0-7146-8182-2 (pbk.)
1. Electronic intelligence – Congresses. 2. Electronic
surveillance–Congresses. 3. Cold War–Congresses. I. Aid, Matthew
M., 1958– . II. Wiebes, Cees. III. Series.
UB255 .S43 2001
327.12--dc21   2001028241

This group of studies first appeared in a Special Issue on
'Secrets of Signals Intelligence during the Cold War and Beyond' of
Intelligence and National Security 16/1 (Spring 2001)
published by Frank Cass (ISSN 0268-4527).

All rights reserved. No part of this publication may be reproduced, stored in or introduced into a retrieval system or transmitted in any form, or by any means, electronic, mechanical, photocopying, recording, or otherwise, without the prior written permission of the publisher of this book.

Digitally printed in Great Britain by
Butler and Tanner, Burgess Hill, West Sussex

CASS SERIES: STUDIES IN INTELLIGENCE
(Series Editors: Christopher Andrew and Michael I. Handel;
Wesley K. Wark and Richard J. Aldrich)
ISSN 1368-9916

Secrets of Signals Intelligence during the Cold War and Beyond

Also in this series

American-British-Canadian Intelligence Relations, 1939–2000 edited by David Stafford and Rhodri Jeffreys-Jones

The Clandestine Cold War in Asia, 1945–65 edited by Richard J. Aldrich, Gary Rawnsley and Ming-Yeh Rawnsley

Allied and Axis Signals Intelligence in World War II edited by David Alvarez

The Norwegian Intelligence Service 1945–1970: Northern Vigil by Olav Riste

British Military Intelligence in the Crimean War, 1854–1856 by Stephen Harris

Intelligence and the Cuban Missile Crisis edited by James G. Blight and David A. Welch

Knowing Your Friends: Intelligence Inside Alliances and Coalitions from 1914 to the Cold War edited by Martin S. Alexander

Eternal Vigilance? 50 Years of the CIA edited by Rhodri Jeffreys-Jones and Christopher Andrew

Nothing Sacred: Nazi Espionage Against the Vatican, 1939–1945 by David Alvarez and Revd Robert A. Graham

Intelligence Analysis and Assessment edited by David Charters, A. Stuart Farson and Glenn P. Hastedt

Intelligence and Imperial Defence: British Intelligence and the Defence of the Indian Empire 1904–1924 by Richard J. Poppelwell

Espionage: Past, Present, Future? edited by Wesley K. Wark

Codebreaker in the Far East by Alan Stripp

Contents

Foreword	Christopher Andrew	vii
List of Illustrations		ix
Preface		xi

1. Introduction: The Importance of Signals Intelligence in the Cold War — **Matthew M. Aid and Cees Wiebes**   1

2. The National Security Agency and the Cold War — **Matthew M. Aid**   27

3. GCHQ and Sigint in the Early Cold War 1945–70 — **Richard J. Aldrich**   67

4. Canada's Communications Security Establishment from Cold War to Globalization — **Martin Rudner**   97

5. The Bundesnachrichtendienst, the Bundeswehr and Sigint in the Cold War and After — **Erich Schmidt-Eenboom**   129

6. France, Sigint and the Cold War — **Roger Faligot**   177

7. Scandinavia, Sigint and the Cold War — **Alf R. Jacobsen**   209

8. Dutch Sigint during the Cold War, 1945–94 — **Cees Wiebes**   243

9. Dutch Sigint and the Conflict with Indonesia, 1950–62 — **Wies Platje**   285

10. Conclusions — **Matthew M. Aid and Cees Wiebes**   313

Abstracts 333

About the Contributors 338

Index 341

Foreword

Signals intelligence (Sigint) was the best kept secret of the Cold War – so well preserved that most histories of that era do not even mention it. Though the Central Intelligence Agency (CIA) quickly became a household name, the National Security Agency (NSA), which runs American Sigint operations, remained almost invisible throughout the Cold War. The small circle of those in the know in Washington joked that NSA stood for 'No Such Agency'. NSA, however, has a far bigger budget than the CIA, employs far more people, and generates far more intelligence.

Since the end of the Cold War NSA has become a little less mysterious. One veteran of the era of total secrecy complained to me a few years ago that NSA now stood for 'Nothing Sacred Anymore'. It still, however, attracts far less public and scholarly interest than the CIA. Though Sigint is a basic fact of modern international relations, even the word remains largely unknown.

Part of the reason why most scholars ignore Sigint is the continued classification of most of the huge historical archive which it has generated. The main reason for its neglect, however, has been what psychologists call cognitive dissonance – the difficulty all of us have in grasping new concepts which disturb our existing view of the world. Sigint is just such a concept. Most scholars working on the international relations of the twentieth century have been unable to come to terms with it.

From 1945 onwards, for example, almost all histories of the Second World War mentioned the American success in breaking the main Japanese diplomatic cipher over a year before the attack on Pearl Harbor. But, until the revelation of the Ultra Secret in 1973, it occurred to almost no historian (save for former intelligence officers who were forbidden to mention it) that there might have been major Sigint successes against Germany as well as Japan. Even after the disclosure of Ultra's important role in British and American wartime operations in the West, it took another 15 years before any historian raised the rather obvious question of whether there was a Russian Ultra on the Eastern Front as well.[1]

Many of the historians who now acknowledge the significance of Sigint in the Second World War still ignore it completely in their studies of the

Cold War. This sudden disappearance of Sigint from the historical landscape immediately after VJ Day has produced a series of eccentric anomalies even in some of the leading studies of policy-makers and international relations. Thus, for example, Sir Martin Gilbert's massive and mostly authoritative multi-volume official biography of Churchill acknowledges his passion for Sigint as war leader but includes not a single reference to his continuing interest in it as peacetime prime minister from 1951 to 1955. There is even less about Sigint in biographies of Stalin. Indeed, it is difficult to think of any history of the Soviet Union which devotes as much as a sentence to the enormous volume of Sigint generated by the KGB and GRU.[2] Studies of the presidency of George Bush (the first) invariably ignore his candid admission that Sigint was a 'prime factor' in his foreign policy – just as they neglect the use of Sigint by other post-war presidents.[3]

Hence the importance of this path-breaking collection edited by Matthew M. Aid and Cees Wiebes. They and the other contributors to this study have brought together a wider and more innovative range of material on the role of Sigint since the Second World War than has ever been published before. Sigint's importance, as they demonstrate, extended far beyond the Cold War superpowers. They are right to argue that intelligence studies need to become more 'internationalist' and take more account of intelligence in middle-ranking and minor powers, some of whom have highly significant liaison arrangements with the major players.

This volume confronts all historians of the Cold War and international relations specialists with a major challenge which most so far have ducked: either to seek to take account of the role of Sigint since the Second World War or to explain why they do not consider it necessary to do so.

CHRISTOPHER ANDREW
Cambridge, May 2001

NOTES

1. Geoff Jukes, 'The Soviets and "Ultra"', *Intelligence and National Security* 3/2 (April 1988) pp.233–47. Though Jukes's conclusions are debatable, his article remains a path-breaking study.
2. Christopher Andrew and Vasili Mitrokhin, *The Mitrokhin Archive: The KGB in Europe and the West* (London: Penguin 1999) Chapter 21.
3. Christopher Andrew, *For The President's Eyes Only: Secret Intelligence and the American Presidency from Washington to Bush* (London: HarperCollins 1995) p.5 and Chapter 13.

Illustrations

Figure 1.1	US Army 'Adventurer' border intercept site	8
Figure 1.2	US Army 'Hippodrome' space collection facility	8
Figure 2.1	US Army radio intercept operators	48
Figure 2.2	AN/FRD-10 antenna array	48
Figure 2.3	Antenna tower	49
Figure 4.1	Canadian Communications Security Establishment HQ	98
Figure 4.2	Enigma cipher machine	100
Figure 4.3	Canadian Army Corps of Signals Sigint intercept station	102
Figure 4.4	Canadian satellite interception dish	110
Map 5.1	Listening, direction-finding and radar installations of the Bundeswehr 1989	142
Map 5.2	Reconnaissance installations of the BND 1989	149
Figure 8.1	The HQ of the Dutch Defence Intelligence Agency	246
Figure 8.2	Secret targets for Dutch Sigint, January 1981	267–9
Figure 8.3	Dutch satellite intercept station	270
Figure 9.1	Captain Henri Koot	289
Figure 9.2	Marid 6 at Hollandia	296
Figure 9.3	Marid 6 at Biak	297
Figure 9.4	Numbers of intercepted Indonesian messages	303
Figure 9.5	Hollandia staff buildings	307

Preface

The Cold War passed into history in 1989–91 with the dissolution of the Soviet Union and its allies bound together in the Warsaw Pact. The American, Canadian and European intelligence communities had to change their focus and attention shifted to different emerging threats like rogue states, money laundering, drug kingpins, organised crime and terrorists but also to environmental disasters and large-scale refugee problems. Partly because of this transformation in the world of intelligence more and more students of the Cold War begin to realise that the Western intelligence communities played an important role between 1945 and 1990. In recent years in particular the importance of Signals Intelligence (Sigint) has been emphasised and especially the capabilities and possibilities of reading and deciphering diplomatic, military, commercial and other communications of foreign nations.

This growing awareness of the importance of intelligence applies not only to the activities of the big services but also to those of the smaller nations like for example the Netherlands. For this exact reason a couple of years ago the Netherlands Intelligence Studies Association (NISA) was established in which academics and (former and still active) members of the Netherlands intelligence community work together in order to promote research into the history of Dutch intelligence communities. This growing interest had led in Holland to publications dealing with the history of the Dutch internal security service (1995), the Dutch Navy intelligence (1997) and the Netherlands foreign intelligence service (1998).

While the NISA hosts an international conference every two years it was this time decided to organise a congress dealing with 'The Importance of Sigint in Western Europe during the Cold War 1945–1999'. This conference took place on Saturday 27 November 1999 in Amsterdam. The speakers came from the United States, Norway, Germany, Great Britain, the Netherlands and altogether six papers were presented. This Sigint work is a spin-off of this conference, which was a great success with more than 100 participants from 11 different countries. The readers can not only find the expanded version of the papers presented at this Sigint conference but also additional contributions on the topic of Sigint in Canada and France. This study also contains an introduction on the topic of the importance of Sigint and will end with conclusions regarding the eight different contributions.

Any questions regarding the activities of the NISA should be addressed to P.O. Box 18 210, 1001 ZC Amsterdam, the Netherlands. The editors of this work can be contacted at: Matthew M. Aid <mmaid@starpower.net> and Cees Wiebes <wiebes@pscw.uva.nl>.

1
Introduction: The Importance of Signals Intelligence in the Cold War

MATTHEW M. AID and CEES WIEBES

Today, our knowledge of about the role and importance of Signals Intelligence (Sigint) in the years after the end of World War II can only be described as an inventory of ignorance. The distinguished British historian Christopher Andrew has written that 'The biggest gap in our knowledge of United States intelligence collection during the Cold War concerns the role of Sigint. No history of the Second World War nowadays fails to mention the role of the Anglo-American codebreakers in hastening victory over Germany and Japan. By contrast, most histories of the Cold War make no reference to Sigint at all.'[1] By the same token, our lack of knowledge about role played by Sigint in countries outside the United States is deeper and even more profound.

Part of the problem stems from the heavy shroud of secrecy that has covered this immensely important subject for so long. Too many academics, researchers and journalists in the US, Europe and elsewhere still speak about the subject *sotto voce*, fearful of the strictures of the Official Secrets Act and similar laws that effectively bar public discussion of this subject. Another factor is that because of its technical nature, Sigint is an extremely difficult subject for the layman to understand, which has deterred academics and journalists from examining the subject in any depth, and what coverage there has been remains focused on the trials and tribulations of World War II.[2]

It is certainly true that Sigint lacks the glamour and sex appeal that surrounds the exploits and derring-do of secret agents, which have dominated the post-war literature on intelligence. One writer has put it thus: 'For many, Sigint conjures up images of grey men eavesdropping on conversations, cracking codes, and installing large-dish antennas. Compared with human intelligence, Sigint can seem rather boring and, frankly, a little grubby.'[3]

Moreover, partisans of the more traditional art of Human Intelligence (Humint), have been less than kind to Sigint in the past. For years the authors of this study have been told horror stories about the failings of Sigint from past or present practitioners of Humint. Former CIA officials, seeking to enhance the reputation of the CIA's clandestine service, have been particularly harsh in their public criticism of the National Security Agency (NSA) and other forms of technical intelligence gathering. In the best-selling novel *Tinker, Tailor, Solder, Spy*, George Smiley, John le Carré's master spy, voiced the widely-held opinion of many Humint professionals about their more technically-oriented counterparts in the intelligence services, saying: 'We all have our prejudices and radio men are mine. They're a thoroughly tiresome lot in my experience, bad fieldmen and overstrung, and disgracefully unreliable when it comes down to doing the job.'[4]

As a result, for more than 50 years Sigint professionals around the world have been forced to fight in complete secrecy an uphill battle arguing the value of radio intelligence. Oftentimes, the adherents of Sigint lost these bureaucratic battles against the numerous and usually more powerful partisans of Humint. A former senior Indian intelligence officer recalled that 'We dithered in creating an integrated set-up for signal interception... because of the pressures from our sprawling network of spies and human analysts, led by a technically illiterate bureaucracy.'[5]

And so, despite the latent prejudice and immense secrecy surrounding the subject, the question has been asked: why was Sigint so important during the Cold War? This contribution and the others that follow suggest that the history of post-World War II intelligence must be radically rewritten to take into account the important contributions made to the security of the United States and the nations of Western Europe by this arcane and difficult to understand intelligence discipline.

WHAT IS SIGNALS INTELLIGENCE?

An US Army publication defines Sigint as intelligence derived from the intercept, analysis, and parametric exploitation of foreign communications and non-communications radio-electronic emissions.[6] An US Marine Corps manual defines Signals Intelligence (Sigint) as 'intelligence gained by exploiting an adversary's use of the electromagnetic spectrum with the aim of gaining undetected firsthand intelligence on the adversary's intentions, dispositions, capabilities, and limitations'.[7]

Sigint is composed of three separate but interrelated intelligence collection techniques: communications intelligence (Comint), electronics intelligence (Elint), and foreign instrumentation signals intelligence

INTRODUCTION   3

(Fisint).[8] Communications Intelligence (Comint) is intelligence information derived from the intercept and processing of voice, Morse code, radioteletype, facsimile, multichannel (or microwave radio relay), and video signals. Comint does *not* include the interception of unencrypted written communications (mail), the monitoring of foreign public media or propaganda broadcasts, the interception of communications obtained during counterintelligence investigations, or wartime censorship activities.[9]

For example, during the 1950s and 1960s NSA intercept operators around the world spent most of their time monitoring and transcribing radio traffic concerning the day-to-day routine activities at foreign military bases around the world, such as communications from airfield control towers or ground stations directing aircraft movements, the radio traffic of ground forces manoeuvring in the field, ship-to-ship and ship-to-shore naval radio traffic, foreign military and civilian weather broadcasts, and air-to-ground civilian airline communications.[10] During the Cold War, a typical American Comint target was the routine activity at Soviet airfields in East Germany and elsewhere. NSA voice intercept operators monitored the early morning radio checks from the air base, followed by radio traffic among the control tower, the firing range controller, the taxi strip monitor, the bombing range controller, the weather station, the aerial intercept controller, the ground safety crews, and the radar operators. The intercept operators then tracked the routine training flights of the base's combat aircraft as they practised aerial intercepts or bombing attacks at ranges near the airfield. This required listening to hours of mundane air-to-air and/or air-to-ground radio chatter, which in turn required further hours to transcribe and process every day.[11]

Electronics Intelligence (Elint) is concerned with the interception and analysis of emissions from foreign electronic devices. The most common Elint targets are the wide variety of radar systems used around the world for early warning, missile detection, ground control intercept, missile targeting, fighter target vectoring, and altitude determination.[12] Through Elint, these radar systems can be identified by their function and type, their range and capabilities assessed, and their locations precisely fixed.[13] This intelligence information is principally of interest to the military because, as a recently declassified US Air Force document put it: 'By counting radars, specifying their precise location, determining their ranges, and evaluating their operational systems, analysts and engineers could develop countermeasures capable of jamming offensive surface-to-air missile radars and other defensive radars.[14] Other Elint targets include navigation aids and radio beacons which provide geographic position information to ships, aircraft and other vehicles; air-to-air and air-to-ground identification signals, such as Identification, Friend or Foe (IFF) transponders, repeaters and interrogators; emissions from countermeasures equipment and radio jamming devices; radiation from missile guidance systems and artillery fuses; and emissions from meteorological devices, diathermy, radio heating, and research and development laboratories and field testing stations working on electronic devices.[15]

Fisint is defined as the collection and processing of emissions associated with the testing and operational deployment of aerospace, surface, and subsurface systems, which may have either military or civilian application. Fisint includes but is not limited to monitoring telemetry from ballistic missiles as well as manned and unmanned space vehicles, beaconry, electronic interrogators, tracking/fusing/arming/command systems, and video data links which relay data to a ground station concerning performance of space vehicles or weapons systems. As such, Fisint is the Sigint collection discipline primarily associated with the monitoring of foreign weapons research and development activities, including but not limited to ballistic missile testing.[16]

Finally, in the last decade Sigint has become deeply involved with a new kind of electronic communications medium: digital data communications signals, which refers to the transmission of vast amounts of digital data among and between computer systems and networks. A good example of the traffic passing along this medium is electronic bank transfer data. NSA and its English-speaking Sigint partners refer to data traffic by the covername 'Proforma'.[17]

THE IMPORTANCE OF SIGINT

Since the dawn of time, all governments have wanted to know what their friends and allies were doing. In justifying the continuing need for the huge Russian radio intercept station at Lourdes, Cuba, in December 2000 the Russian newspaper *Izvestia* wrote that 'Not a single state has yet been able to deny itself the temptation to learn more about other states (especially those it sees as rivals) than they would like to tell.'[18] And that the easiest way to do this is to listen to the secret communications of foreign governments. The former head of the US Navy Communications Intelligence Organisation, Captain (later Vice Admiral) Joseph N. Wenger, wrote that 'The ambition of every nation has been to develop unbreakable ciphers for its own use and to solve every cipher in use by its actual or potential enemies.'[19]

By its very nature, Sigint has certain intrinsic qualities, which make it a particularly effective intelligence-gathering tool.

The first is that Sigint is a passive intelligence collection technique that generally is conducted without the target's knowledge. Moreover, Sigint collects information against communications targets that are oftentimes thousands of miles away, thus negating the need for the intercept sites to be

INTRODUCTION 5

near the targets being monitored. This means, generally speaking, that Sigint involves relatively little political or physical risk.[20] There was one exception to this rule, however. This was the peripheral aerial and maritime reconnaissance missions conducted by all sides during the Cold War, which resulted in a number of reconnaissance platforms either being destroyed or captured. A total of 146 NSA military and civilian personnel were killed in the line of duty during the Cold War, 60 of whom were killed in Vietnam. The single worst loss occurred during the Israeli attack on the NSA spy ship USS *Liberty* in June 1967, which resulted in the death of 34 Navy, Marine and NSA civilian cryptologists.[21]

By comparison, Humint collection during the Cold War was a particularly risky proposition. For example, the CIA, MI6, as well as the Norwegian, French, West German, and Turkish intelligence services lost more than 300 agents during attempts to infiltrate the Soviet Union between 1949 and 1955; plus several hundred more operatives who were lost trying to establish agent networks in Eastern Europe during the 1950s.[22] Between 1951 and 1953, 212 Chinese agents trained by the CIA were parachuted into northern China. According to declassified CIA documents, 101 of the agents were killed and 111 were captured.[23] Of the 49 CIA agents parachuted into Chinese-occupied Tibet between 1957 and 1960, only 12 survived. Of the remainder, 37 were killed or committed suicide, one surrendered, and one was captured by the Chinese.[24] Between 1958 and 1966, the CIA and Taiwanese lost 120 agents who were parachuted into mainland China as part of an operation called 'Grosbeak'.[25] Finally, according to a recent South Korean news report, between 1950 and 1972 a staggering 7,726 Korean agents working for American or South Korean intelligence were killed or disappeared while spying inside North Korea.[26]

Second, the objectivity and reliability of Sigint is great, but far from perfect. Former CIA Director Vice Admiral Stansfield Turner wrote in 1991 'electronic intercepts may be even more useful [than agents] in discerning intentions. For instance, if a foreign official writes about plans in a message and the United States intercepts it, if he discusses it and we record it with a listening device, those verbatim intercepts are likely to be more reliable than second-hand reports from an agent.'[27] A retired senior CIA officer opined that Humint can never be free from the biases and perceptions of its sources, that the information is oftentimes deemed tainted because it came from traitors motivated by greed or personal grievances, or that it was obtained by corrupting or seducing vulnerable human beings.[28] But in its raw form, Sigint reproduces exactly what it records in an unvarnished, unbiased and undistorted fashion. This historically has given Sigint tremendous credibility with intelligence consumers, particularly since paranoia on both sides of the Iron Curtain created an atmosphere that led analysts inherently to question the credibility of every secret agent that came forward. A former CIA intelligence analyst was quoted as saying of Sigint's reliability that 'You know the origin and you know that this is genuine. It's not like a clandestine [Humint] report where you don't know if this is a good agent or a weak agent or a bad agent or a double agent.'[29] But another CIA officer was recently quoted as saying of Sigint that 'Electronic intercepts are great, but you don't know if you've got two idiots talking on the phone', suggesting that the reliability of a particular intercept is largely dependent on the seniority of the sender and receiver of the transmission.[30]

For example, some senior CIA intelligence analysts questioned the reliability of the information provided by Colonel Oleg Penkovskiy, in large part because the CIA's Clandestine Service would not tell them where the information came from.[31] A Top Secret 1976 study of CIA estimates on the Soviet military found that 'Because the Soviet Union remains a uniquely closed society, human contacts, traditionally the principal source of foreign intelligence, play a distinctly subordinate role in the preparation of these documents: not only is such information exceedingly scarce, but it is always suspect of being the product of a deliberate disinformation effort in which the Soviet government engages on a massive scale. Furthermore, information obtained from sensitive human sources often has such limited distribution that it does not play a significant part in the preparation of NIEs [National Intelligence Estimates].'[32]

Third, unlike other sources, *some but certainly not all* Sigint intercepts can stand on their own without the need for analysis or correlation with other sources, although practitioners of the Sigint craft and 'all-source' intelligence analysts screech in dismay whenever this occurs. This led to the practice during the Cold War of the President of the United States and senior White House officials getting each morning a Top Secret intelligence summary from the CIA and an even more highly classified publication called the Black Book, containing the most important decrypts produced by NSA during the previous 24 hours, along with the Agency's commentary.[33] The same was true in the Soviet Union, where every day the KGB sent a selection of key intercepts in a bound volume called the Red Book to the top six members of the Politburo, although the KGB did not forward any materials contained in the decrypts which ran contrary to the prevailing political trends of the time within the Kremlin.[34] In Britain, the daily Sigint digest produced by GCHQ for the Prime Minister and senior Cabinet officials was called the Blue Book.[35] Individual Sigint reports produced by GCHQ were called Blue Jackets (BJs) because of the distinctive blue-coloured file folders that the reports came in.[36] The highly classified daily collection of diplomatic decrypts that was sent daily to the Dutch Cabinet was called the Green Edition. There were also special daily reports containing Sigint materials on

INTRODUCTION 7

the Middle East (Red Edition), Latin America (Yellow Edition), and on economic intelligence matters (Blue Edition).[37]

Fourth, because of its reliability and the high-level attention that intelligence derived from Sigint received on both sides of the Iron Curtain, it proved to be (with apologies to U-2 and spy satellite aficionados) the premier source of information for national security officials and foreign policymakers during the Cold War. In 1966, Senator Milton Young of North Dakota stated that 'As far as foreign policy is concerned, I think the National Security Agency and the intelligence it develops has far more to do with foreign policy than does the intelligence developed by the CIA.'[38] In October 1998, John Millis, the late staff director of the House Permanent Select Committee on Intelligence, described Sigint as 'the INT of choice of the policy maker and the military commander'.[39] For example, Sigint has been used publicly by the US government in a number of instances, such as the Tonkin Gulf incidents (1964), the North Korean seizure of the USS *Pueblo* in 1968, and the C-130 (1958), EC-121 (1969), KAL 007 (1983) and Brothers to the Rescue (1996) shootdown incidents. President Ronald Reagan publicly justified the 1986 air strikes on Libya using NSA intercepts that reportedly linked Libya to the La Belle Disco bombing in West Berlin. By contrast, rare indeed have been the instances where major policy decisions have been significantly influenced or determined solely by information received from a human intelligence source.

Fifth, Sigint was usually the fastest source of current intelligence information available to consumers. A congressional intelligence committee official said of Sigint that 'it's there quickly when needed'.[40] Lieutenant General Daniel O. Graham, the former Director of the Defense Intelligence Agency (DIA), was quoted as saying 'Most collection agencies give us history. The NSA is giving us the present.'[41] During the Cold War, NSA emphasised the speedy delivery of finished Sigint to its customers because of the perishable nature of the product. Today, thanks to improvements in communications and data processing technologies, NSA can get the results of its Sigint collection efforts to its consumers in the field in near real-time.[42] NSA had its own dedicated communications system just for transmitting intercepts to Fort Meade, and another special distribution network called the SSO system for getting the finished product from NSA headquarters to its consumers. Between 1962 and 1965, NSA placed into operation the Critical Intelligence Communications (CRITICOMM), which allowed each of NSA's 150 Sigint collection and processing units around the world to bypass normal communications channels and transmit especially important intelligence information to Washington DC within 15 minutes.[43] NSA even had its own direct communications link with the White House, which was established shortly after the Cuban Missile Crisis, to feed the

FIGURE 1.1

US Army 'Adventurer' border intercept site in South Korea.
US Army Intelligence and Security Command

FIGURE 1.2

US Army 'Hippodrome' space collection facility at Sinop, Turkey.
US Army Intelligence and Security Command

President of the United States and the National Security Council with key decrypts, thus bypassing the analysts at the CIA and DIA.[44]

As a result, Sigint intercepts usually found their way to consumers in far less time than imagery interpretation reports from satellites, and usually arrived weeks before Humint reports found their way to the hands of intelligence analysts.[45] For example, during the Cuban Missile Crisis in 1962, most CIA agent reporting from inside Cuba was accomplished by secret writing, which took a week or more to get to the CIA analysts from inside Cuba. Information derived from refugee interrogations was oftentimes months old by the time the Cubans were questioned by the CIA at Opa Locka, Florida.[46]

The Soviet experience was also the same. During the war in Afghanistan during the 1980s, the Russian Army depended on radio intercepts as their primary means of locating major units of Afghan guerrillas. The principal reason for the dependence on Sigint was that Russian field commanders found that the Humint reports that they received from the Afghan secret service – the Khad – almost always arrived to late to be of any tactical value.[47] This has meant that one of Sigint's most important functions was to provide forewarning of an enemy attack, something which slower intelligence sources, especially Humint, could not provide.

Sixth, Sigint produces more intelligence information on a broader range of subjects than any other intelligence source. In 1964 alone, NSA sent out approximately 150,000 finished intelligence reports and translations to its consumers in Washington, or more than 400 Sigint reports a day. By the end of the 1960s, this figure more than doubled to almost 400,000 finished intelligence reports being produced annually, which is the equivalent of more than 1,000 Sigint reports going to NSA's consumers every day.[48] By comparison, in 1960 the KGB deciphered 209,000 diplomatic cables sent by 51 countries, or the equivalent of 572 decrypts a day going to consumers.[49] In 1967, the KGB deciphered 188,400 diplomatic messages sent by 72 countries, or 516 decrypts a day.[50]

Seventh, Sigint never sleeps. Agents and their handlers must sleep (we are after all only human!), and darkness or adverse weather could shut down imagery collection systems for weeks at a time. But Sigint collects and produces intelligence 24 hours a day, 365 days a year, regardless of the weather or other environmental conditions.[51]

Eighth, Sigint is flexible and more responsive to consumer tasking than most other intelligence sources. A 1998 congressional report stated that 'much of NSA's past strength has come from its localised creativity and quick-reaction capability'.[52] You can quickly retarget Sigint, assuming that you possess the appropriate collection platforms and the manpower with the requisite skills to perform the mission. This made Sigint the source of choice during fast moving world crises during the Cold War. For example, NSA was able to react to a series of Cold War crises faster than the rest of the US intelligence community, such as the Korean War, the Cuban Missile Crisis, and the Vietnam conflict.[53] You cannot, however, quickly change the tasking of agents nor build a new agent network overnight (unless of course you are a reporter for CNN); and the cost of retasking a spy satellite (and thus shortening its orbital life) was prohibitively expensive.

Ninth, intelligence insiders argue that Sigint's potential as an intelligence source is greater than all other intelligence collection disciplines. One successful solution of a major foreign cryptographic system can generate more intelligence information in a day than all other sources combined. A former American defence and intelligence official has written that a 'break' is the 'equivalent not of one but of a thousand spies, all ideally placed, all secure, and all reporting instantaneously'.[54] Even such a stalwart believer in Humint as the late Allen W. Dulles, the Director of the CIA from 1953 to 1961, opined that Sigint was 'the best and "hottest" intelligence that one government can gather about another'.[55]

And tenth, relative to all other intelligence disciplines, many intelligence 'insiders' consider Sigint to be one of the most cost effective means of gathering intelligence. American intelligence observers believe, for example, that despite NSA's huge workforce and budget, it produces on a dollar-for-dollar basis more 'bang for the buck' than any other intelligence source, with perhaps the exception of the National Reconnaissance Office's spy satellites.[56] This has been an extremely contentious issue within the US intelligence community for decades because of NSA's huge budget. The authors estimate that NSA and its predecessors have spent about $100 billion since 1945, 75 per cent of which was spent on Sigint and the rest on communications security.[57] More importantly, throughout the Cold War the US government spent four to five times as much money on Sigint than they did on Humint collection.[58]

There are no hard figures available for how much the Soviet Union and its allies spent on Sigint, but according to a former KGB official, by the late 1980s, Sigint collection was eating up 25 per cent of the KGB's annual budget.[59] But what is now clear is that by the early 1980s, the Soviet intelligence community, in particular the GRU, had come to depend to a greater degree on Comint because it proved to be a more productive source for strategic intelligence than the more traditional Humint sources.[60] According to a 1993 statement by Cuban Defence Minister Raul Castro, Russia got about 75 per cent of its strategic military intelligence information from the huge listening post at Lourdes, Cuba.[61]

The East German Sigint organisation, Hauptverwaltung III, was particularly successful during the Cold War. Throughout the 1970s and

INTRODUCTION 11

1980s, the East German Sigint service eavesdropped on the telephone conversations of almost every important West German politician, including the sensitive conversations of former West German chancellor Helmut Kohl. The former head of the East German Sigint organisation, Major General Horst Männchen, told an interviewer that during the early 1980s, his service was intercepting approximately 40,000 West German telephone conversations per year, including those of the most senior members of the West German government.[62]

The result is that, as will be demonstrated throughout this work, Sigint was arguably the most important intelligence source for the US and its European allies during the Cold War. And since the demise of the Soviet Union, one can argue that the relative importance of Sigint has only increased. In the late 1950s, Sigint and U-2 imagery were producing about 75 per cent of the hard information available to the US intelligence community about the Soviet military.[63]

Dependence on Sigint by many Western European nations was even greater. For example, during the 1980s the vast majority (80–90 per cent) of the raw intelligence information reaching the British Joint Intelligence Committee in London every day came from Sigint.[64] In May 1999, British Foreign Secretary Robin Cook stated that 'GCHQ's work is vital in supporting our foreign and defence policies.'[65] The 2000 Annual Report of the British Parliament's Intelligence and Security Committee revealed that 'The quality of the intelligence gathered [by GCHQ] clearly reflects the value of the close co-ordination under the UKUSA agreement.'[66]

Sigint also accounted for the majority of the intelligence information generated by the Canadian intelligence community.[67] One informed observer has written that Canada's national Sigint organisation, the Communications Security Establishment (CSE), is that country's premier intelligence producer, adding that CSE was Canada's 'single-most important contributor to allied intelligence sharing agreements'.[68]

A 1952 French report stated that 'The study of enemy radio is by far our best source of intelligence.'[69] A 1973 memorandum to the Dutch Prime Minister described that nation's Sigint organisation as '[T]he most valuable asset we have to collect an intelligence product that is valuable to all interested parties.'[70]

PROBLEMS AND LIMITATIONS OF SIGINT

The following is a short summary of some of the basic weaknesses and limitations of Sigint, some of which apply to many of the other intelligence disciplines as well:

Secrecy of Sigint: Historically, because of the need to protect sensitive sources, Sigint intercepts were given extremely limited distribution with the highest levels of government and the military, and even then, only on a need-to-know basis.[71] A declassified 1952 US Army memorandum states: 'It is fully realised that enemy communications are probably the most sensitive of all intelligence sources, and that every precaution must be taken to protect the security of our efforts to exploit them.'[72] Each Comint report coming out of NSA during the 1950s and 1960s stated on its cover that 'This document is to be distributed and read by only those persons who are officially indoctrinated in accordance with communications intelligence security regulations and who need the information in order to perform their duties.'

There are numerous examples of the negative ramifications stemming from the decision to keep Sigint under wraps. The first was that few government officials ever came to fully appreciate Sigint. For example, a former senior NSA official has written that this high degree of secrecy 'preserved the anonymity (but also limited the appreciation) of the source' within the US intelligence community.[73] This also meant that, in many instances, government officials and military commanders who needed the information were denied access to Sigint because someone had determined that they did not have the 'need-to-know'.

A 1951 report concerning Comint support to the US Army during the Korean War found that the effectiveness of this crucial intelligence source was limited, in large part because the high security classification level of the intercepts prevented all but a few senior Army officers in Korea from seeing them. In addition, Army intelligence officers in Korea were barred by Comint security regulations from merging Comint with other forms of intelligence (such as Photint or Humint) into an 'all-source' intelligence product; and the sensitivity of the source oftentimes prevented Army field commanders from using the Comint information, leading some senior Army commanders in Korea to refer to Comint as a 'wasting asset'.[74] A 1951 report by a US Air Force inspection team found that Comint intercepts being generated by listening posts in Japan were so highly classified that they could not be distributed to most USAF intelligence consumers in Korea.[75] US Navy commanders in the Far East were also complaining that few combat commanders ever saw either tactical or strategic Comint during the Korean War because they did not possess the requisite security clearances.[76]

During the Vietnam War, NSA's Sigint coverage of North Vietnamese MiG flight activity was excruciatingly detailed and accurate, but because of security concerns NSA refused to give this highly perishable intelligence to the American pilots flying combat missions over North Vietnam. When senior US Air Force and Navy commanders in Southeast Asia found out that NSA was collecting intelligence that could save the lives of American pilots, but was not distributing the intelligence because of security concerns, they were understandably furious.[77] Voicing the opinion of many soldiers in Southeast Asia, Command Sergeant Major John Martin, who served with the US Army Special Forces in Vietnam, recalled 'Because they [the US Army Sigint units and personnel] were so "special" you could never get them to work for you; you could only hope they would share something if it could save your ass. Other than those unusual situations, their info was just "too special" for us average boonie rats.'[78]

In the mid-1980s, American officials frequently hinted that they had 'indisputable evidence' [i.e. Sigint] demonstrating Nicaraguan and Cuban support for guerrilla forces operating in El Salvador. Despite calls for the evidence to be made public, the Reagan administration refused to release the materials because it would jeopardise sensitive intelligence sources and methods. This led a former CIA intelligence analyst to remark that 'Radio intercepts are not so novel, or so critical. They won't jeopardise anything... I can't believe that in the past years they wouldn't have been produced – so critical is this matter to US policy.' The analyst added that 'You don't save these expensive intelligence sources for the junior prom.'[79]

The United States was not the only country that applied these stringent security measures to protect their Sigint product. In the Soviet Union, Sigint intercepts were deemed so sensitive that they were delivered directly to a very small number of specially cleared consumers in the Politburo, the KGB and the GRU, thus bypassing regular intelligence analysis channels. Sigint sharing with the Soviet Union's allies in Eastern Europe was also specifically forbidden.[80]

In the British Foreign Office, a special messenger from the FO's Permanent Under Secretary's Department would periodically deliver copies of the Top Secret Blue Jacket reports from GCHQ to those few officials cleared to see them. After reading the reports, the GCHQ material was given back to the messenger, who returned them to the Permanent Under Secretary's office for storage. The rule was that there was never to be any mention of decrypts in official Foreign Office papers.[81]

In October 1975, senior officials of the Australian Sigint organisation, now known as the Defence Signals Directorate (DSD), apparently decided not to send to the Australian Prime Minister a sensitive intercept which revealed that the Indonesian military intended to kill five Australian journalists in East Timor covering the Indonesian invasion of that country. The DSD officials feared that Prime Minister Gough Whitlam would act on the information, thus revealing DSD's ability to read sensitive Indonesian communications traffic. All five of the journalists were murdered by Indonesian special forces troops, and their bodies burned.[82]

Diminished Utility: The ability of consumers to use Sigint was strictly limited because of pervasive security considerations.[83] For example, during the Korean War, American Top Secret intelligence reports derived from Comint carried the following caveat emptor: 'Certain restrictions prohibit the further dissemination of this information either direct or paraphrased. Pertinent order of battle information included herein, that is not confirmed by other sources will be passed to divisions on a "need to know basis" only and will not be included in any routine intelligence reports or summaries.'[84]

During the 1950s and 1960s, every NSA Comint report carried the following edict on its cover: 'No action is to be taken on information herein reported, regardless of temporary advantage, if such action might have the effect of revealing the existence and nature of the source.'[85] Because of these limitations, US government officials and military commanders oftentimes found themselves severely circumscribed in how they could act on the Sigint that they received, which naturally diminished the utility of the intelligence to its consumers.[86]

Failure to believe Sigint: Cold War history is replete with many examples of government officials, military commanders, and intelligence analysts who chose not to believe the Sigint they received. In part, this was because the reader did not understand the information they received, or trust the reliability of the Sigint source. More often then not, the Sigint was misused or ignored because it did not fit some preconceived notion already held by the reader. Field-Marshal Lord Montgomery and General Douglas MacArthur, for example, were two commanders who did not particularly trust Sigint unless it confirmed their own personal assessments.[87]

During the Chinese Civil War, American intelligence analysts failed to heed information contained in decrypted Soviet clandestine radio traffic between Moscow and Mao Tse-tung's headquarters in Yenan, China, which revealed that as of October 1945, Mao's People's Liberation Army consisted of almost 1.1 million men under arms. A year later, in mid-1946, the US Army G-2 still was estimating that Mao's forces consisted of only 600,000 men, despite the fact that the PLA had by that time probably grown to almost two million men under arms.[88] During the war in Indochina in the early 1950s, some senior French commanders chose not to believe the information contained in decrypts of high-level Viet Minh military radio traffic because it did not match their assessments of the strength and capabilities of Ho Chi Minh's forces.[89]

Over-Reliance on Sigint. There are numerous examples of intelligence officials and military commanders placing undue reliance on Sigint to the exclusion of other sources of intelligence information. For instance, by the

INTRODUCTION   15

late 1950s the US intelligence community was relying almost exclusively on Comint to provide warning of a Soviet military attack. Former senior CIA official Lyman Kirkpatrick stated that 'If the Soviets ever decided to go for broke, they wouldn't put anything on electronic communications or do anything visible by satellite. All the orders would go by officer couriers, which was what Hitler did at the Battle of the Bulge and caught us totally unprepared. We were relying too heavily on communications intelligence.'[90]

By the late 1970s, the US intelligence community had become so dependent on Sigint that in 1978, President Jimmy Carter said 'Recently... I have been concerned that the trend that was established about 15 years ago to get intelligence from electronic means might have been over-emphasised.'[91] A declassified 1976 CIA intelligence assessment confirmed that the Soviets also largely depended on Sigint as their primary source for strategic warning of a nuclear or conventional attack.[92]

The results of over-reliance on Sigint can be disturbing. The system backfired in November 1983, when Soviet Sigint stations in East Germany and Czechoslovakia detected the sudden cessation of radio traffic coming from American nuclear weapons units in West Germany, particularly among the units of the US Army's 56th Artillery Brigade, which was armed with Pershing nuclear missiles. This was followed by a change of ciphers and radio frequencies throughout the US Seventh Army. The Soviets interpreted these moves as indicative of an imminent nuclear attack by the US. In fact, it was only a nuclear release exercise, called 'Able Archer 83'.[93] Nevertheless, the Soviets were so alarmed that they placed many of their forces in Eastern Europe on alert. Between 2 and 11 November 1983, American and British Sigint stations in West Germany detected Soviet ground forces in East Germany and the Baltic Military District going to a heightened readiness status; Soviet air units in East Germany and Poland were placed on alert and routine training flights suddenly stopped; Soviet nuclear-capable fighter bombers were placed on runway alert on airfields in East Germany; and the Soviets suddenly ceased broadcasting weather reports throughout the Soviet Union and Eastern Europe.[94]

More recently, on 11 May 1998, the government of India tested three nuclear devices at the Pokhran nuclear test site in western India. Not surprisingly, the Indian nuclear tests dramatically heightened the state of tension between India and neighbouring Pakistan. Two weeks later, on 27 May 1998, the Signals Intelligence Directorate of the Indian Army intercepted and decrypted a message from the Pakistani Foreign Ministry in Islamabad to the Pakistani High Commission in New Delhi, which reported that Pakistan had 'credible information' that India was ready to mount a pre-emptive strike against Pakistani nuclear installations. The following day, 28 May 1998, Pakistan conducted its own series of nuclear tests. India publicly denied that it had any intention of attacking Pakistani military installations, but the military forces of the two countries remained on hair-trigger alert for many weeks afterwards.[95]

Sigint Snobbery. As the importance of Sigint grew during the years after World War II, followed by the introduction of spy satellites in the 1960s, the value of Humint was rapidly marginalised within the American and British intelligence communities.[96] This led to a pervasive sense of snobbery and self-infatuation by the denizens of the Sigint community in the West. Intelligence insiders referred to this elitism as the 'Green Door' syndrome. This led Humint partisans to complain that greater credence was almost always given to Sigint over Humint.[97] A recently declassified NSA history noted that during the 1950s, many high-level intelligence reports referred to Humint and other collateral intelligence as 'unconfirmed information'. Only Sigint was deemed reliable enough to be described as 'a usually reliable source'.[98]

Talking about his experiences with the US Army Sigint organisation, the Army Security Agency (ASA), during the Vietnam War, one US Army officer wrote:[99]

> ASA came very close to total alienation from the [US Army] combat arms. This seemed to result partly from the avid devotion [by ASA] to the strategic [Comint] mission and its highly classified and controlled product, as well as a self-imposed snobbery and self-infatuation. Every combat arms officer from the Vietnam War I've met has his own story of the 'green door' syndrome and hyper-classification of signal intelligence from ASA. While ASA served the combat troops better than they realised during the conflict, the superior and separatist attitude of even tactical ASA units has left a bad taste in the mouths of many commanders even to this day. Those who don't dread our attachment to their commands in the future look forward to 'straightening us out' and making 'real' soldiers of us.

The Fragmentary Nature of Sigint: Sigint usually will provide hundreds if not thousands of pieces of a complex puzzle, but rarely will it yield the entire puzzle. Much of the information obtained by Sigint is fragmentary and indirect, requiring that analysts patiently sift through hundreds or thousands of intercepts in order to piece together the pieces of a puzzle. Even then, the puzzle more often than not remains largely incomplete, as in the case of the much-touted Venona decrypts. The fragmentary nature of most decrypts make them extremely difficult to understand, much less use.[100] A senior American intelligence official was quoted as saying that 'You rarely get a Sigint smoking gun. It's usually very fragmentary... Very often you don't even know who you're listening to.'[101] Voicing a feeling often heard from intelligence consumers around the world trying to understand Sigint, in 1976 the Dutch Prime Minister complained that the Sigint that he was receiving was raw, unfinished materials that were sometimes unbalanced or fragmentary.[102]

Sigint Does Not Provide All the Answers. Generally, Sigint cannot measure a nation's political will or morale, or detail the innermost workings of foreign governments. Even Ultra and Magic during World War II failed to yield this kind of information, but it should be pointed out that the much vaunted Humint effort during the war did not either. Current and former senior intelligence officials in the US, Canada and Europe interviewed by the authors have all emphasised that Sigint is only useful when it is combined with intelligence obtained from other sources into an 'all-source' product.[103]

Lack of Timeliness: Although Sigint insiders pride themselves on being fast, sometimes they are not fast enough because of the time and effort required to process, analyse and report to consumers the results of Sigint collection.[104] For example, an intelligence community post-mortem of 1968 Czechoslovakian Crisis found that Sigint did not find its way to consumers in Washington until days after it had been intercepted. A small consolation for NSA was the fact that the CIA's Humint and Imint reporting was much slower.[105]

Too Much Information: Experience during the Cold War showed that NSA often did drown intelligence analysts in a sea of paper, such as during the 1968 Czech crisis and before the 1973 Middle East War. After the 1973 Mid-East War, the CIA blamed NSA in part for the failure to predict the war, claiming that CIA intelligence analysts were swamped by hundreds of Comint summaries every week pertaining to Egyptian and Syrian military activities. A post-mortem study done after the war concluded, however, that the overworked CIA and DIA intelligence analysts responsible for the Middle East were not trained to understand or effectively evaluate the information contained in the NSA Comint summaries.[106]

When Sigint satellites were placed in orbit, the vast amount of information that they generated swamped the available analytic resources. For instance, the intercept tapes generated by the US Navy's first Grab Elint satellite, which was launched from Cape Canaveral, Florida on 22 June 1960, so thoroughly saturated NSA's analysts that it took months to process a few weeks of intercepts.[107]

Smaller nations, such as India, are also suffering from an information glut from their Sigint collection operations. A former Indian Cabinet official recently wrote that 'We are in the throes of a similar crisis in our technical collection capability as we are also not able to utilise a major portion of intercepted traffic.'[108]

Lack of Sigint: In some instances during the Cold War, good operational and communications security by the Soviet Union and its allies 'blacked out' Sigint, although these instances were fewer than previously believed. For example, in 1959 the Algerian National Liberation Front (FLN), which was fighting for the independence of Algeria from France, changed all of its codes, making them impossible for the French Comint service to decrypt.[109]

Deniability: Access to Sigint data can be denied by the use of encryption and other secure forms of communications, such as landline telephone and telegraph circuits, or more recently fibre-optic cables.[110] For example, NSA lost much of its access to high-level Soviet communications traffic in the late 1940s and early 1950s when the Russian military shifted much of its high-level communications traffic to landlines.[111] Recently introduced complex communications technologies, such as frequency hopping radio systems, have made the job of the Sigint intercept operator far more difficult then in the past.[112] In recent years, Pakistani-backed guerrillas operating in the Indian state of Kashmir, calling themselves the Hizbul Mujahideen, have begun using frequency-hopping radios, burst transmission technology, citizen-band radios, satellite telephones, even sophisticated encryption technology, which has made it increasingly difficult for the Indian government's Sigint services to monitor their communications traffic.[113]

Fragility of the Source: Because it is dependent on extremely fragile and sensitive sources and methods, Sigint is particularly vulnerable to damage caused by treason, defections, news leaks, or poorly considered public statements by government officials.[114] A KGB operative named William Wolfe Weisband, who worked inside the US Army Sigint organisation, the Army Security Agency (ASA), during the late 1940s single handedly destroyed three years of successful work against Soviet cipher systems. After Weisband told the KGB about ASA's successes against Soviet systems, on 29 October 1948, known within NSA as 'Black Friday', the Russians executed a massive change of all of their cryptographic systems and communications operating procedures, shifting all of their mainline systems to unbreakable one-time pads. It would take six years before NSA could solve another high-level Soviet cipher system.[115]

The June 1960 defection of two NSA civilian employees, William H. Martin and Bernon F. Mitchell, caused immense damage to NSA's Sigint effort against the Soviet Union. According to one source, Martin and Mitchell's defection resulted in a 'partial dimout of United States communications intelligence', requiring that NSA work in double shifts for months trying to fix the damage done to the US Sigint effort.[116]

In 1969, President Richard M. Nixon revealed at a press conference that the US had the ability to read Soviet and North Korean communications. After making this statement, the Soviets, Chinese and North Koreans

INTRODUCTION   19

changed many of their cryptographic systems. It took NSA months to repair the damage caused by Nixon's off-the-cuff remarks.[117]

In the 1970s, the East German Secret Service, the STASI, was able to infiltrate the French listening post in West Berlin. This led the East Germans to change their communications procedures and manipulate their radio traffic so as to deceive the French.[118]

In 1989, the transcript of an intercepted telephone conversation involving Colombian drug lord Pablo Escobar was leaked to Bogotá newspapers, which revealed to the members of the Medellin Cartel that the US was eavesdropping on their telephone calls.[119]

Communications Deception: Sigint is vulnerable to communications deception, although this is a very difficult and dangerous game to play.[120] For example, the KGB decided not to play communications deception games with the Berlin Tunnel in order to protect their source inside MI6, George Blake.

Because of it is often compartmentalised away from all other intelligence sources, Sigint is particularly vulnerable to political manipulation by those senior government officials who control access to the intelligence reports. Henry Kissinger, President Richard Nixon's National Security Advisor, reportedly ordered that certain sensitive NSA intercepts not be shared with the Secretaries of State and Defense.[121] In 1986, NSA refused a request by Secretary of Defense Caspar Weinberger for access to NSA intercepts concerning the Iran-Contra affair, stating that the Pentagon had no 'need-to-know'. What Weinberger and Secretary of State Schultz did not know was that on orders from the White House, the State and Defense Departments had been specifically barred from access to these intercepts. Senior Pentagon officials were later outraged to learn that while they were not allowed to see the intercepts, NSA was providing copies of these reports to Richard Secord, who although a retired Air Force General, was not a government official and held no security clearance.[122]

Lack of a Co-ordinated Sigint Effort: Competing bureaucracies were the bane of the American and Soviet Sigint efforts during the Cold War, resulting in massive duplication of effort and wasted resources. Declassified documents clearly demonstrate that during the 1940s and 1950s, the intelligence components of the three US military services pursued independent intelligence collection and processing efforts that were 'conducted with a minimum of service co-ordination'.[123] For instance, by 1951 senior CIA officials had become so frustrated by the continuing internecine fighting between the three American military Comint organisations in the Far East, as well as the inability of NSA's predecessor organisation, the Armed Forces Security Agency (AFSA), to impose order over its quarrelling subordinates, that the CIA came to view AFSA as unresponsive to civilian authority or national intelligence requirements. By mid-1951 CIA Director General Walter Bedell Smith was so unhappy with how the American Comint system was working that he threatened to stand on his right as the head of the US intelligence community and reorganise the entire American Comint system unless the State and Defense Departments stepped in and did something to fix the problems.[124]

During the manhunt for Colombian drug lord Pablo Escobar in 1992–93, US Army and CIA airborne Sigint units operated independently of each other in order to prove that their personnel, equipment and equipment were superior to their counterparts. US Army intelligence officers believed that the CIA station chief in Bogotá was taking credit for information that they had collected.[125]

In the Soviet Union there was also little cooperation or coordination of effort between the Sigint units of the KGB and the intelligence organisation of the Soviet military, the GRU, throughout the Cold War.[126]

But the superpowers were not the only ones who suffered from this afflication. During the Cold War the West Germans fielded five Sigint organisations, four military and one civilian unit run by the West German foreign intelligence services, the Bundesnachrichtendienst (BND) For almost 20 years, from 1970 until the end of Cold War in 1989, the West German military services and the BND fought a series of bitter internecine battles over mission and resource allocation, as well as control of the West German government's Sigint effort.[127]

During the 1950s there was no coordination of effort among the Dutch army, navy and air force Sigint units, resulting in tremendous duplication of effort. The Dutch Army Sigint unit, for example, did not even bother to tell the Dutch naval Sigint organisation, the TIVC, which targets it was copying or what results it was obtaining. A proposed merger of the three service Sigint organisations collapsed because each of the services feared that their requirements would not be met in a unified Sigint organisation. A former Dutch Sigint official later admitted that Sigint cooperation with foreign services was better than between the three Dutch Sigint organisations.[128]

Technical Issues: Sigint's ability to perform effectively is subject to the vagaries of atmospheric conditions and solar flare activities. For example, in the mid-1950s, the Canadian intercept site at Churchill in Manitoba was forced to shut down its operations for days at a time because atmospheric anomalies, which are common in the northern climes, prevented the station's operators from hearing any high-frequency signals.[129] Terrain is also a significant limiting factor. For example, Sigint intercept operators have historically experienced great difficulty copying radio signals emanating from urban areas, densely wooded terrain or in mountainous regions.[130]

INTRODUCTION

Finally, radio interference coming from major urban areas or industrial activities in the vicinity of the listening post can wreak havoc with radio intercept operations.[131] For example, the Canadian listening post at Inuvik in northern Canada had to be closed in April 1970 because radio interference from nearby oil exploration activity significantly affected the station's ability to monitor HF radio signals coming from the Soviet Union.[132]

NOTES

1. Christopher Andrew, 'Conclusion: An Agenda for Future Research', *Intelligence and National Security* 12/1 (Jan. 1997) p.228.
2. A more detailed discussion of these matters is contained in Matthew M. Aid, 'Not So Anonymous: Parting the Veil of Secrecy About the National Security Agency', in Athan G. Theoharis (ed.) *A Culture of Secrecy: The Government Versus the People's Right to Know* (Lawrence: UP of Kansas 1998) pp.65–7.
3. William Rosenau, 'A Deafening Silence: US Policy and the Sigint Facility at Lourdes', *Intelligence and National Security* 9/4 (Oct. 1994) pp.730–1.
4. John le Carré, *Tinker Tailor Soldier Spy* [1974] (NY: Coronet Books 1994) p.206.
5. Maj. Gen. Yashwant Deva (Ret.), 'Of Tapes and Tapping: Technical Intelligence Scores Over Human Intelligence', 2 July 1999, located at http://www.ipcs.org/issues/articles/217-ip-deva.htm.
6. Department of the Army, Field Manual FM 34-2, *Collection Management*, Oct. 1990, pp.2–5.
7. US Marine Corps, Marine Corps Warfighting Publication (MCWP) 2-15.2, *Signals Intelligence*, June 1999, p.1-1.
8. US House of Representatives, Permanent Select Committee on Intelligence, *Annual Report by the Permanent Select Committee on Intelligence*, 95th Congress, 2nd Session, 1978, p.50.
9. US Senate, Report No. 94-755, *Final Report of the Select Committee to Study Governmental Operations With Respect to Intelligence Activities*, 94th Congress, 2nd Session, 1976, Book III, p.737; US House of Representatives, Permanent Select Committee on Intelligence, Report No. 95-1795, *Annual Report by the Permanent Select Committee on Intelligence*, 1978, pp.31, 58; Department of Defense Directive S-5100-20, *The National Security Agency and the Central Security Service*, 23 Dec. 1971, p.2, DoD FOIA; Air University Extension Career Institute, *Electronic Signals Intelligence Exploitation Craftsman Career development Course AFSC 1N571* (Goodfellow AFB, 17th Training Wing, 21 March 1995) Vol.II, pp.34–5, USAF FOIA; *Naval Cryptology in National Security*, 1985, p.30, COMNAVSECGRU FOIA; David L. Christianson, 'Signals Intelligence', in Gerald W. Hopple and Bruce W. Watson (eds.) *The Military Intelligence Community* (Boulder, CO: Westview Press 1986) p.41.
10. Patrick J. McGarvey, *CIA: The Myth and the Madness* (NY: Saturday Review Press 1972) pp.42–3.
11. Ibid., pp.74–6.
12. Headquarters Strategic Air Command History Division, *SAC Reconnaissance History: January 1968 – June 1971*, 7 Nov. 1973, p.72, via Dr Jeffrey T. Richelson.
13. Air University Extension Career Institute, *Electronic Signals Intelligence Exploitation Craftsman Career development Course AFSC 1N571* (Goodfellow AFB, 17th Training Wing, 21 March 1995) p.19, USAF FOIA.
14. Headquarters Strategic Air Command, History Division, *SAC Reconnaissance History* (note 12) p.3, via Dr Jeffrey T. Richelson.
15. US House of Representatives, *House Intelligence Committee 1978 Annual Report*, p.36; NSGTP 69304-B, *Naval Cryptology in National Security* (note 9) pp.30–1, COMNAVSECGRU FOIA; NSGTP 68322, *Cryptologic Operators Manual for Noncommunications Operations*, 1980,

pp.6–12, 40–4, COMNAVSECGRU FOIA; FM 34-2, *Collection Management*, Oct. 1990, pp.2–5; Charles A. Kroger Jr, 'ELINT: A Scientific Intelligence System', *Studies in Intelligence* (Winter 1958) p.72, RG-263, Entry 37, Box 1, Folder 6, US National Archives, College Park, Maryland (hereafter 'NA, CP'); Memo with attachment, Radford to Secretary of Defense, 20 July 1954, RG-330, Entry 200B OSD 1954, Box 47, 334 Joint Electronics Analysis Group, NA, CP.
16. FISINT was formerly known as Telemetry Intelligence (Telint). Christianson, 'Signals Intelligence' (note 9) p.40; *Naval Cryptology in National Security* (note 9) p.48.
17. National Air Intelligence Center, *Draft, Technical Requirements Document (TRD) for the Sigint System Integration Contract*, 24 July 2000, located at www.pixs.wpafb.af.mil/pixslibr/DATAEX/sigsoo_drft1.doc.
18. 'Paper Looks at Continuing Value of Russian Tracking Station Near Havana', *BBC Monitoring*, 18 Dec. 2000.
19. SRH-264, *A Lecture on Communications Intelligence by Captain J.N. Wenger, USN*, 14 Aug. 1946, p.8, RG-457, NA, CP.
20. US Marine Corps, MCWP 2-15.2, *Signals Intelligence*, June 1999, pp.1-4 – 1-5.
21. *NCVA Cryptolog*, Summer 1996, p.3.
22. Confidential interviews.
23. Evan Thomas, *The Very Best Men* (NY: Simon & Schuster 1995) pp.52–3, 360.
24. John Kenneth Knaus, *Orphans of the Cold War: America and the Tibetan Struggle for Survival* (NY: BBS Public Affairs 1999) p.233; William M. Leary, 'Secret Mission to Tibet', *Air & Space*, Jan. 1998, pp.62, 70.
25. Department of State, *Foreign Relations of the United States 1964–1968: Vol. XXX, China* (Washington DC: GPO 1998) pp.476–7, 495.
26. 'ROK Spies Who Died After Infiltrating NK Number 7,726', *Korea Times*, 27 July 1999.
27. Stansfield Turner, 'Intelligence for a New World Order', *Foreign Affairs*, Fall 1991, p.158.
28. Confidential interview.
29. Transcript, National Public Radio, Morning Edition, National Security Agency, 14–16 March 2000.
30. Bob Drogin, 'At CIA School, Data Outweigh Derring-do', *Los Angeles Times*, 27 Aug. 2000, p.A7.
31. John M. Maury, *Memorandum for the Record: Conversations With Messrs. Ed Proctor and Jack Smith Re Use of CHICKADEE Material for NIE 11-8-61*, 7 June 1961, CIA FOIA Page at www/odci.gov/. Another former CIA official stated that Penkovskiy gave 'little information of major importance that had any significant effect on our intelligence estimates. He was primarily useful in... confirming data from other sources, and adding confidence to existing assessments', for which see Herbert Scoville Jr, 'Is Espionage Necessary For Our Security', *Foreign Affairs*, April 1976, p.488.
32. *Intelligence Community Experiment in Competitive Analysis: Soviet Strategic Objectives: An Alternative View: Report of Team 'B'*, Dec. 1976, p.9, RG-263, NA, CP.
33. David Kahn, *The Codebreakers* (NY: Macmillan 1967) p.727.
34. Christopher Andrew and Oleg Gordievsky, *KGB: The Inside Story* (NY: HarperCollins 1990) p.455; Christopher Andrew and Vasili Mitrokhin, *The Sword and the Shield: The Mitrokhin Archive and the Secret History of the KGB* (NY: Basic Books 1999) pp.352–3.
35. Mark Urban, *UK Eyes Alpha* (London: Faber 1996) p.8.
36. Aldrich contribution.
37. Wiebes contribution, p.254.
38. Harry Rowe Ransom, *The Intelligence Establishment* (Cambridge, MA: Harvard UP 1970) p.127.
39. 'Address at the CIRA Luncheon, 5 Oct. 1998; John Millis' Speech', *CIRA Newsletter*, Winter 1998/1999, p.6.
40. Ibid.
41. David Kahn, 'Big Ear or Big Brother?', *New York Times Magazine*, 16 May 1976, p.62.
42. US Marine Corps, MCWP 2-15.2, *Signals Intelligence*, June 1999, p.1-5; Penelope S. Horgan, *Signals Intelligence Support to US Military Commanders: Past and Present* (Carlisle Barracks, PA: US Army War College 1991) p.88.

INTRODUCTION 23

43. DOD Directive S-5100.19, *Implementation of National Security Council Intelligence Directive No. 7*, 19 March 1958, pp.1–4, DOD FOIA; JCS 2010/143, *Fulfillment of Proposed NSCID No. 7 and COMINT Communications Requirements*, 18 July 1958, RG-218, CCS 334 NSA, Section 22, NA, CP; NSA/CSS Manual No. 22-1, *National Security Agency Central Security Service Organization Manual*, 21 Jan. 1974, p.2, NSA FOIA; McGarvey, *CIA* (note 10) pp.45–6; Central Intelligence Agency, *A Consumer's Guide to Intelligence*, Sept. 1993, pp.5, 38; NWP-5, *Naval Cryptologic Operations*, p.3-2, ONI FOIA; NSGI C3211.1D, *CRITIC (Critical) Information*, 2 Aug. 1990, COMNAVSECGRU FOIA; ESCR 200-40, *The CRITIC Test and Evaluation Program*, 12 May 1987, AIA FOIA; Tad Szulc, 'The NSA-America's $10 Billion Frankenstein', *Penthouse*, Nov. 1975; 'US Electronic Espionage: A Memoir', *Ramparts*, Aug. 1972, p.43.
44. Paul Bracken, *The Command and Control of Nuclear Forces* (New Haven, CT: Yale UP 1983) p.26; Bruce D. Berkowitz and Allan E. Goodman, *Strategic Intelligence for American National Security* (Princeton UP 1989) p.35.
45. *CIA: The Pike Report* (London: Spokesman Books 1977) p.140; Roy Godson (ed.) *Intelligence Requirements for the 1980s: Clandestine Collection* (Washington DC: National Strategy Information Center 1982) p.119.
46. *Report to the President'as Foreign Intelligence Advisory Board on Intelligence Community Activities Relating to the Cuban Arms Build-Up: 14 April through 14 October 1962*, undated, p.24, National Security Files: Countries: Cuba, Box 61, Kennedy Library, Boston, Massachussetts.
47. Lester W. Grau, *Road Warriors of the Hindu Kush: The Battle for the Lines of Communication in the Soviet-Afghan War* (Ft Leavenworth, KS: Foreign Military Studies Office, Aug.1996) located at http://call.army.mil/call/fmso/fmsopubs/issues/roadwar/road war.htm.
48. *Annual Historical Report US Army Security Agency FY 1964*, p.72, INSCOM FOIA and Confidential interviews.
49. Vladislav M. Zubok, 'Spy vs. Spy: The KGB vs. the CIA, 1960-1962', *Cold War International History Project Bulletin*, No. 4, Fall 1994, pp.22–33.
50. Raymond Garthoff and Amy Knight, 'New Evidence on Soviet Intelligence: The KGB's 1967 Annual Report', *Cold War International History Project Bulletin*, No. 10, March 1998, p.214.
51. Department of the Army, DA Field Manual 34-40-12, *Morse Code Intercept Operations*, 26 Aug. 1991, p.2-4, INSCOM FOIA; US Marine Corps, MCWP 2-15.2, *Signals Intelligence*, June 1999, p.1-5; SSgt Regina Mason, 'Flight Commanders: Moral and Mission Are Key Responsibilities', *Spokesman*, Jan. 1991, p.9.
52. US House of Representatives, Permanent Select Committee on Intelligence, Report 105 508, *Intelligence Authorization Act for Fiscal Year 1999*, 105th Congress, 2nd Session, 5 May 1998, p.10.
53. See the case studies detailed in USAFSS Historical Office, *A Special Historical Study of USAFSS Response to World Crises, 1949–1969*, 22 April 1970, AIA FOIA.
54. Angello Codevilla, *Informing Statecraft: Intelligence for a New Century* (NY: The Free Press 1992) pp.14–15.
55. David Kahn, 'Cryptology', *The Encyclopedia Americana*, 1987, Vol.8, p.276.
56. David A. Fulghum, 'Sigint Aircraft May Face Obsolescence in Five Years', *Aviation Week & Space Technology*, 21 Oct. 1996, p.54.
57. A discussion of this contentious issue can be found in Godson, *Intelligence Requirements for the 1980s* (note 45) p.119.
58. 'Address at the CIRA Luncheon, 5 October 1998; John Millis' Speechs', *CIRA Newsletter*, Winter 1998/1999, p.6.
59. Rosenau, 'A Deafening Silence' (note 3) p.726.
60. Confidential information.
61. US Defense Intelligence Agency Testimony to US Senate Select Committee on Intelligence, 'Worldwide Threat to US National Security', Aug. 1996, located at www.securitymanagement.com/library/000255.html.

62. Peter Conradi, 'Stasi Phone Taps Sound Alarm in Kohl's Inner Circle', *Sunday Times*, 9 April 2000, p.26; Roger Boyes, 'Stasi's Spies Bugged 100 Kohl Phone Lines', *London Sunday Times*, 29 June 2000. On the East German Sigint service in general, see Ben B. Fischer, 'One of the Biggest Ears in the World: East German Sigint', *International Journal of Intelligence and Counterintelligence* 12/2 (Summer 1998) pp.142–53.
63. Gene Poteat, 'Stealth, Countermeasures, and ELINT, 1960–1975', *Studies in Intelligence*, 42/1 (1998) p.52, CIA FOIA.
64. Mark Urban, *UK Eyes Alpha* (London: Faber 1996) p.5.
65. Press Release, 'GCHQ Accomodation Project Site Announced', 7 May 1999, located at www.fco.gov.uk/news/newstext.asp?2391.
66. CM 4897, Intelligence and Security Committee, *Annual Report 1999–2000*, 2 Nov. 2000, www.official-documents.co.uk/document/cm48/4897/4897-02.htm.
67. Rudner contribution, p.103.
68. Stuart Farson, 'Accountable and Prepared? Reorganizing Canada's Intelligence Community for the 21st Century', *Canadian Foreign Policy* 1/3 (Fall 1993) p.49.
69. Alexander Zervoudakis, 'Nihil Mirare, Nihil Contemptare, Omni Intelligere: Franco-Vietnamese Intelligence in Indochina, 1950–1954', *Intelligence and National Security* 13/1 (Spring 1998) pp.203–25.
70. Wiebes contribution, p.243.
71. Horgan, *Signals Intelligence Support to US Military Commanders* (note 42) p.84.
72. Memo, Col. T.F. Van Natta, Assistant Chief of Staff, G2, Eighth US Army to Assistant Chief of Staff, G-2, Department of the Army, *Clearance for Access to Communications Intelligence*, 24 Oct. 1952, RG-338 Records of GHQ Far East Command, Entry 34015A AC of S, G-2 Executive (Coordination) Division, Box 70, File: 333.5, NA, CP.
73. Oliver Kirby, 'Louis Tordella Led by Example', *Cryptolog*, Spring 1996, p. 8.
74. *Agenda Prepared by Army Field Forces Observer Team No. 5*, August 1951, p. 167, MISC 333 AFF-FECOM, OCMH, Washington DC; Memo for the NSA/CSS Representative Defense, *NSA Transition Book for the Department of Defense*, 9 Dec. 1992, p.2. The author is grateful to Dr. Jeffrey T. Richelson for making a copy of this document available. Ed Evanhoe, *Dark Moon: Eighth Army Special Operations in the Korean War* (Annapolis MD: Naval Institute Press 1995) pp.86–7; Private information.
75. Eduard Mark, *Aerial Interdiction in Three Wars* (Washington DC: Center for Air Force History 1994) p.277.
76. US Naval Security Group, *US Naval Communications Supplementary Activities in the Korean War: June 1950–August 1953*, p.80, COMNAVSECGRU FOIA; Confidential interview.
77. Marshall L. Michel III, *Clashes: Air Combat Over North Vietnam, 1965–1972* (Annapolis, MD: Naval Institute Press 1997) pp.114–15.
78. 'Good Times With Bad-Eye: The Adventures of John 'Bad-Eye' Martin, undated, located at www.vietvet.org/badeye.htm.
79. Joan Edwards, 'Reagan's Charges "Total Untruths", ex-CIA Man Says', *Toronto Globe and Mail*, 29 June 1984, p.12.
80. Victor Sheymov, *Tower of Secrets* (Annapolis, MD: Naval Institute Press 199?) p.15.
81. Aldrich contribution, p.88.
82. Marian Wilkinson, 'Our Spies Knew Balibo Five at Risk', *Sydney Morning Herald*, 13 July 2000, located at www.smh.com.au/news/0007/13/pageone/pageone13.html
83. Horgan (note 42) p.85.
84. See for example Memo, Tarkenton to Assistant Chief of Staff, G2, I Corps *et al.*, *Classified Information for Limited Use*, 30 Dec. 1950, RG-500 Records of Eighth US Army 1946–1956, AcofS, G-2, Box 53, File: Classified Information for Limited Use, NA, CP.
85. The NSA COMINT report language is taken from NSA, 3/0/__/R26-63, *COMINT Report*, 23 Nov. 1963 2103Z, JFK Assassination Files.
86. Memo, Belmont to Boardman, 1 Feb. 1956, pp.1, 9, FBI Venona Files; SRH-123, *Brownell Committee Report*, pp.103–4, 108, RG-457, NA, CP; Dr Ray S. Cline, *The CIA Under Reagan, Bush and Casey* (Washington DC: Acropolis Books 1981) p.130; Woodrow J. Kuhns (ed.) *Assessing the Soviet Threat: The Early Cold War Years* (Washington DC:

Center for the Study of Intelligence: 1997) p.5; Russell Jack Smith, *The Unknown CIA: My Three Decades with the Agency* (NY: Berkley Books 1989) p.42; RADM Tom Brooks, 'The Y1 Story: Continued', *Naval Intelligence Professionals Quarterly* (Winter 1999) p.3.
87. Horgan (note 42) p.86.
88. 'COMINT and the PRC Intervention in the Korean War', *Cryptologic Quarterly*, date unknown, p.7, NSA FOIA.
89. Faligot contribution, p.9.
90. Phillip Knightley, *The Second Oldest Profession* (NY: Norton 1986) p.376.
91. Godson (note 45) p.118.
92. NIE 11-3/8-76, *Soviet Forces for Intercontinental Conflict Through the Mid-1980s*, 1976, p.59. RG-263, NA, CP.
93. Robert M. Gates, *From the Shadows* (NY: Simon & Schuster 1996) pp.270–3; Confidential interview.
94. Gates, *From the Shadows* (note 93) p.272; Jeffrey T. Richelson, *A Century of Spies* (NY: Oxford University Press 1995) p.386; Christopher Andrew, *For the President's Eyes Only* (NY: HarperCollins 1995) p.475.
95. Raj Chengappa and Zahid Hussain, 'Bang for Bang', *India Today*, 8 June 1998, www.indiatoday.com/itoday/08061998/cover.html.
96. Percy Kemp, 'The Fall and Rise of France's Spymasters', *Intelligence and National Security* 9/1 (Jan. 1994) p.20.
97. Tony Geraghty, *BRIXMIS* (London: HarperCollins 1997) p.284.
98. 'COMINT and the PRC Intervention in the Korean War', *Cryptologic Quarterly*, date unknown, p.10, NSA FOIA.
99. Captain James D. Pearson, 'On Going Tactical', *The Hallmark*, June 1976, p.2.
100. Memo, Ladd to Director, 28 Feb. 1951, p.1, FBI Venona Files.
101. Bob Drogin, 'Crash Jolts US e-Spy Agency', *Los Angeles Times*, 21 March 2000, p.1.
102. Wiebes contribution, p.274.
103. Horgan (note 42) pp.86–7.
104. Ibid. p.85.
105. *CIA: The Pike Report* (London: Spokesman Books 1977) p.140.
106. US House of Representatives, Select Committee on Intelligence, *US Intelligence Agencies and Activities: The Performance of the Intelligence Community*, 94th Congress, 1st Session, 1975, Part 2, pp.658–9, 680–1; *The Pike Report* (note 105) pp.143, 147.
107. Naval Research Laboratory, NRL Press Release 32-00r, 'A Tribute to the Father of US Electronic Warfare: Howard Otto Lorenzen', 22 May 2000, located at www.pao.nrl.navy.mil/rel-00/32-00r.html.
108. 'Intelligible Intelligence: An Alchemy of Collation and Coordination', *The Times of India*, 21 Sept. 2000.
109. Faligot contribution, p.188.
110. Horgan (note 42) p.87.
111. DA Pamphlet No. 30-2, *The Soviet Army*, July 1949, p.41, RG-6, Box 107, MacArthur Memorial Library (MACL); SRH-123, *Brownell Committee Report*, p.29, RG-457, NA, CP; SRH-277, *A Lecture on Communications Intelligence by Rear Admiral E.E. Stone, DIRAFSA*, 5 June 1951, p.34, RG-457, NA, CP; S/ARU/C735, *Developments in Soviet Cypher [sic] and Signals Security, 1946–1948*, Dec. 1948, RG-38, Translations of Intercepted Enemy Radio Traffic, 1940–1946, Box 2739, NA, CP; David E. Murphy, Sergei A. Kondrashev and George Bailey, *Battleground Berlin* (New Haven, CT: Yale UP 1997) p.208.
112. US Marine Corps, MCWP 2-15.2, *Signals Intelligence*, June 1999, p.1-5.
113. Ramesh Vinayak, 'Wireless Wars', *India Today*, 14 Sept. 1998, www.india-today.com/itoday/14091998/war.html.
114. Vice Admiral B.R. Inman, 'The NSA Perspective on Telecommunications Protection in the Nongovernmental Sector', *Cryptologic Spectrum* (Summer 1979) p.5.
115. National Cryptologic School, *On Watch* (Ft Meade, MD: NSA/CSS, Sept. 1986) pp.19–20, NSA FOIA; TI Item #137, NT-1 Traffic Intelligence, *Unprecedented Coordinated Russian Communications Changes*, 4 Nov. 1948, RG-38, Box 2742, NA, CP.

116. US House of Representatives, Committee on Un-American Activities, *Security Practices in the National Security Agency*, 88th Congress, 2nd Session, 1962, pp.7–9; Harry Howe Ransom, *The Intelligence Establishment* (Cambridge, MA: Harvard UP 1970) p.129; Richard J. Aldrich, *Espionage, Security and Intelligence in Britain: 1945–1970* (Manchester UP 1998) pp.57–8; Christopher Andrew, *The Sword and the Shield* (NY: Basic Books 1999) p.179; Confidential interview.
117. Seymour M. Hersh, *The Price of Power: Kissinger in the Nixon White House* (NY: Summit Books 1983) pp.73–4.
118. Faligot contribution, pp.194–5.
119. Mark Bowden, 'Killing Pablo: Martinez Pushes Ahead with the Hunt', *Philadelphia Inquirer*, 3 Dec. 2000, located at www.killingpablo.com/content/killingpablo/philly/1047343243.htm.
120. For one unsubstantiated allegation of Soviet communications deception, see Edward Jay Epstein, *Deception: The Invisible War Between the KGB and the CIA* (NY: Simon & Schuster 1989) pp.162–76.
121. Hersh, *The Price of Power* (note 117) p.207.
122. These NSA intercepts were highly classified and given very limited distribution, with copies going only to National Security Advisor Robert McFarlane, Lt. Col. Oliver North, CIA Director William Casey, and Vice Admiral Arthur S. Moreau Jr, assistant to the Chairman of the Joint Chiefs of Staff. Lawrence E. Walsh, *Iran-Contra: The Final Report* (NY: Times Books 1994) pp.3, 207; Stephen Engelberg, '3 Agencies Said to Have Received Data About Iran Money Transfers', *New York Times*, 27 Nov. 1986, p.A1.
123. Mats R. Berdal, *The United States, Norway and the Cold War 1954–60* (London: Macmillan Press 1997) p.29.
124. Memo, Director of Central Intelligence to Executive Secretary, National Security Council, *Potential Soviet Attack on Our Essential Communications Systems and Organizations*, 25 Jan. 1951, RG-59, State Department Participation in the National Security Council, Box 124, File: NSC – Miscellaneous, 1950–1952, NA, CP; Jack E. Ingram, 'The Origins of NSA', *American Intelligence Journal*, Spring/Summer 1994, p.40; Christopher Andrew, *For The President's Eyes Only* (NY: HarperCollins 1995) p.196.
125. Mark Bowden, 'A Rivalry Grows Between Spy Units', *Philadelphia Inquirer*, 27 Nov. 2000, located at www.killingpablo.com/content/killingpablo/philly/1047343226.htm.
126. Confidential interview.
127. Schmidt-Eenboom contribution, p.145 and *passim*.
128. Wiebes contribution, *passim*.
129. N.C. Gerson, 'Collaboration in Sigint; Canada – U.S.', *NCVA Cryptolog*, Spring 1999, p.2.
130. US Marine Corps, MCWP 2-15.2, *Signals Intelligence*, June 1999, p.1-5.
131. DoD Directive C-3222.5, *Electromagnetic Compatibility (EMC) Management Program for Sigint Sites*, 22 April 1987, DOD FOIA.
132. Gerson, 'Collaboration in Sigint' (note 129) p.2.

2
The National Security Agency and the Cold War

MATTHEW M. AID

Since its formation in November 1952, the National Security Agency (NSA) has managed and directed all US government Signals Intelligence (Sigint) collection and processing activities. Today, NSA is the largest and arguably the most powerful intelligence organization in the world. It is the sole collector and processor of Communications Intelligence (Comint), the primary processor of Foreign Instrumentation Signals Intelligence (Fisint), and the coordinator of the US government's national Electronics Intelligence (Elint) program since 1958. Per its charter, NSA does not produce finished intelligence reports. This is the responsibility of NSA's consumers within the US intelligence community. Since its inception, NSA has also been responsible for overseeing the security of the US government's communications and data processing systems (referred to within NSA as Information Security or Infosec), and since the mid-1980s NSA has also managed the US government's Operations Security (Opsec) program.[1]

NSA's principal consumers are: the White House, the National Security Council, the Secretary of Defense, the Secretary of State, the Secretary of the Treasury, the Secretary of Energy, the Secretary of Commerce, the Federal Bureau of Investigation, the CIA, the Defense Intelligence Agency (DIA), the Joint Chiefs of Staff, the three service intelligence agencies, all unified and specified military commands, military operational commanders (as required), and the intelligence agencies of collaborating foreign nations.[2]

Although it is currently impossible because of secrecy constraints to empirically demonstrate how important NSA's role was during the Cold War, comments by former senior American intelligence officials suggest that the value of the Agency's Sigint product was great. For example, former NSA Director Vice Admiral Bobby Ray Inman stated in 1979 that 'In the signals intelligence field, [NSA] has, since its inception in 1952, provided a vast quantity of intelligence information of inestimable value in the conduct of the Nation's defense and foreign policy.'[3]

The secrecy that historically has cloaked the subject of post-World War II Sigint is the hardest obstacle for the historian to overcome. In 1995, the Canadian historian John Ferris wrote 'We know very little about American signals intelligence. Because its product was not circulated widely or referred to explicitly in intelligence summaries, its significance will be unusually hard to trace.' Nonetheless, Ferris added, 'The government may well be able to keep most of the evidence from the public record if it wishes to do so.'[4]

But American intelligence officials have long known that it was inevitable that this curtain of secrecy would eventually have to come down. In a 1946 Top Secret memorandum, one of the founding fathers of NSA, Dr Solomon Kullback, wrote 'It is obvious that we cannot hope to keep the cryptologic achievements of the United States and Great Britain completely secret for an indefinite or large number of years.'[5]

The late Dr Kullback was, of course, correct. As will be detailed below, interviews with former NSA officials and newly released documents reveal for the first time that by the time the Cold War came to an abrupt end with the fall of the Berlin Wall in 1989, Sigint had achieved a preeminent status within the American intelligence community, supplanting (if not actually submerging) the more traditional manpower intensive intelligence sources, such as spies, agents, defectors, diplomats, and low-level military Human Intelligence (Humint) sources, which former NSA director Vice Admiral William O. Studeman obliquely referred to as 'historically less productive intelligence means'.[6]

The reason for Sigint's rise to the top of the American intelligence 'food chain' was simple. Because during the 45 years of the Cold War in Europe, as will be detailed below, Sigint consistently produced where other intelligence sources failed. In fact, it can be argued that Sigint was not only the most important source of intelligence – it was practically the only reliable intelligence source for the United States and its Western European allies about what was going on behind the Iron Curtain. NSA accomplishments, most of which remain classified and buried in NSA's storage vaults, were arguably the most impressive of any American intelligence organization during the Cold War. Former NSA director, Vice Admiral J.M. 'Mike' McConnell, has recently compared NSA's Cold War accomplishments with those achieved during World War II, writing that NSA's 'Sigint successes during the Cold War were no less significant in terms of gravity and magnitude for changing world history and for protecting the interests of the United States, our allies, and democracy.'[7]

THE SOVIET TARGET

This contribution focuses on NSA's work against its main target during the Cold War, the Soviet Union. But it is important to note that the Soviet Union was not NSA's sole target. For example, as of 1949 American and British codebreakers were reading the diplomatic traffic of 52 countries, and were probably reading the communications traffic of at least ten other nations.[8]

But the Soviet Union was by far NSA's largest and most important target during the 45 years of the Cold War in Europe, and as such, the vast majority of NSA's collection resources were devoted to monitoring the Soviet Union and its allies in Eastern Europe. As of June 1949, 71 per cent of all American Comint intercept personnel and 60 per cent of all American Comint processing personnel were working on the 'Soviet Problem', and yet the US was only able to intercept and process a very small portion of the Soviet radio traffic then coursing through the airwaves.[9] This level of effort against the Soviet Union remained relatively constant throughout the Cold War. During the 1970s, NSA's Sigint collection resources dedicated to monitoring the Soviet Union and Eastern Europe closely matched the rest of the US intelligence community. As of 1975, the US intelligence community's target coverage was as follows: the USSR and Eastern Europe: 65 per cent, Asia: 25 per cent, the Middle East: 7 per cent, Latin America: 2 per cent; and the rest of the world: 1 per cent.[10] As of 1980, almost 60 per cent of NSA's collection resources were still dedicated to covering the Soviet Union and its Eastern European allies. By 1989, the year that the Berlin Wall fell, more than 50 per cent of the Agency's assets were still devoted to coverage of the Soviets.[11]

As one can readily imagine, given its sheer size, trying to spy on the Soviet Union during the Cold War was a monumental task. The Soviet Union was the largest country in the world, covering an area of 6.65 million square miles, which made it almost three times the size of the United States. It occupied 17 per cent of the world's land mass, stretching across two continents and 11 time zones, which came to a distance of about 6,000 miles running from east to west. Most of the Soviet Union lay north of the 49th Parallel (i.e. north of Vancouver, Canada). The Soviet Union's two largest cities, Moscow and Leningrad, were on the same latitude as Edmonton, Canada and Anchorage, Alaska respectively.[12]

The second major obstacle was that obtaining reliable intelligence information using traditional human intelligence (Humint) means inside the Soviet Union and Eastern Europe was an extremely difficult proposition. The Soviet Union was hermetically sealed from the outside world, and paranoia was the norm in Stalin's Russia. The Russian government strictly limited travel visas to the Soviet Union to a very small number of foreign diplomats and a few reporters and officially approved visitors. Photography of anything but the most mundane subject inside the Soviet Union was strictly prohibited. Some 220,000 KGB border guards made it extremely difficult to sneak in or out of the Soviet Union. Travel inside the Soviet Union was severely restricted for both Russians and foreigners by an internal passport system, which coupled with the KGB's large and extremely efficient counterintelligence service and a massive network of informers at all levels of Russian society, meant that operating agent networks in the country was virtually impossible. Foreigners were followed by the KGB whenever they ventured onto the streets of Moscow or other Eastern European cities, their phones were tapped, their mail read, their apartments and cars were bugged, and those 'normal' Russians that they spoke to were closely questioned by the KGB. Because of the extreme difficulty of conducting covert Humint operations in this region of the world, the US intelligence community came to refer to the Soviet Union and Eastern Europe as 'denied areas'.[13]

The historical record, such as it exists today, clearly shows that throughout the Cold War, the spies of the CIA and its European partners produced relatively little in the way of hard intelligence about what was going on inside the Soviet Union.

For example, conveniently forgotten in the recent historiography produced by CIA historians is the fact that repeated attempts by the CIA and its European allies to infiltrate agents into the Soviet Union in the late 1940s and early 1950s were a complete failure. Between 1950 and 1954, the Norwegian Intelligence Service infiltrated a small number of agents from neutral Finland into the strategically important Kola Peninsula in northern Russia. But these missions yielded little intelligence.[14]

Attempts by the CIA, MI6 and the West German Gehlen Organization to infiltrate agents by air and sea into the Soviet-controlled Baltic states of Latvia, Lithuania and Estonia between 1946 and 1954 were smashed by the KGB. Newly available KGB internal historical documents reveal that the Soviets compromised these operations at a very early stage, luring dozens of American, British and Swedish agents into the hands of the KGB. More than 30 of the British agents alone were lost during the operation, and those operatives who survived were almost all captured and 'turned' by the KGB.[15]

Between September 1949 and the end of 1954, the CIA, MI6 and the Gehlen Organization parachuted some 150 operatives into the Ukraine, Belorussia and Moldavia to try and establish agent networks that could provide early warning of a Soviet attack on Western Europe, as well as support anti-Communist Ukrainian guerrilla forces operating in the region. None of these operatives ever came back from their mission. Most were killed or disappeared without a trace shortly after landing. KGB documents confirm that those agents that did survive were captured by the KGB and 'played back', feeding the CIA and MI6 with false or misleading information.[16]

CIA attempts to infiltrate agents into the southern Soviet Union from Turkey during the early 1950s were, according to a former CIA officer, 'spectacularly unsuccessful'.[17] Starting in the early 1950s, Britain's MI6 parachuted dozens of agents trained on a Turkish island into the Soviet Caucasus. The operation was terminated in 1956 because of a 100 per cent loss rate among the agents.[18]

The level of frustration felt by CIA and MI6 officers about their inability to infiltrate the Soviet Union was enormous. A CIA historian has only recently admitted that during the early postwar years 'the intelligence collection arm of the CIA was finding it impossibly difficult to penetrate Stalin's paranoid police state with agents'.[19] In 1952 a CIA 'Murder Board' reviewed the Humint operations in the Soviet Union and Eastern Europe, and concluded that 'Our operations have failed and there is no alternative to offer.'[20] Peer de Silva, who was a senior official in the CIA's Soviet Russian Division until 1955, described the parachute insertion program as 'wasteful and, in many ways, tragic'. He added that the operation only produced 'calamitous results'.[21] Anthony Cavendish, who ran many of the agent parachute operations into the Soviet Baltic states for MI6, wrote that 'The operations must now seem nothing more than a catalogue of disasters.'[22]

There is no question that the CIA's early Humint efforts against the Soviet Union were poorly managed and largely ineffectual. A Top Secret 1955 report to President Dwight D. Eisenhower bluntly stated that 'we obtain little significant information from classical covert operations inside Russia'.[23] Dr George Kistiakowsky, President Dwight D. Eisenhower's science advisor and a key player in postwar US intelligence, characterized the American clandestine Humint effort against the Soviet Union in the postwar years as '[P]itiful. The time of Mata Hari had passed.'[24]

With no other sources available (photo reconnaissance satellites would not become available until the early 1960s), out of sheer necessity the US intelligence community had to rely to a very large degree on Sigint because at the time, it was the only reliable source producing intelligence information about what was going on behind the Iron Curtain, particularly for intelligence on Soviet military strength and capabilities.[25]

SIGINT IN THE EARLY COLD WAR YEARS

On the day that Japan surrendered in September 1945, the combined US Army and Navy Comint organizations, the Signal Security Agency (SSA) and the Naval Communications Intelligence Organization (OP-20-G), consisted of more than 36,000 military and civilian personnel manning 37 strategic listening posts and direction finding stations and dozens of tactical mobile Comint intercept units that were equipped with more than 2,500 radio intercept receivers, although most of these assets were in the Pacific where they had been targeted against Japan.[26]

The end of World War II led to a restructuring of the Army and Navy cryptologic organizations. On 15 September 1945, the SSA was redesignated as the Army Security Agency (ASA), which unlike its predecessor was given complete control over all US Army Comint activities.[27] On 10 July 1946, OP-20-G was deactivated, and all US Navy Comint intercept and processing units around the world were merged into a new organization headquartered in Washington DC called the Communications Supplementary Activities (CSA).[28]

Postwar demobilization and deep cuts in the US defense budget badly hurt the Army and Navy Comint organizations. By the summer of 1946, ASA's personnel strength had dropped from 27,000 at the end of World War II to less than 4,000 military and civilian personnel.[29] CSA's manpower strength fell from 9,100 in September 1945 to about 3,000 military and civilian personnel by mid-1946.[30] In the late 1940s, two more American cryptologic organizations were created. On 20 October 1948, a third American Comint organization was created when the US Air Force activated its own Comint organization, the US Air Force Security Service (USAFSS).[31] By the end of 1949, USAFSS consisted of a 150-man headquarters staff in San Antonio, Texas, plus 600 personnel assigned to five field units.[32]

On 20 May 1949, Secretary of Defense Louis Johnson issued a directive creating the Armed Forces Security Agency (AFSA), under which was placed the responsibility for the direction and control of all US Communications Intelligence (Comint) and Communications Security (Comsec) activities *except* for tactical cryptologic activities, which remained under the control of the Army, Navy and Air Force.[33] AFSA's headquarters staff in Washington DC consisted of slightly more than 4,500 personnel (1,240 military and 3,279 civilians). Almost all of AFSA's headquarters staff would come from the Army and the Navy, with the Navy providing the majority of AFSA's military personnel, and the Army contributed the majority of AFSA's civilian staff.[34]

By the time of the Korean War began in June 1950, the US Sigint community had rebounded somewhat, with AFSA and the three Service Cryptologic Agencies (SCAs) together consisting of about 16,000 military and civilian personnel.[35]

Geographical factors made the task of intercepting Russian radio traffic far more difficult than German and Japanese communications during World

War II. This meant that the US and Great Britain had to reformulate the agreements signed during World War II in order to reflect the difficulties associated with collecting intelligence on the new Soviet target. On 5 March 1946, a Top Secret 25-page document called the British-United States Communications Intelligence Agreement (BRUSA) was signed, which renewed the 1943 bilateral agreement between the two countries to cooperate in the field of Comint.[36] In June 1948, the US and Great Britain signed a new agreement, called the UK-USA Communications Intelligence Agreement (UKUSA), which in essence delineated the division of Comint effort by the parties to the agreement against the Soviet Union and its communist allies. Under the terms of the UKUSA agreement, scarce Comint intercept and processing resources were allocated and targets carefully chosen by the US, Britain, and key member nations of the British Commonwealth, specifically Canada, Australia and New Zealand.[37]

Throughout the Cold War, the relationship between NSA and GCHQ, which was the core of the UKUSA Agreement, was constantly evolving with the time. And yet, there was a fundamental doctrine which bound the two organizations together despite the political, military and economic sea changes that marked the relationship between the US and Great Britain. The man whom many NSA officials credit with being the 'father' of the UKUSA, former NSA deputy director Dr Louis Tordella, believed that NSA 'had a partnership with GCHQ and that we shared tasking and operations but could function independently of one another'. According to Tordella, 'We are each mindful of our national interests but can still find common ground.'[38]

Recently declassified decrypts and materials found by Western researchers in the KGB archives in Moscow show that the UKUSA partners were remarkably successful in their early efforts against the codes and ciphers of the Soviet Union and its allies. These new materials reveal that despite the fact that the Americans and British cryptanalysts apparently could not solve most of the high-level Soviet diplomatic ciphers in use during the late 1940s, they were getting a great deal of good intelligence information from intercepted Soviet communications about the strength and capabilities of the Soviet armed forces, the production capacity of Soviet industry, and even some useful intelligence about Soviet work in the field of atomic energy.[39]

Declassified documents show that the Americans and British were able to collect this information because following the end of World War II, the Soviet military continued to use a large number of foreign-made cryptographic systems, which the Russians had either acquired under Lend Lease from the US and Britain, or which they had captured from the Germans during World War II. It was the Russian military's use of these systems that the American and British cryptanalysts initially exploited.[40]

The first Russian machine cipher system that was solved was a Russian version of the Hagelin B211 cipher machine, which presumably had been captured by the Russians from German forces during World War II. By September 1945, American cryptanalysts had reconstructed the internal wiring of the machine's rotors, and a copy of the machine was made, which was called Sauterne Mark I. Beginning on 4 April 1946, a regular supply of Sauterne decrypts began being produced by the US Navy cryptanalysts.[41] In late 1945, ASA cryptanalysts began attacking a more sophisticated Enigma-type cipher machine system used by the Russian Army, which was given the codename 'Albatross'. According to available information, it took ASA's cryptanalysts years to figure out the internal workings of the machine, and it was not until the late 1940s that 'Albatross' was reportedly solved.[42]

In March 1946, the British began sending to ASA and CSA intelligence information derived from their solution of another Russian cipher machine system that the British called Coleridge, which was used on Soviet Army, Navy and Air Force mainline teletype nets. The Coleridge decrypts generated important information about the Soviet military's order of battle, training activities, and logistics matters.[43]

The Russian Army cipher system called RUMUA-1 by the American and British cryptanalysts was solved in the fall of 1948. ASA engineers managed to reconstruct the RUMUA cipher machine, which was known as the RUMUA Tan Analogue. Other recently declassified documents revealed that in October 1949, the ASA scored what was then deemed to be a significant cryptologic success against a Soviet military cipher system being used in the Far East, but no further details have been released concerning this particular success.[44]

In 1947, ASA cryptanalysts solved the RUARA-1 operational cipher system which was then being used by the Soviet Air Force headquarters in Moscow to communicate with its subordinate commands throughout the Soviet Union and Eastern Europe.[45] By 1948, US Army cryptanalysts in Japan had solved the operational ciphers used by the Soviet 9th and 10th Air Armies at Vozdvizhenka and Khabarovsk respectively in the Soviet Far East.[46]

Some significant cryptanalytic successes were also achieved against Soviet naval cipher systems. In late 1945, CSA and GCHQ were trying to solve three Soviet naval code and cipher systems, known as RUNLB-1, RUNLC-1, and RUNMD-1.[47] CSA's cryptanalysts solved the cipher system used by Soviet Navy headquarters in Moscow in September 1948, followed in November 1948 by the solution of some of the cryptographic systems used by the Russian Navy in the Far East. There was also some success in reading the cipher systems used by the two fleets operating in the Baltic Sea, as well as those cryptographic systems used by the Black Sea Fleet and the Caspian Sea Flotilla.[48]

American and British cryptanalysts also managed to solve a significant number of high-level Eastern European diplomatic, military, intelligence and internal security systems during the early days of the Cold War. Decrypts of Eastern European diplomatic traffic revealed the details of the complicated and sometimes strained relationship between the Soviet Union and its Eastern European allies, such as the degree of Soviet influence over the governments of Eastern Europe, the military and political response of the Soviet Union to the formation on NATO, and the political reaction of the Eastern European states to the withdrawal of Marshal Tito's Yugoslavia from the Cominform in 1948.[49]

Many (but not all) of these Anglo-American cryptologic successes came to an abrupt end on the morning of 29 October 1948, when the Russians executed a massive worldwide change of their cryptographic systems and communications operating procedures. This day is still known as 'Black Friday' within NSA. This move by the Russians was prompted by the treachery of William W. Weisband, a Russian linguist and cryptanalyst working inside the Army Security Agency, who sold to the Russians information about ASA's codebreaking efforts. By July 1949, Weisband was able to tell his KGB handler in Washington that ASA had 'suddenly lost the ability to read Russian cables, and the head of the [ASA] deciphering service was concerned that a Communist might have penetrated its ranks'.[50]

Despite the Black Friday disaster, in the decades that followed NSA and its UKUSA partners not only survived, but prospered. As will be detailed below, perseverance and adaptation are recurring themes in the story of Sigint during the Cold War.

SIGINT IN THE 1950S

Starting with the North Korean invasion of South Korea on 25 June 1950, the US Sigint System (USSS) underwent dramatic change and growth as the Cold War heated up.

By 1951, American military and civilian intelligence officials had become extremely unhappy about the performance of AFSA and American Sigint during the Korean War. For example, in mid-1951 the commander of US forces in Far East, General Matthew B. Ridgway, was furious about the poor quality of the Comint support that he was receiving in Korea.[51] The director of the CIA, General Walter Bedell Smith and his deputies were also extremely unhappy with the condition of the American Comint community, warning that 'existing means of control over, and coordination of, the collection and processing of Communications Intelligence have proved ineffective'.[52]

On 24 October 1952 President Harry S. Truman signed a Top Secret eight-page directive entitled 'Communications Intelligence Activities', which made Comint a national responsibility, and directed that all American Comint activities 'must be so organized and managed as to exploit to the maximum the available resources of all participating departments and agencies and to satisfy the legitimate intelligence requirements of all such departments and agencies.' The directive abolished AFSA and transferred all of its responsibilities and resources to the newly created National Security Agency (NSA).[53]

The Korean War rejuvenated the American Sigint effort. By late 1952, NSA consisted of more than 30,000 military and civilian personnel.[54] By 1960, eight years after the Agency had been created, NSA had more than doubled in size to approximately 65,000 military and civilian personnel, of whom about 10,000 worked at NSA headquarters at Fort Meade, Maryland.[55]

Between 1950 and 1960, NSA constructed at a cost of hundreds of millions of dollars a multi-layered network of 70 strategic intercept stations and an equal number of tactical Comint units around the world. This network of American Sigint intercept stations, together with the 35 or so radio intercept stations operated by Britain, Canada, Australia and New Zealand, stretched completely around the periphery of the Soviet Union, Eastern Europe, the People's Republic of China and North Korea, and was able to copy all important shortwave radio traffic coming from inside these countries.

Add to this mix the not inconsiderable Sigint collection resources of NSA's Third Party partners in Norway, Denmark, West Germany, Austria, Italy, Greece, Turkey, Pakistan, Thailand, Nationalist China, Japan, and South Korea, and one can make the argument that the UKUSA partners possessed the largest and geographically best situated Sigint collection system in the world, even if the Soviet Sigint system was larger in terms of personnel.

NSA's success rate built slowly during the 1950s as the agency grew and became more capable, but the learning process was slow and at times painful. A former senior NSA official recalled that 'We [NSA] were on probation in those early years. We were constantly being examined and reexamined. We owed our success in part to [President Dwight D.] Eisenhower, who remembered how important Comint had been to him during the war, and wasn't going to jeopardize that.'[56] Despite its slow start, the US intelligence community's dependence on the Sigint product coming out of NSA was admitted by virtually every senior American intelligence official who testified in 1952 before a review panel headed by New York attorney George Brownell, which was tasked with examining the American

Comint effort. According to the Brownell panel's final report, America's intelligence 'Mandarins' unanimously stated their belief that Comint was the United States' single most important source of intelligence information.[57] A Top Secret 1955 study commissioned by the White House found that Comint was 'the timeliest, most complete, and most reliable source of intelligence' available to the US intelligence community.[58]

Solving high-level Russian cipher traffic was an extremely difficult proposition during the 1950s. Russian radio operators made fewer mistakes than they had during the 1940s, and the Russian traffic became increasing more difficult to decipher as newer and more sophisticated encipherment techniques became available.[59] But there is some evidence that by the mid-1950s, NSA and its allies had solved some high-level Soviet machine cipher systems with the help of the huge computers in NSA's basement. A recently declassified NSA historical study revealed that NSA may have solved a high-level Soviet machine cipher system in 1954, which would have been the first significant American cryptanalytic success against Soviet cryptologic systems since 'Black Friday' on 29 October 1948.[60] By 1958, the West Germans had solved a Russian cryptographic system called PT-53, although it is not known how important this system was or who within the Russian military used it.[61] A former US Air Force cryptologist has indicated that by 1959, NSA had solved additional high-level Soviet cryptographic systems.[62]

Sigint was by far the best source for hard intelligence about the strength, capabilities and activities of the Soviet armed forces, which was the primary focus of the US intelligence effort during the 1950s.

Comint was one of the principal sources for information about Russian strategic bomber activities. Sigint helped debunk the myth that the Russians had more strategic bombers than the US by tracking the delivery flights of Russian bombers from the factories where they had been built to their operational bases throughout Russia. Sigint also revealed that the combat readiness levels of the Russian bomber force was low, that Soviet bombers did not have the range of their American counterparts, that the flying skills of the Soviet crews were deficient because of a lack of flying time, and that the Russians had no mid-air refueling capability, which made it virtually impossible for the Soviets to strike targets in the Continental United States.[63]

Sigint was virtually the only viable source for reliable intelligence about the strength and capabilities of the Soviet national air defense command, the Protivovozdushnaya Oborona Strany (PVO). Elint was the Strategic Air Command's principal source of hard intelligence information about the locations and detection capabilities of the early warning radars positioned around the periphery of the Soviet Union and Eastern Europe.[64] On 16 May 1955, a Secret directive entitled National Security Council Intelligence

Directive No. 17, *Electronics Intelligence*, was issued. NSCID 17 established for the first time a unified National Elint Program that brought together the disparate Elint collection and processing efforts of the three US military services and the CIA, and focused their efforts almost exclusively against the Soviet Union and its allies. In addition, the directive ordered the creation of a new national Elint processing and analysis organization called the National Technical Processing Center (NTPC), which was situated on the grounds of the Naval Security Station in Washington DC. NTPC was run by the US Air Force as Executive Agent for the Department of Defense, but was staffed by personnel from all three military services and the CIA.[65]

From Comint intercepts of PVO radio traffic, American intelligence analysts were able to accurately measure the alertness, timeliness and accuracy of the PVO's radars and its command and control system; the reaction times, weapons capabilities and altitude limitations of the Soviet air defense fighter force; as well as map the order of battle of the PVO, all of which was essential information for planning strategic bombing attacks on the Soviet Union.[66] A former US Air Force intercept operator who monitored PVO radar tracking communications has confirmed how important the intelligence information from this source was: 'During my tenure our main job was to identify Russian and Chinese strengths by determining where their bases were, how many and what type of aircraft they had, and how long it would take them to scramble and intercept a hostile [aircraft]. We did this by monitoring their radar stations.'[67]

Comint generated vitally important intelligence information on the movements and activities of Russian Army combat units in Eastern Europe and the Soviet Union; the results of Soviet nuclear weapons tests on Novaya Zemlya in the Soviet Arctic; and Soviet military logistics and transportation activities.[68]

Comint coverage of the radio traffic of the Soviet tactical air force (Frontal Aviation) allowed American intelligence analysts to ascertain by 1957 the location of the FA's 320 Soviet fighter and 61 bomber regiments and identify what kind of aircraft each regiment flew. Comint also tracked Soviet aircraft movements, followed FA training exercises, and determined the level of experience and capabilities of Soviet fighter and bomber crews. Comint coverage of the SAF also yielded information about Russian weapons production, capabilities of weapons systems, data on fuel consumption and storage levels, and other logistical information.[69]

Monitoring the activities of the Russian Navy was particularly difficult because at the time it rarely left home waters. One of the solutions was the use of US Navy submarines to perform reconnaissance missions designed to monitor Soviet naval activities, including extensive Sigint collection of Soviet naval radio traffic. In February 1957, the American diesel attack submarine USS *Tirante* (SS 420) commenced the first of a series of US Navy submarine reconnaissance patrols off the Soviet port of Murmansk, which duplicated similar operations then being undertaken by American submarines monitoring the activities of the Soviet Navy in the Pacific. In August 1957, for example, the attack submarine USS *Gudgeon* (SS 567) conducted a reconnaissance mission off Vladivostok, but was detected by Soviet naval forces and depth charged for 30 hours by Soviet naval forces until finally it was forced to surface because of a lack of oxygen. The Russians allowed the *Gudgeon* to flee the area rather than destroy it. In early 1958, the diesel attack submarine USS *Wahoo* (SS 565) was caught by Russian naval units inside their territorial waters in the Far East and forced to flee after being depth charged. On board the submarine was a young American naval officer who would later become the Chairman of the Joint Chiefs of Staff and then Ambassador to London, William J. Crowe Jr.[70]

The data obtained from NSA's 24 hour-a-day Comint monitoring of these Soviet military targets allowed American intelligence analysts to follow even minute changes in Soviet military strength and capabilities, as well as spot variances from the day-to-day norm which might be indicative of a potential Soviet invasion. NSA analysts watched for any sign of anomalous Soviet military behavior; including dramatic increases in the volume of Soviet military radio traffic; the changing of major code and cipher systems; the formation of new military units and headquarters; the 'fleshing out' of Red Army combat divisions; intensified or more realistic training exercises; large scale redeployment of combat forces and support units from the Russian hinterland to Eastern Europe or the border with China and North Korea; surge deployments of Soviet naval units and submarines from their home ports; the movement of strategic bombers and tactical air units to forward bases and dispersal fields in the Arctic; the activation of the Russian emergency command posts and related emergency communications networks; the withdrawal of Aeroflot civilian airliners from regularly scheduled service to augment the military air transport service; increased rail and commercial shipping traffic, and other logistical preparations for war.[71]

It is true that Sigint failed to provide any warning prior to the North Korean invasion of South Korea on 25 June 1950.[72] Sigint had detected Soviet military forces in the Far East having been placed on a higher degree of alert just prior to the North Korean invasion, including increased radio traffic among Soviet air, air defense, and submarine forces in the region.[73] After the invasion, Sigint monitoring of Soviet communications revealed that Russian forces in the Far East had been moved to a higher state of alert and that some 20,000 to 25,000 troops of the Russian 25th Army had been moved to the North Korean border. But Sigint found no further signs of any

Soviet military preparations indicative of intent to intervene in Korea.[74] Sigint also provided warnings of the Soviet intention to crush the popular uprising in Budapest, Hungary by military means in November 1956. The month before, October 1956, Comint intercepts tracked 40 Russian tank battalions converging on the Hungarian capital to crush the civilian uprising there.[75]

The introduction of the U-2 spyplane by the CIA in 1956 did not materially affect the importance of Comint within the US intelligence community. By the late 1950s, Sigint and U-2 imagery were deemed to be the most credible and productive sources of hard intelligence available to the West about what was going on inside the Soviet Union, accounting for 75 per cent of the information available about the Soviet military. And not surprisingly, Comint and U-2 imagery provided the bulk of the material used by the CIA in drawing up the Top Secret National Intelligence Estimates (NIE) on Soviet military capabilities that went to the White House and other high-level intelligence consumers in Washington.[76]

It should be noted, however, that the much-touted U-2 overflight program of the Soviet Union had its fair share of problems. First, it does not take a genius to realize that an average of five overflights a year between 1956 and 1960 of the Soviet Union was not nearly enough to adequately follow Soviet military and technological developments. Nor was the U-2 capable of providing any warning of an impending Soviet attack. Worse still, in order to protect the source of the U-2 imagery, CIA photo interpreters were ordered to convert the photos into detailed drawings before they could be distributed to analysts with no reference to what was the source of the information. As a result, many analysts openly questioned the veracity of the information contained in these drawings, with some analysts claiming that they were fabrications.[77]

Sigint also had non-military applications when it came to the Soviet target. For instance, the so-called Venona decrypts were a crucial weapon in allowing the FBI and its counterparts in Britain and Australia to identify and neutralize several Soviet spy rings during the 1950s, although it should be noted that only 15 of the 206 Soviet agents identified as having spied for the Soviet Union were ever prosecuted, in large part because the secrecy of these decrypts prevented them from being used in an American court of law.[78]

By 1954, NSA appears to have penetrated the communications system of the Soviet national civil defense network.[79] NSA was able to gather information about the performance of the Soviet economy, such as information about Soviet foreign trade, internal consumer goods policies, gold production, petroleum shipments, shipbuilding activities, as well as military and civilian aircraft production.[80] Comint was also used to track the movements of officials belonging to the Soviet ministry responsible for arms sales, thus giving advanced warning of where Russian weapons were about to be exported.[81]

From the 1950s onwards, the US and its UKUSA partners became increasingly dependent on clandestine means for many of their Sigint successes as cryptographic systems became increasingly difficult to solve by traditional cryptanalytic means. In 1953, a very small and highly secret unit was created within the CIA called Staff D (renamed Division D in the 1960s), whose mission was to steal foreign cryptographic materials, recruit foreign government code clerks, tap communications lines, and plant audio surveillance devices in the code rooms of foreign government offices overseas.[82] In the US, specially trained teams of FBI agents, all graduates of the FBI Laboratory's 'Sound School' at Quantico, Virginia, broke into foreign diplomatic establishments in Washington DC and New York City to obtain cryptologic material for NSA. In one instance, FBI agents drove a garbage truck into the central courtyard of the Czech embassy in Washington DC in the middle of the night and spirited away one of the embassy's cipher machines for study.[83] In 1953, MI6 created its own cryptologic unit, called Section Y, which was responsible for intercepting through surreptitious means Soviet and other nations' military and diplomatic communications, including the use of microphone eavesdropping devices planted in Soviet and Warsaw Pact embassies.[84] MI5, the British internal security service, also formed its own clandestine unit in the mid-1950s, which specialized in bugging foreign embassies in London.[85]

Few details of the successes of these units have made their way into the public realm. The joint CIA-MI6 Berlin Tunnel cable tapping operation, which ran from 1954 to 1956, produced reams of invaluable high-level intelligence information concerning the strength and capabilities of Soviet military forces in East Germany; data on the reduction in size of the Soviet armed forces and Soviet military cooperation with the other Eastern European nations; Soviet political intentions in Berlin; details about the Soviet Union's relationship with the newly created East German government; the first indications that East Germany was creating its own army; details of the Soviet Union's nuclear weapons program; and information about Soviet and East German intelligence operations.[86]

SIGINT IN THE 1960S

In many respects, during the 1960s NSA reached the peak of its power. The growth of NSA continued unabated during the 1960s, in large part because of the demands of the Vietnam War. By 1969, NSA and the other components of the US Sigint System (USSS) had grown to more than

95,000 military and civilian personnel, 15,000–16,000 of whom worked at NSA headquarters at Fort Meade, Maryland.[87]

The 1960s also marked the nadir of the American Humint program against the Soviet Union, and the ascendancy of high-tech spying to the top of the American intelligence community. During the 1960s, the CIA's Humint effort against the Soviet Union was decimated by a molehunt that devastated the Agency's operations inside Russia. By one account, as of 1968 the CIA had only five relatively low-level agents reporting from inside the Soviet Union.[88] For example, the CIA's Humint assets inside the Soviet Union, including Colonel Oleg Penkovskiy, failed to detect the Soviet military's movement of men and weaponry to Cuba prior to the 1962 Cuban Missile Crisis. Rather, these missiles were discovered by CIA operatives in Cuba well after the missiles had arrived in that country.[89] It should be noted that MI6's operations against the Soviet Union during the 1960s were also paralyzed by a massive internal molehunt. It was not until the 1971 recruitment of a Soviet trade official named Oleg Lyalin that MI6 once again began recruiting Russian agents.[90]

As the quantity and quality of intelligence produced by Humint sources about what was going on behind the Iron Curtain declined, American intelligence analysts out of necessity came to increasingly rely to an even greater degree on Sigint and the newly developed photo reconnaissance satellites that were being put into space beginning in the early 1960s by the newly created National Reconnaissance Office (NRO).[91]

There were three major problems, however, with the satellite reconnaissance photos that CIA historians have perhaps conveniently forgotten to mention in the recent spate of writings about the origins of the American reconnaissance satellite program: (1) the satellite photographs were so highly classified that only a few intelligence analysts were allowed access to them; (2) the CIA's reconnaissance satellites had an extremely limited focus, namely, Soviet strategic military and industrial targets; and (3) it usually took weeks from the time the photos were taken until the moment they made their way to the intelligence analysts, which meant that satellite imagery had limited utility during fast-moving crises. Not surprisingly, these factors severely limited the utility of the product coming from NRO's reconnaissance satellites. It was not until the late 1970s that the security classification of satellite imagery was downgraded and made more available to intelligence consumers.[92]

These limitations on the use of imagery obtained from the photo reconnaissance satellites further enhanced the importance of Sigint, which remained the premier source of timely and reliable intelligence available to the US intelligence community during the 1960s, despite its rapidly escalating cost in terms of both human and material resources.[93]

Sigint performed particularly well during a series of international crises involving the Soviet Union during the 1960s. In July 1962, NSA's Comint coverage of Soviet merchant marine communications traffic revealed a dramatic surge in the number of Soviet merchant ships arriving in Cuba with undisclosed cargos.[94] Two months later, in September 1962, the first hard indications of the presence of Soviet military forces in Cuba were detected by the NSA spyship USS *Oxford*.[95] Two months prior to the June 1967 Arab–Israeli War, Comint detected the movement of a division of Soviet airborne troops and their supporting transport aircraft from the Transcaucasus Military District in the Soviet Union to Bulgaria as well as other military preparations.[96]

During the spring and summer of 1968, Sigint detected the movement of 34 Russian combat divisions into Czechoslovakia or to positions along its border, the deployment of 400 Soviet military aircraft to airfields in Poland and East Germany, the call up thousands of reserve military personnel in the Baltic, Belorussian and Carpathian Military Districts in the western Soviet Union, and the activation of a new theater headquarters in the Carpathian Military District. On 20 August 1968, the Soviets invaded Czechoslovakia.[97]

During the August 1969 fighting between Soviet and Chinese forces along the Amur Ussuri Rivers, Sigint detected that Russian strategic bomber regiments deployed in Central Asia and the Far East had been placed on alert, as well as accelerated Soviet photo reconnaissance satellite activity, transport aircraft bringing additional Russian troops and supplies to reinforce Soviet forces in the Far East; the deployment of additional fighter and fighter bomber units to air bases in the Far East and Mongolia; a heightened alert status of Soviet air defense forces in the Far East; and the activation of a special Soviet Army headquarters complex near Khabarovsk.[98]

There is some evidence that NSA made some important strides during the 1960s against high-level Soviet cryptographic systems. According to press reports, in April 1964 NSA broke the Silver cipher system used by Russian military forces in Cuba during the 1962 missile crisis.[99] Until at least 1973, NSA was reportedly able to read the communications traffic between Moscow and the Russian ambassador in Washington, Anatoly Dobrynin.[100] NSA was also apparently able to read some of the Soviet diplomatic traffic passing between Moscow and its embassies in North Vietnam and Laos during the Vietnam War.[101] In February 1967, NSA and GCHQ were able to intercept the telephone conversations between Soviet Premier Alexei Kosygin and President Leonid Brezhnev while Kosygin was staying in London trying to negotiate a settlement to the Vietnam War.[102] As of the spring of 1967, NSA was reportedly reading Soviet diplomatic radio traffic between Moscow and its representatives in Egypt, which showed that the Soviets were convinced that war between Israel and Egypt was imminent.[103]

The Soviet target was obviously not the only subject of NSA's attentions in Eastern Europe. During the 1960s and 1970s, the NSA-GCHQ listening post at Teufelsberg in West Berlin was reading East German Communist Party (SED) communications traffic, the communications of the East German Army and Air Force, as well as the most secret communications traffic of the East German Ministry of State Security (STASI).[104] By analyzing hundreds of intercepted East German commercial telegrams, in the late 1960s the US Army Security Agency's Target Exploitation (TAREX) office in Frankfurt, Germany was able to identify the complete order of battle of the East German Army.[105]

Sigint performed particularly well on strategic military subjects during the 1960s. In 1961, American engineers and analysts made their first break into the Soviet missile telemetry signals that had been intercepted since 1955 by NSA, allowing intelligence analysts for the first time to extract performance data on Soviet strategic missiles then being tested as well as on Soviet space flights. For the next 15 years, NSA and the CIA were able to read virtually all Soviet telemetry signals, and thus understand the intricacies of all Soviet missile systems being developed, as well as Soviet activities in space.[106]

The launch of the first 'Ferret' satellites in the early 1960s revolutionized Elint collection against the Soviet Union. Between 1963 and 1967, American Ferret satellites completely mapped the locations and ascertained the capabilities of virtually every Soviet radar site in Eastern Europe and the Soviet Union, a mammoth task by any measure, as well as all Chinese, North Korean and North Vietnamese radar systems. By 1967, the Elint database was so detailed that the CIA was able to issue the first truly comprehensive National Intelligence Estimate on the state of Soviet air defenses, based largely on Sigint.[107]

Sigint was crucial in allowing the US intelligence community to follow Soviet Anti-Ballistic Missile (ABM) activities during the 1960s. In the early 1960s, American Elint sensors mapped the locations and capabilities of the first Soviet ABM early warning radars called 'Hen House' that were being built around the periphery of the Soviet Union. In 1968, Sigint provided the first detailed information about the capabilities and weaknesses of three new Soviet ABM radars deployed around Moscow.[108]

At the operational level, NSA's Sigint coverage of the Soviet military during the 1960s gave American intelligence analysts an incredibly detailed view of what was going on deep inside the Soviet Union, as well as allowed the US intelligence community to track the increasing tempo of Russian military activities outside the Soviet Union. For instance, in the 1960s a new worldwide ocean surveillance Sigint system called 'Classic Bullseye' was activated, which merged and modernized the naval Sigint intercept and high frequency direction finding (HFDF) resources of all five UKUSA members nations. 'Classic Bullseye' was an automated HFDF system that was larger, faster and more capable than the manual HFDF systems previously used by the UKUSA Sigint organizations, and allowed the US Navy and its English-speaking partners to track in near realtime the movements and activities of Soviet warships and submarines around the world. By the early 1970s, the US Naval Security Group Command was operating 21 'Classic Bullseye' HFDF stations in the US and around the world. The American stations were integrated with eight modernized HFDF stations operated by NSA's UKUSA partners.[109]

At about the same time, Britain's GCHQ began operating a highly sensitive direction finding system called 'Sambo', which was used to locate and record the burst transmissions from Soviet 'Yankee' and 'Delta'-class ballistic missile submarines that were patrolling in the North Atlantic.[110]

In 1972, a former US Air Force intelligence analyst described the extent of NSA's intelligence coverage of the Soviet military as follows:[111]

> As far as the Soviet Union is concerned, we know the whereabouts at any given time of all its aircraft, exclusive of small private planes, and its naval forces, including its missile-firing submarines. The fact is that we're able to break every code they've got, understand every type of communications equipment and enciphering device they've got. We know where their submarines are, what every one of their VIPs is doing, and generally their capabilities and the disposition of all their forces. This information is constantly computer correlated, updated, and the operations go on twenty-four hours a day.

By the late 1960s, satellite imagery and Sigint were rated by the CIA as the two most 'essential sources' needed to verify any arms control treaty with the Soviet Union.[112] In Vietnam, American military commanders became so dependent on Sigint that an American intelligence officer told a congressional committee that American military commanders in Vietnam 'were getting Sigint with their orange juice every morning and have now come to expect it everywhere'.[113]

SIGINT IN THE 1970s

Following the end of American troop withdrawal from Vietnam in 1972, NSA's budget was severely cut, and the Agency's manpower fell from 95,000 to approximately 52,000 military and civilian personnel by 1980, of whom 19,900 personnel worked at NSA headquarters.[114] Despite the cuts in its size, NSA remained a behemoth within the US intelligence community. A former senior NSA official recalled that by the early 1970s, NSA had '... become a several billion dollar a year corporation, with thousands of people operating a global system'.[115]

According to some sources, the overall importance of Comint within the US intelligence community declined in the 1970s. In part, this was because a GCHQ employee named Geoffrey Prime compromised the few remaining high-level Soviet systems that the US and Britain could still read during the mid-1970s.[116] But the main reason for the decline of Comint in the 1970s was the advent of microelectronics and new computer-based encryption technologies, all of which made it increasingly difficult for NSA and its allies to solve foreign cryptographic systems. A presidential review panel was forced to report in 1975 that 'The reason for the decline in importance [of Comint] is that the senders have come out ahead of the interceptors in the never-ending struggle to encrypt messages so that they cannot be deciphered.'[117]

Despite the doomsayers, in fact the 1970s was the era when Sigint and Imint came to truly dominate the US intelligence community. A 1976 study of US intelligence reporting on the Soviet Union found that virtually all of the material contained in the CIA's National Intelligence Estimates about Soviet strategic and conventional military forces came from Sigint and satellite imagery.[118] Conversely, a 1975 study of the US intelligence community found that less than 5 per cent of the finished intelligence being generated by it came from Humint.[119]

NSA and its partners adapted to the changing technological environment during the 1970s. A series of new high-tech Sigint collection systems came online in the 1970s, which gave the US and its allies increased access to high-level Soviet and Warsaw Pact communications traffic. During the 1970s, several generations of Sigint satellites were put into space, including the 'Canyon', 'Jumpseat' and 'Chalet' Comint satellites, which gave NSA access to high-level microwave communications traffic inside the Soviet Union that it had never had before.[120] The level of detail obtained from these satellites was so high that a former American intelligence officer stated that 'we could hear their teeth chattering in the Ukraine'.[121] The CIA's 'Rhyolite' Telint satellite revolutionized the US intelligence community's knowledge of Soviet strategic weapons development by intercepting previously unheard telemetry coming from Soviet strategic ballistic missile and bomber testing deep inside the Soviet Union. A new US Navy Elint ocean surveillance satellite called 'White Cloud' allowed the US Navy to track the movements of Russian Navy warships to a degree that heretofore had not been possible. And a new, fourth-generation Elint ferret satellite called 'Farrah' gave the US intelligence community exceptional detail about the new Soviet ABM radars and weapons systems then being developed.[122]

When the Soviet Union began shifting a significant part of its communications traffic to communications satellites in the 1970s, NSA and its UKUSA partners established a worldwide network of stations to intercept this new source of raw traffic. Dozens of new covert listening posts were established inside American, British, Canadian and Australian embassies around the world, especially behind the Iron Curtain, to intercept internal radio and telephone traffic.[123]

As before, NSA continued to produced high-grade intelligence information. In 1972, intercepts from the NSA listening post inside the US embassy in Moscow revealed an important part of the Soviet negotiating position during a summit meeting in Moscow leading up to the signing of the first Strategic Arms Limitation Talks (SALT) agreement. A senior US intelligence official who read the intercepts was quoted as saying 'That's the sort of thing that pays NSA's wages for a year.'[124] In the fall of 1973, decrypts of Soviet traffic revealed that the Soviet Union had become concerned about the likelihood of war between Egypt and Israel, and was concerned about the safety of the Russian citizens and dependents living in Egypt.[125]

In the 1970s, improved Sigint coverage of Soviet ballistic missile submarine communications, together with the Sound Surveillance System (SOSUS) underwater acoustic tracking network, allowed American intelligence analysts to accurately track the movements of Soviet submarines as they increasingly ventured further out into the Atlantic and Pacific Oceans. The coverage was reportedly so good that US Navy analysts were able to pinpoint the location of the patrol 'boxes' off the North American coastline frequented by Soviet ballistic missile submarines.[126]

On 22 December 1979, three days before the first of 85,000 Soviet troops crossed the Soviet-Afghan border on 25 December 1979, NSA was able to report to the White House that the Russians intended to invade Afghanistan.[127] In the late 1970s, NSA helped the FBI determine which US government telephone calls the Soviets were intercepting from inside their diplomatic establishments in Washington, New York and San Francisco.[128]

SIGINT IN THE 1980s

During the 1980s, NSA once again grew dramatically in size. As of 1989, the year the Berlin Wall came down, there were 75,000 American military, civilian and contractor personnel performing the Sigint mission in the US and overseas. Of the 75,000 personnel working for the Agency, almost 25,000 worked at NSA headquarters at Fort Meade, Maryland.[129]

FIGURE 2.1

US Army radio intercept operators at the listening post at Misawa, Japan, in 1982.

US Army Intelligence & Security Command

FIGURE 2.2

AN/FRD-10 antenna array at the US Navy listening post near Homestead Air Force Base, Florida.

US Army Intelligence & Security Command

FIGURE 2.3

Antenna tower at the US Army border intercept site at Wobeck, West Germany.
US Army Intelligence & Security Command

Continuing the trend that began in the 1960s, NSA's traditional Comint collection programs continued to produce less information because of a dramatic increase in worldwide telecommunications traffic volumes, which NSA had great difficulty coping with; and the growing availability and complexity of new telecommunications technologies, such as cheaper and more sophisticated encryption systems.[130] By the late 1980s, the number of intercepted messages flowing into NSA headquarters at Fort Meade had increased to the point that the Agency's staff and computers were only able to process about 20 per cent of the incoming materials.[131]

But despite these obstacles, the 1980s was a boomtime for NSA and its UKUSA partners, with the Agency enjoying some unheralded successes. Former CIA director Robert M. Gates recently stated that 'The truth is, until the late 1980s, US signals intelligence was way out in front of the rest of the world.'[132] The former Director of Intelligence of the US Marine Corps described Sigint as 'the success story of the 1980s'.[133] According to former senior American intelligence officials, on some days during the mid-1980s, Sigint accounted for over 70 per cent of the material contained in President Ronald Reagan's daily intelligence report that he received from the CIA.[134]

The depth of NSA's Sigint coverage of Soviet Union deepened during the 1980s, although NSA's best known successes were in parts of the world outside the Soviet Union and the Warsaw Pact.[135] In late 1980, NSA and American spy satellites detected a sizeable Soviet military buildup along the Polish border. The military buildup was prompted by the rising popularity of the Solidarity movement in Poland, and the inability of the Polish government to suppress the movement.[136]

In November 1983, NSA listening posts detected a dramatic increase in Soviet military radio traffic during a NATO nuclear weapons release exercise called 'Able Archer 83', reflecting the fact that Soviet forces throughout the Soviet Union and Eastern Europe had been placed on a heightened state of alert because of the exercise.[137]

During the *Achille Lauro* crisis in 1985, NSA listened in on what the Russians thought was a secure telephone line between the KGB listening post at Lourdes, Cuba and the KGB Rezidentura in Washington DC. The intercepts of these conversations revealed which US government communications circuits the KGB was listening to during the crisis.[138] In the late 1980s, NSA's Sigint satellites were used to monitor the radio circuits used by Russian SS-24 mobile ICBM batteries in the field to communicate with their operating bases. From these intercepts, NSA was reportedly able to pinpoint the locations of the SS-24 main operating bases as well as the deployment sites that the Soviets were using for their missiles.[139]

The best intelligence coverage of the April 1986 disaster at the Russian Chernobyl nuclear reactor came from intercepts supplied by a 'Vortex' Sigint satellite. In May 1988, a 'Vortex' satellite picked up radio traffic indicating that an explosion had taken place at a Russian missile propellant plant at Pavlograd, which made fuel components for the Soviet SS-24 ICBM.[140]

Sigint was an essential tool for monitoring Soviet forces during the war in Afghanistan from 1979 to 1989. Sigint kept tabs on the flow of reinforcements and supplies being sent from Russia to Afghanistan; the effectiveness of Soviet bombing raids; the movements of Soviet and Afghan combat units, and the mistrust that existed between the Soviets and their Afghan allies. Sigint helped the CIA design tactics to counter Soviet fighter bombers and helicopter gunships with new weapons, such as the Stinger surface-to-air missiles. Sigint also revealed that the Pakistani military and Afghan middlemen stole as much as 30 per cent of the aid that the CIA was sending to the Mujahidin resistance forces.[141]

On the basis of Sigint data, detailed post-mortem assessments of the 1 September 1983 KAL-007 incident undertaken by NSA and the US Air Force concluded that the massive Soviet PVO national air defense system had not performed well. Radar tracking information had not made its way in a timely manner from radar stations to the PVO regional command centers or to the PVO national command center at Balashikha outside Moscow; there was no lateral sharing of information among various PVO command centers and operating units in the region; PVO radar tracking data had been inaccurate and poorly assessed, with the Korean airliner and other targets flying in the area being repeatedly misidentified and not properly located. PVO fighter interceptors did not respond quickly, failing to intercept the lumbering Boeing 747 airliner as it slowly traversed the Kamchatka Peninsula and Sakhalin Island. Apparently Soviet fighters never got closer than 20 miles to the 747 as it cruised across Kamchatka. It took PVO fighters on Sakhalin two hours to intercept the lumbering 747 because of faulty directions from ground control intercept operators. The normally staid and disciplined PVO command and control system degenerated into chaos, with the Soviet fighter pilot who shot down the airliner, Lt Colonel Gennadi Osipovich, having to be ordered six times to destroy the target because of incessant radio chatter on the overloaded radio frequencies used by the fighter pilots.[142]

These findings were confirmed less than four years later in May 1987, when a 19-year-old West German amateur pilot named Mathias Rust managed to fly his Cessna 172 light plane 400 miles from Finland, through the heart of the PVO air defense system, and managed to successfully land his aircraft in the middle of Red Square without a scratch. To put it mildly, Soviet PVO officers were mortified by this embarrassing breach of their much-vaunted air defense system. Equally shocked were Reagan administration intelligence officials, who had loudly proclaimed the strength of the Soviet air defense system throughout the 1980s. NSA intercepts showed that Russian fighter interceptor pilots sent to intercept Rust initially identified his aircraft as Finnish, causing tremendous confusion within PVO air defense system as to how to deal with an errant aircraft belonging to a neutral nation.[143] Two days after Rust landed on Red Square, 30 May 1987, Soviet Defense Minister Marshal Sergei L. Sokolov and the chief of the PVO were fired by the Politburo. Rust was convicted by a military tribunal and sentenced to four years imprisonment, but was released after serving only one year of his sentence.

By the late 1980s, clandestine Sigint programs reportedly produced almost 50 per cent of the 'useful' Comint coming out of NSA.[144] For example, in October 1971, the US Navy submarine USS *Halibut* emplaced a tap on a Russian communications cable that ran under the surface of the Sea of Okhotsk which linked the naval base at Petropavlovsk on Kamchatka with the naval base at Magadan on the Soviet mainland. The operation was called 'Ivy Bells'. This tap generated superb intelligence for ten years until it was betrayed to the Russians in the fall of 1981 by a former NSA employee named Ronald W. Pelton. But unbeknownst to Pelton, a second tap had been installed by the US Navy in August 1979 on a cable in the Barents Sea that linked Russian naval bases on the Kola Peninsula. This cable was still producing intelligence as late as 1990. American intelligence officials later told Congress that the tap on the Soviet communications cable was one of the most valuable source of information available to NSA during the latter half of the Cold War, particularly communications relating to the operations of the Soviet Navy.[145]

In 1980, CIA agents tapped an underground communications cable that carried traffic between KGB headquarters in Moscow and a KGB communications center at Troitsk, 25 miles southwest of Moscow. This operation, designated Operation 'TAW', reportedly produced stellar results for five years until it was betrayed to the Soviets in 1985 by a KGB agent inside the CIA, Aldrich H. Ames.[146]

NSA was reportedly able to break into some Soviet military computer systems during the 1980s, yielding vast amounts of highly classified Soviet military data. A press report suggested that some of the penetrations were made possible by 'Black Bag' jobs of Soviet defense installations.[147]

The Sigint information coming from NSA clearly impressed consumers in the White House. President Reagan's first National Security Advisor, Richard V. Allen, was quoted as saying that NSA 'never ceases to amaze me. They have incredible capabilities.'[148] A former Reagan administration intelligence official was recently quoted as saying that 'There was a constant stream of incredibly good stuff' coming from NSA during the 1980s.[149] In September 1986, President Ronald Reagan himself told NSA that 'The simple truth is: Without you I could not do my job; nor could Secretary [of State George P.] Schultz conduct diplomacy; nor could Secretary [of Defense Caspar] Weinberger, nor [JCS chairman] Admiral [William J.] Crowe, muster the forces that defend us.'[150] In a speech to NSA staff at Fort Meade on 1 May 1991, President George W. Bush stated 'As President and Commander-in-Chief, I can assure you, signals intelligence is a prime factor in the decisionmaking process by which we chart the course of this nation's foreign affairs.'[151]

CONCLUSIONS

So, how to sum up. Just as historians are rewriting the history of World War II on the basis of newly released intelligence records, perhaps it is time that we begin rewriting the history of the Cold War on the basis of the new information that has been released concerning the important role played by intelligence, especially Sigint and the other technical intelligence disciplines.

Clearly, there is no shortage of evidence to suggest that Sigint played a vital role in helping to keep American policymakers and defense officials informed about what was going on behind the Iron Curtain during the Cold War. The late Senator Milton Young of North Dakota was quoted as saying that 'As far as foreign policy is concerned, I think the National Security Agency and the intelligence it develops has far more to do with foreign policy than does the intelligence developed by the CIA.'[152] Canadian historian John Ferris noted that NSA 'probably had more influence on American diplomacy and strategy' than the CIA.[153]

Recent statements by US government officials confirm what outsiders have long suspected about the importance of Sigint. For instance, in May 1994, Defense Secretary William J. Perry told NSA staff members that 'I don't make a significant decision without taking into account the products from this agency... NSA's work will be a critical factor in our ability to deal with this particularly complex and particularly uncertain world.'[154] And former NSA director John M. McConnell has written that 'Just as during World War II, robust world-class signals intelligence contributed significantly to winning the Cold War.'[155]

But what is presented above about the importance of NSA's intelligence product to US national security is clearly just the tip of the iceberg. Because of the paucity of records released by NSA about its Cold War accomplishments, it is difficult indeed to correctly gauge the importance of Sigint in the Cold War. Sadly, we must await the release of additional information from the archives of the National Security Agency and the other Sigint agencies around the world before a more definitive historical appreciation of the value of Sigint during the Cold War can be made. This writer, for one, can only hope that NSA will begin releasing more post-World War II cryptologic records in the near future.

NOTES

1. Executive Order No. 12333, *United States Intelligence Activities*, 4 Dec. 1981, in *Federal Register*, Vol. 46, No. 235, 8 Dec. 1981, pp.59947–8; National Security Council Intelligence Directive No. 6, *Signals Intelligence*, 17 Feb. 1972, National Security Council FOIA; Department of Defense Directive No. S-5100.20, *The National Security Agency and the Central Security Service*, 23 Dec. 1971, DoD FOIA; Department of Defense Directive No. S-3115.7, *Signals Intelligence (SIGINT)*, 25 Jan. 1973, DoD FOIA; United States Signals Intelligence Directive (USSID) 1, *SIGINT Operating Policy*, 29 June 1987, p.1, NSA FOIA; DIAM 56-3, *Defense Intelligence Organization, Operations and Management*, 8 July 1979, p.21, DIA FOIA; Memo for the NSA/CSS Representative Defense, *NSA Transition Book for the Department of Defense*, 9 Dec. 1992, Top Secret Edition, pp.1–2, NSA FOIA.
2. CIA, *A Consumer's Guide to Intelligence*, Sept. 1993, p.17; Penelope S. Horgan, *Signals Intelligence Support to US Military Commanders: Past and Present* (Carlisle Barracks, PA: US Army War College 1991) p.113.
3. Vice Admiral B.R. Inman, 'The NSA Perspective on Telecommunications Protection in the Nongovernmental Sector', *Cryptologic Spectrum*, Summer 1979, p.5.
4. John Ferris, 'Coming in from the Cold War: The Historiography of American Intelligence, 1945–1990', *Diplomatic History* 19/1 (Winter 1995) p.102.
5. Memo, *The Necessity for Continuance of Research and Development in the Cryptologic Field*, undated, p.2, in *Army Security Agency Summary Annual Report, Fiscal Year 1946*, Tab 5, INSCOM FOIA.
6. Memo, W.O. Studeman to All NSA Employees, *Farewell*, 8 April 1992, NSA FOIA.
7. J.M. McConnell, 'The Future of Sigint: Opportunities and Challenges in the Information Age', *Defense Intelligence Journal*, Summer 2000, p.42.
8. Copies of gists and full text decrypts of foreign military, diplomatic and other decrypts for the period 1945–46 can be found in RG-38, Records of the Office of the Chief of Naval Operations, No Entry Number, Translations of Intercepted Enemy Radio Traffic and Miscellaneous Materials, 1940–1946, Boxes 2739-2747, National Archives, College Park, Maryland (hereafter 'NA, CP'); RG-457, Historic Cryptologic Collection, Box 192, File: Red Intelligence Summaries, NA, CP; RG-457, Historic Cryptologic Collection, Box 521, File: Decrypted Diplomatic Traffic from World War II, T3101 – T3200, NA, CP.
9. Memo, USCIB to Secretary of Defense, 12 May 1949; Memo for the Secretary of Defense from Admiral Louis Denfield, USN, *Atomic Energy Program of the USSR*, 30 June 1949; and Memo for the Secretary of Defense, *Atomic Energy Program of the USSR*, 23 June 1949, all in RG-330, Box 61, CD 11-1-2, NA, CP.
10. US Senate, Select Committee to Study Governmental Operations With Respect to Intelligence Activities, *Final Report*, 94th Congress, 2nd Session, 1976, Book I, p.348.

11. H.D.S. Greenway and Paul Quinn-Judge, 'CIA Chief Voices Final Hopes and Fears', *Boston Globe*, 15 Jan. 1993, p.B17; Confidential interviews.
12. J.P. Cole, *Geography of the Soviet Union* (London: Butterworths 1984) p.3; Harold Fullard (ed.) *Soviet Union in Maps* (London: George Philip & Son 1972) p.1.
13. *Technological Capabilities Panel Report*, Feb. 1955, p.145, Dwight D. Eisenhower Library, Abilene, Kansas; Victor Marchetti and John D. Marks, *The CIA and the Cult of Intelligence* (NY: Laurel Books 1980) pp.21-2; Harry A. Rositzke, *The CIA's Secret Operations: Espionage, Counterespionage and Covert Action* (Boulder, CO: Westview Press 1977) p.20; Harry A. Rositzke, *The KGB: The Eyes of Russia* (Garden City, NY: Doubleday 1981) pp.80-3; Peer de Silva, *Sub Rosa: The CIA and the Uses of Intelligence* (NY: Times Books 1978) pp.23-36; Angelo Codevilla, *Informing Statecraft: Intelligence for a New Century* (NY: Free Press 1992) p.81; Bradley F. Smith, *Sharing Secrets With Stalin* (Lawrence: UP of Kansas 1996) p.254; Gregory W. Pedlow and Donald E. Welzenbach, *The Central Intelligence Agency and Overhead Reconnaissance: The U-2 and OXCART Programs, 1954-1974*, April 1992, declassified by the CIA in Sept. 1998.
14. Olav Riste and Arnfinn Moland, *Strengt Hemmelig: Norsk Etterretningsteneste 1945-1970* (Oslo: Universitetsforlaget 1997) pp.96-101; Rolf Tamnes, *The United States and the Cold War in the High North* (Oslo: ad Notam 1991) p.75.
15. The best description of the abortive MI6-CIA operations in the Baltic is Tom Bower, *The Red Web* (London: Aurum Press 1989). For a description of the Gehlen Org infiltration operations in the Baltic States, see Heinz Hohne and Hermann Zolling, *The General Was a Spy* (NY: Coward, McCann & Geoghegan 1972) pp.147-52. The KGB side of the story is told in *History of the Soviet State Security Organs*, 1977, pp.466, 473n, 481.
16. Sanche de Gramont, *The Secret War* (NY: Putnam's 1962) pp.184-8; Hohne and Zolling, *The General Was a Spy* (note 15) p.153; Rositzke, *The CIA's Secret Operations* (note 13) pp.18-20; Rositzke, *The KGB* (note 13) pp.82-3; Christopher Simpson, *Blowback* (NY: Weidenfeld 1988) p.173; Bower, *The Red Web* (note 15) p.120; Burton Hersh, *The Old Boys* (NY: Scribner's 1992) p.249; Evan Thomas, *The Very Best Men* (NY: Simon & Schuster 1995) p.36; Tom Bower, *The Perfect English Spy* (NY: St Martin's Press 1995) pp.205-6; Michael Smith, *New Cloak, Old Dagger* (London: Gollancz 1996) pp.114-15; Anthony Cavendish, *Inside Intelligence: The Revelations of an MI6 Officer* (London: HarperCollins 1997) p.51. For the KGB perspective, see *History of the Soviet State Security Organs*, 1977, p.473n.
17. Wilbur Crane Eveland, *Ropes of Sand* (NY: Norton 1980) p.263.
18. Bower, *The Perfect English Spy* (note 16) p.207.
19. Woodrow J. Kuhns, *Assessing the Soviet Threat: The Early Cold War Years* (Washington DC: Center for the Study of Intelligence 1998) p.7.
20. Bower, *The Red Web* (note 15) p.165.
21. Peer de Silva, *Sub Rosa: The CIA and the Uses of Intelligence* (NY: Times Books 1978) p.84.
22. Smith, *New Cloak, Old Dagger* (note 16) p.114.
23. *Technological Capabilities Panel Report*, Feb. 1955, p.145, Dwight D. Eisenhower Library, Abilene, Kansas.
24. William E. Burrows, *Imaging Space Reconnaissance Operations During the Cold War: Cause, Effect and Legacy*, undated, located at http://wbster.hibo.no/asf/Cold_War/report1/williame.html.
25. SRH-277, *A Lecture on Communication Intelligence by RADM E.E. Stone, DIRAFSA*, 5 June 1951, p.35, RG-457, NA, CP.
26. Included in the figure for the 37,000 Army COMINT personnel were approximately 17,000 personnel assigned to dozens of tactical Comint collection units stationed overseas. SRH-277, *A Lecture on Communications Intelligence* (note 25) p.12.
27. Memo, Adjutant General to Commanding Generals, *Establishment of the Army Security Agency*, 6 Sept. 1945; Memo, Adjutant General to Chief, Military Intelligence Service, *Establishment of the Army Security Agency*, 19 Sept. 1945; Memo, Adjutant General to

Commanding General, Army Service Forces, *Transfer of Signal Security Agency to Army Security Agency*, 21 Sept. 1945; Memo, Assistant Chief of Staff, G-2 to Commanding Generals, *Establishment of the Army Security Agency*, 7 Nov. 1945; Memo for Record, *General Provisions of the Army Security Agency*, 17 May 1946, all in RG-165, Entry 421 ABC Files, Box 269, File: ABC 350.05 (8 Dec. 1943), Sec. 1, NA, CP; John Patrick Finnegan, *Military Intelligence: An Overview, 1885–1987*, INSCOM History Office, Jan. 1988, p.116, INSCOM FOIA.
28. SRH-355, Part II, pp.6, 11, RG-457, NA, CP; SRMN-084, *The Evolution of the Navy's Cryptologic Organization*, p.9, RG-457, NA, CP.
29. *ASA Summary Annual Report: FY 1946*, p.7, INSCOM FOIA; *ASA Summary Annual Report: FY 1947*, Feb. 1950, p.2, INSCOM FOIA; Letter, Clayton to Patterson, 17 Sept. 1946, RG-107, Entry 108, Patterson General Decimal File 1946–1947, Box 4, File: 311.5 Communications, Crypto Security, NA, CP; Oral History, Arnold I. Dumey, 9 Oct. 1984, p.18, Charles Babbage Institute, Center for the History of Information Processing, University of Minnesota, Minneapolis, Minnesota.
30. SRH-197, J.N. Wenger, Captain, USN, *US Navy Communication Intelligence Organization, Liaison and Collaboration, 1941–1945*, 8 Oct. 1945, Chart A, pp.7, 34, RG-457, NA, CP; SRMN-084, *The Evolution of the Navy's Cryptologic Organization*, p.10, RG-457, NA, CP; OH, Arnold Dumey, 9 Oct. 1984, Charles Babbage Institute, Center for the History of Information Processing, University of Minnesota, Minneapolis, Minnesota.
31. HQ USAF, AFOIR-SR 322, *Functions of the USAF Security Service*, 20 Oct. 1948, p.1, AIA FOIA; Dickey, pp.5–6; *Insight*, p.8; '35 Years of Excellence', *Spokesman*, Oct. 1983, p.9.
32. Memo for USAF Deputy Chief of Staff/Operations from Major General C.P. Cabell, USAF Director of Intelligence, *Change in Personnel and Equipment Priorities for US Air Force Security Service*, 14 Dec. 1949, RG-341, Entry 214, File 2-10500 – 2-10599, NA, CP; USAFSS, *Organizational Development of the USAFSS, 1948–1962*, 15 Feb. 1963, p.122, AIA FOIA.
33. JCS 2010/1, p.12; Memo, Secretary of Defense to Secretaries of the Army, Navy and Air Force, *Organization of Cryptologic Activities Within the National Military Establishment*, 20 May 1945, with attachment, RG-330, Entry 199 OSD 1947–1950, Box 97, CD 22-1-23, NA, CP; JCS 2010, *Organization of Cryptologic Activities Within the National Military Establishment*, 20 May 1949, p.1, RG-341, Entry 214, File 2-8100 – 2-8199, NA, CP.
34. Memo, Denfield to Secretary of Defense, 26 Oct. 1949, RG-330, Entry 199 OSD 1947-1950, Box 97, CD 22-1-23, NA, CP; JCS 2010/19, *Expanded Requirements of the Armed Forces Security Agency in View of Current World Situation*, dated 27 July 1950, RG-218, CCS 334 NSA, Section 4, NA, CP.
35. As of 30 June 1950, AFSA headquarters in Washington consisted of 4,500 personnel; ASA consisted of 6,380 military and civilian personnel; NSG of approximately 3,500 military and civilian personnel; and USAFSS had 3,232 military and civilian personnel. *History, Army Security Agency and Subordinate Units, FY 1950*, p.1, INSCOM FOIA; Headquarters USAF, *United States Air Force Statistical Digest, Fiscal Year 1951*, pp.426, 429, Decimal File K134.11-6, AFHRA, Maxwell AFB, AL; USAFSS, *Organizational Development of the USAFSS, 1948–1962*, 15 Feb. 1963, pp.130–1; USAFSS, *Organizational Development of the USAFSS, 1948–1966*, 1 Feb. 1967, p.71.
36. SRMA-011, *SSS/SSA/ASA Staff Meeting Minutes*, pp.293, 321, RG-457, NA, CP; *ASA Summary Annual Report FY 1947*, pp.22–3, INSCOM FOIA; *Army-Navy Communication Intelligence Board Organizational Bulletin No. 1*, June 1945, RG-457, HCC, Box 1364, NA, CP; Meilinger, p.71; NSA-OH-03-83, Oral History Interview with Earl E. Stone, RADM, USN (Ret.), 9 Feb. 1983, pp.34–5, NSA FOIA.
37. Christopher Andrew, *For The President's Eyes Only* (NY: HarperCollins 1995) p.163; Nicky Hager, *Secret Power* (Nelson, NZ: Craig Potton 1996) pp.61–2; Christopher Andrew, 'The Growth of the Australian Intelligence Community and the Anglo-American Connection', *Intelligence and National Security* 4/2 (April 1989) p.224; Christopher Andrew, 'The Making of the Anglo-American SIGINT Alliance', p.106, and John

Ranelagh, 'Through the Looking Glass: A Comparison of United States and United Kingdom Intelligence Cultures', both in Hayden B. Peake and Samuel Halpern (eds.) In *The Name of Intelligence: Essays in Honor of Walter Pforzheimer* (Washington DC: NIBC Press 1994) pp.411, 435–6.
38. Gene Becker, 'If Lou Tordella Says It's OK, That's Good Enough for Me', *Cryptolog*, Spring 1996, pp.26–7.
39. Allen Weinstein and Alexander Vassiliev, *The Haunted Wood* (NY: Random House 1998) pp.291–2; David Alvarez, 'Behind Venona: American Signals Intelligence in the Early Cold War', *Intelligence and National Security* 14/2 (Summer 1999) p.180.
40. National Cryptologic School, *On Watch* (Ft Meade: NSA/CSS, Sept. 1986) p.19, NSA FOIA.
41. *War Diary Report OP-20-G-4A: 1 September to 1 October 1945*, 1 Oct. 1945, p.3, RG-38, CNSG 5750/160, NA, CP; *War Diary Summary of G4A for December 1945*, 4 Jan. 1946, p.1, RG-38, CNSG 5750/160, NA, CP; *War Diary Summary of G4A for March 1946*, 9 April 1946, p.1, RG-38, CNSG 5750/160, NA, CP; *War Diary Summary of G4A for April 1946*, 6 May 1946, p.1, RG-38, CNSG 5750/160, NA, CP. The author is grateful to Ralph Erskine for making copies of these documents available. See also Alvarez, 'Behind Venona' (note 39) pp.181, 185n4.
42. Alvarez, 'Behind Venona' (note 39) p.180.
43. *War Diary Summary of G4A for April 1946*, 6 May 1946 (note 41) p.1; Alvarez, 'Behind Venona' (note 39) p.181.
44. *Army Security Agency Summary Annual Report, FY 1948*, p.68, and *History, Army Security Agency and Subordinate Units, FY 1950*, p.6, both INSCOM FOIA. For examples of RUMUA decrypts, see RUMI-0622, Riga to Moscow, RUMUA-1A, intercepted 28 Dec. 1946, solved 12 Oct. 1948; RUMI-0625, Tbilisi to Moscow MVS, RUMUA-1, intercepted 8 Jan. 1948, solved 12 Oct. 1948; RUMI-0705, Vienna to Mukachevo, RUMUA-1A, intercepted 3 Dec. 1947, solved 23 Dec. 1948; RUMI-0712, Vienna to Mukachevo, RUMUA-1A, intercepted 3 Dec. 1947, solved 23 Dec. 1948, all in RG-38, Box 2742, NA, CP.
45. See for example RUM-12083, Moscow VVS VS to Vienna 2nd Air Army, RUARA-1, intercepted 6 Oct. 1947, solved 20 Oct. 1948; RUM-12375, Dairen 7 Air Corps to Vozdvizhenka 9th Air Army, RUMUC-2, intercepted 1 Dec. 1947, solved UNK, all in RG-38, Box 2742, NA, CP; RUM-10828, Vozdvizhenka 9th Air Army to Moscow VVS VS, intercepted 4 May 1947, solved 6 Aug. 1948, RG-38, Box 2744, NA, CP; RUMI-0505, Tbilisi 11 Air Army to VVS VS, intercepted 30 April 1948, solved 31 Aug. 1948, RG-38, Box 2745, NA, CP.
46. Michael Haydock, *The Cold War in Asia*, *VFW Magazine*, located at www.asa.npoint.net/cldwarasia.htm.
47. Memo, *Miscellaneous*, undated but circa late 1945, RG-457, Historic Cryptologic Collection (HCC), Box 522, File: List of Russian Codes Captured by Finns, NA, CP.
48. For examples of Soviet Navy decrypts, see NI-1-#14928, CinC 5th Fleet to Moscow Naval Headquarters, RUNRA-1, intercepted 18 April 1948, solved 16 March 1949; NI-1-#23815, Vladivostok to Moscow, RUNY, intercepted 8 Dec. 1948, solved 21 April 1949; RUN-16971, Petropavlovsk Naval Base to Sovetskaya Gavan Naval Base, RUNRA-1, intercepted 16 Jan. 1948, solved 18 Nov. 1948, all in RG-38, Box 2739, NA, CP.
49. See for example, DA-76354, Belgrade to Athens, Greek, intercepted 21 Oct. 1948, solved 1 Nov. 1948, RG-38, Box 2741, NA, CP; DA-96549, Warsaw to Ankara, Turkish, intercepted 18 Feb. 1949, solved 15 March 1949; ZA-1716, Sofia to Rome, Italian, intercepted 24 Sept. 1948, solved 16 Dec. 1948; ZA-2335, Bucharest to Rome, Italian, intercepted 27 Dec. 1948, solved 28 Feb. 1949; DN-69432, Prague to Nanking, Chinese, intercepted 11 Sept. 1948, solved 21 Sept. 1948, all in RG-38, Box 2744, NA, CP.
50. National Cryptologic School, *On Watch* (Ft Meade: NSA/CSS, Sept. 1986) p.19, NSA FOIA; TI Item #137, NT-1 Traffic Intelligence, *Unprecedented Coordinated Russian Communications Changes*, 4 Nov. 1948, RG-38, Box 2742, NA, CP; Weinstein and Vassiliev, *The Haunted Wood* (note 39) pp.291–3.

51. *Annual Report for G-3 ASA FY 1951*, p.76, INSCOM FOIA; *Agenda Prepared by Army Field Forces Observer Team No. 5*, Aug. 1951, p.167, MISC 333 AFF-FECOM, OCMH, Washington DC; Memo for the NSA/CSS Representative Defense, *NSA Transition Book for the Department of Defense*, 9 Dec. 1992, p.2. The author is grateful to Dr Jeffrey T. Richelson for making a copy of this document available.
52. Ludwell Lee Montague, *General Walter Bedell Smith as Director of Central Intelligence, October 1950–February 1953*, Dec. 1971, Vol. V, p.54, RG-263, NA, CP; Memo for Executive Secretary, National Security Council from Walter B. Smith, Director of Central Intelligence, *Proposed Survey of Communications Intelligence Activities*, 10 Dec. 1951; *AHR Army Security Agency FY 1953*, p.14, INSCOM FOIA.
53. Memo, President Truman to Secretaries of State and Defense, 24 Oct. 1952, pp.1–4, SRH-271, Presidential Memorandum on Communications Intelligence Activities, RG-457, NA, CP; *History of the Army Security Agency and Subordinate Units: Fiscal Year 1953*, Vol. I, pp.3–4, INSCOM FOIA; *National Security Agency Organization Manual*, 19 April 1954, Chapter III, p.1, NSA FOIA.
54. Confidential interview.
55. As of 22 Jan. 1963, there were assigned to NSA Headquarters at Ft George G. Meade 718 officers, 1,726 enlisted men and 7,475 civilian personnel, with another 82 officers, 78 enlisted men and 405 civilians attached to NSA for 'support'. US House of Representatives, Committee on Appropriations, *Military Construction Appropriations for 1964*, Part 2, p.487. The 10,000 figure at Ft Meade in the early 1960s is confirmed by the former Assistant Chief of NSA for Production, Army Major General John Davis, in Senior Officers Oral History Program, *Oral History, Lt. General John J. Davis, USA (Ret.)*, 1986, p.136, OCMH, Washington DC. For additional data on the size of NSA, see JCS 2010/104, *General Guidelines for FY 1956 and 1957, and NSA Civilian Manpower and Space Requirements*, 29 Nov. 1955, RG-341, Entry 355, OPD 311, Section 9, Box 107, NA, CP; *Report of the President's Board of Consultants on Foreign Intelligence Activities to the President*, 20 Dec. 1956, p.7, DDRS 1992 #0487; John C. Schmidt, 'America's Supersecret Watchdog', *The Baltimore Sun*, 8 Jan. 1961, p.A1.
56. 'CRYPTOLOG Interviews NSA Employee Gene Becker', *Cryptolog*, Spring 1996, p.19.
57. SRH-123, *The Brownell Committee Report*, pp.24–5, RG-457, NA, CP.
58. TS Control #55-040, *Supplement to Part II of the Report of the Technological Capabilities Panel*, Feb. 1955, p.2. The author is grateful to Dr Jeffrey T. Richelson for a copy of this document.
59. SRH-123, *Brownell Committee Report*, pp.29, 135–6, RG-457, NA, CP; SRH-277, *A Lecture on Communications Intelligence by Rear Admiral E.E. Stone, DIRAFSA*, 5 June 1951, p.34, RG-457, NA, CP.
60. National Cryptologic School, *On Watch* (Ft Meade: NSA/CSS, Sept. 1986) p.20, NSA FOIA.
61. Michael van der Meulen, 'German Air Force Signal Intelligence 1956: A Museum of Comint and Sigint', *Cryptologia*, July 1999, p.242.
62. Jerry Mooney Deposition, p.239, RG-46, Records of the US Senate, Select Committee on POW/MIA Affairs, Depositions Box 14, NA, Washington DC. See also the William F. Friedman Catalog, 5 Sept. 1957, RG-457, Yardley Collection, Box 2, Document 21, NA, CP for indications of NSA's cryptanalytic work in the mid-1950s against Soviet high-level systems.
63. SNIE 11-6-57, *Soviet Gross Capabilities for Attack on the Continental US in Mid-1960s*, pp.5–6, RG-263, NA, CP; SNIE 11-7-58, *Strength and Composition of the Soviet Long Range Bomber Force*, p.3, RG-263, NA, CP; NIE 11-8-59, *Soviet Capabilities for Strategic Attack Through Mid-1964*, p.27, RG-263, NA, CP; NIE 11-8-61, Annex C (Top Secret Codeword), RG-263, Entry 37, Box 2, Folder 22, NA, CP; K.F. Spielmann Jr, *The Evolution of Soviet Strategic Command and Control and Warning: 1945–72*, Institute for Defense Analysis, May 1975, pp.49, 121–2, 126–31, partially declassified and on file at the National Security Archives, Washington DC; Scott D. Breckinridge, *The CIA and the Cold War: A Memoir* (Westport, CT: Praeger 1993) pp.52–3; Riste and Moland, *Strengt*

THE NSA AND THE COLD WAR   59

Hemmelig (note 14) pp.133–4; R. Cargill Hall, 'The Truth About Overflights', *Military History Quarterly* (Spring 1997) p.31.
64. Charles A. Kroger Jr, 'ELINT: A Scientific Intelligence System', *Studies in Intelligence* (Winter 1958) p.77, RG-263, Entry 37, Box 1, Folder 6, NA, CP; 'Electronic Intercept or Technical Search Operations: History and Importance', *ONI Review*, April 1959, pp.149–50, Operational Archives, Naval Historical Center, Washington DC.
65. Memo Approval No. 1400 (NSCID No. 17, Action #1400), RG-273, NA, CP; NSCID No. 17, *Electronic Intelligence (ELINT)*, 16 May 1955, RG-263, HRP 89-2/00286, Box 4, File: NSCIDs, NA, CP; Kroger, 'ELINT' (note 64) pp.77–8, RG-263, Entry 27, Box 1, NA, CP; CIA Historical Staff, *Dulles as DCI*, Vol. II, pp.85–6, RG-263, NA, CP;JCS 990852, JCS to USCINCEUR PARIS FRANCE, 24 Oct. 1955, RG-218, CCS 334 NSA, Section 16, NA, CP; JIC 463/93, *Note by the Secretaries to the Joint Intelligence Committee on Revision of JCS 222/44*, 16 May 1956, pp.4–5, RG-218, CCS 334 JIC, Sec 8, Box 58, NA, CP; National Security Agency/Central Security Service, NSA/CSS Manual 22-1, *NSA Organizational Manual*, 21 Oct. 1986, p.2, NSA FOIA; O.D. Dickey, *The Development of the US Elint Effort*, undated, p.3, AIA FOIA; *The Implementation and Expansion of the USAF Electronics Intelligence Program*, p.21, AIA FOIA.
66. Letter, Ackerman to Fressanges, 21 April 1953, RG-341, Entry 214, File 3-1644, NA, CP; *Reconnaissance*, pp.1–2, and *Soviet Air Defense Capabilities*, circa 1954, pp.7–8, both in RG-341, Entry 337, Box 30, File: WPC 54-1, Intelligence Annex, Vol. I, NA, CP; Gregory W. Pedlow and Donald E. Welzenbach, *The Central Intelligence Agency and Overhead Reconnaissance: The U-2 and OXCART Programs, 1954–1974*, April 1992, pp.87, 101, 106, declassified by the CIA in Sept. 1998; Alfred Price, *The History of US Electronic Warfare: Vol. III: Rolling Thunder Through Allied Force, 1964 to 2000* (Washington DC: Association of Old Crows 2000) p.2; Scott D. Breckinridge, *The CIA and the US Intelligence System* (Boulder, CO: Westview Press 1986) pp.132–3; Chris Pocock, *Dragon Lady: The History of the U-2 Spyplane* (Osceola: Motorbooks International 1989) p.35; Chris Pocock, 'From Peshawar to Bodo: Mission Impossible?', in Svein Lundestad, *U-2 Flights and the Cold War in the High North* (Bodo: Bodo College 1995) pp.4–5; Riste and Moland, *Strengt Hemmelig* (note 14) pp.104–5; Cargill Hall, 'The Truth About Overflights' (note 63) pp.32–3; David Colley, 'Shadow Warriors: Intelligence Operatives Waged Clandestine Cold War', *VFW Magazine*, Sept. 1997.
67. Letter from Ronald D. Schultz, undated, in 1st Radio Squadron Mobile Updates, located at http://www.bobnfumi.com/1radioupdate.html.
68. Memo, Willems to Director of Central Intelligence with enclosures, 15 Sept. 1953, RG-59, Entry 1498, Box 3, File: Sheldon Committee 1953, NA, CP; Breckinridge, *The CIA* (note 66) pp.131–2; Breckinridge, *CIA and the Cold War* (note 63) p.72; Paul Bracken, *The Command and Control of Nuclear Forces* (New Haven: Yale UP 1983) pp.10, 30; Riste and Moland, *Strengt Hemmelig* (note 14) pp.148–9, 243, 247–8; Rolf Tamnes, *The United States and the Cold War in the High North* (Oslo: Ad Notam forlag 1991) p.122.
69. Memo, Ackerman to Control Division, USAF DCS Operations, *Precedence of Units of the USAF Security Service*, 2 Nov. 1951, RG-341, Entry 214, Box 60, File 2-21300 – 2-21399, NA, CP; Memo, Willems to Director of Central Intelligence, *et al.* with enclosures, 15 Sept. 1953, RG-59, Entry 1498, Box 3, File: Sheldon Committee 1953, NA, CP; NIE 11-57, *Sino-Soviet Air Defense Capabilities Through Mid-1962*, 16 July 1957, p.10, RG-263, Entry 37, Box 7, NA, CP; Patrick J. McGarvey, *CIA: The Myth and the Madness* (NY: Saturday Review Press 1972) pp.42–3, 74–6; Patrick J. McGarvey, *Communist Military Strength in Asia: 1956*, undated, at www.netrox.net/-condev/203mcg10.htm; Breckinridge, *The CIA and the US Intelligence System* (note 66) p.131; idem *CIA and the Cold War* (note 63) p.72; Andrew, *For The President's Eyes Only* (note 14) p.219; Tony Geraghty, *BRIXMIS* (London: HarperCollins 1997) p.87; Riste and Moland, *Strengt Hemmelig* (note 14) pp.104–5, 133; William E. Burrows, 'Beyond The Iron Curtain', *Air & Space*, Aug./Sept. 1994, p.31.
70. Tamnes, *The United States and the Cold War in the High North* (note 14) p.122; Sherry

Sontag and Christopher Drew, *Blind Man's Bluff* (NY: Public Affairs 1998) pp.25–39; Norman Polmar and Thomas B. Allen, *Spy Book: The Encyclopedia of Espionage* (NY: Random House 1997) p.539; David Colley, 'Stealth Beneath the Sea: The Wet Cold War', *VFW Magazine*, Aug. 1997.

71. SRH-123, *Brownell Committee Report*, p.29, RG-457, NA, CP; Memo for Record, 24 Aug. 1950, RG-341, Entry 214, File: 2-15900 – 2-15999, NA, CP; Memo, Willems to Director of Central Intelligence, *et al.* with enclosures, 15 Sept. 1953, RG-59, Entry 1498, Box 3, File: Sheldon Committee 1953, NA, CP; NIE 11-6-55, *Probable Intelligence Warning of Soviet Attack on the US Through Mid-1958*, 1 July 1955, p.8, RG-263, NA, CP; *Supplement to Part II of the Report of the Technological Capabilities Panel*, Feb. 1955, p.2, via Dr Jeffrey T. Richelson; Sanche de Gramont, *The Secret War* (NY: Putnam's 1962) p.168; Breckinridge, *The CIA* (note 66) pp.131–2; Breckinridge, *CIA and the Cold War* (note 63) p.72; Paul Bracken, *The Command and Control of Nuclear Forces* (New Haven, CT: Yale UP 1983) pp.10, 30.

72. James E. Pierson, *USAFSS Response to World Crises, 1949–1969* (San Antonio, TX: USAFSS Historical Office, 22 April 1970) p.1, AIA FOIA; SRH-123, *The Brownell Committee Report*, p.29, RG-457, NA, CP.

73. GHQ Far East Command, G-3 Section, *A Study of the Current Situation in the Far East Command*, 17 June 1950, p.47, RG-6, Box 105, MacArthur Library (hereafter 'MACL'), Norfolk, Virginia.

74. *USAFSS Response to World Crises*, p.1, AIA FOIA; Department of Defense, *Soviet Military Power* (Washington DC: GPO, April 1984) p.10; Msg, G-2, Chinese Nationalist Ministry of National Defense (via Chinese Mission, Tokyo) to FECOM G-2 Foreign Liaison Branch, 3 July 1950, RG-6, Box 14, MACL.

75. David Colley, 'Shadow Warriors: Intelligence Operatives Waged Clandestine Cold War', *VFW Magazine*, Sept. 1997.

76. On the general importance of Sigint during the 1950s and 1960s, see *Oral History of Laurence H. Frost*, pp.7–8, John F. Kennedy Presidential Library, Boston, Massachusetts; Victor Marchetti and John D. Marks, *The CIA and the Cult of Intelligence* (NY: Laurel Books 1980) p.22; Bracken, *Command and Control of Nuclear Forces* (note 71) p.13; Patrick Fitzgerald and Mark Leopold, *Stranger on the Line: The Secret History of Phone Tapping* (London: Bodley Head 1987) p.46; and Dino A. Brugioni, *Eyeball to Eyeball: The Inside Story of the Cuban Missile Crisis* (NY: Random House 1990) pp.6–7. For Sigint and U-2 Imint providing 75 per cent of intelligence of information about the Soviet Union, see Gene Poteat, 'Stealth, Countermeasures, and Elint, 1960–1975', *Studies in Intelligence* 42/1 (1998) p.52, CIA FOIA; Pedlow and Welzenbach, *The Central Intelligence Agency and Overhead Reconnaissance: The U-2 and OXCART Programs* (note 66) pp.100ff, declassified by the CIA in Sept. 1998.

77. Philip K. Edwards, 'The President's Board: 1956–60', *Studies in Intelligence* 13/2 (Summer 1969) pp.117–18, RG-263, Box 16, NA, CP; *The Joint Study Group Report on Foreign Intelligence Activities of the United States Government*, 15 Dec. 1960, p.48, Dwight D. Eisenhower Library, Abilene, Kansas; Lt. Gen. Eugene F. Tighe, USAF (Ret.), 'Imagery and Reconnaissance: Reminiscences', *American Intelligence Journal*, Winter/Spring 1992, p.84.

78. Of the 206 Russian spies identified by the FBI, by 1955 101 had left the US and could not be prosecuted, including 61 Russian officials; 11 had died; 14 were cooperating with the FBI; and 15 had been prosecuted. These individuals were Abraham Brothman, Judith Coplon, Klaus Fuchs, Harry Gold, David Greenglass, Valentine A. Gubitchev (Judith Coplon's KGB handler), Miriam Moskowitz, Julius Rosenberg, Ethel Rosenberg, Alfred Slack, Morton Sobell, Jack Soble, Myra Soble, William Perl, and Alger Hiss. This left 77 individuals whom the FBI had investigated but the US Justice Department could not or would not prosecute for various reasons. Memo, Belmont to Boardman, 27 Nov. 1957, pp.2–3, FBI Venona Files.

79. Confidential interviews.

80. Woodrow J. Kuhns (ed.) *Assessing the Soviet Threat: The Early Cold War Years*

(Washington DC: Center for the Study of Intelligence 1997) p.11 fn39; CIA, IM-323-SRC, *Soviet Preparations for Major Hostilities in 1950*, 25 Aug. 1950, Intelligence File, President's Secretary's File, Box 250, Harry S. Truman Papers, Truman Library, Independence, Missouri.
81. Melvin Beck, *Secret Contenders* (NY: Sheridan Square 1984) pp.43–4.
82. NSA OH 76-1, Oral History, Frank B. Rowlett, 1976, p.374, NSA FOIA; Oral History, Richard Drain, 8 Jan. 1976, p.62, CIA FOIA; Victor Marchetti and John D. Marks, *The CIA and the Cult of Intelligence* (NY: Laurel Books 1974) pp.164, 170–1; Philip Agee, *Inside the Company: CIA Diary* (NY: Bantam Books 1976) p.33; Joseph Burkholder Smith, *Portrait of a Cold Warrior* (NY: Putnam's 1976) p.397; David Martin, *Wilderness of Mirrors* (NY: Harper & Row 1980) p.121; Melvin Beck, *Secret Contenders* (note 81) pp.69, 72; Peter Wright, *Spycatcher* (NY: Viking 1987) pp.145, 148, 151; David E. Murphy, Sergei A. Kondrashev and George Bailey, *Battleground Berlin* (New Haven, CT: Yale UP 1997) p.209; Riste and Moland, *Strengt Hemmelig* (note 14) pp.134–5; Nigel West, *Venona: The Greatest Secret of the Cold War* (London: HarperCollins 1999) pp.34–5.
83. US Senate, Select Committee to Study Governmental Operations With Respect to Intelligence Activities, *Final Report*, 94th Congress, 2nd Session, 1976, Book III, p.359; William C. Sullivan, *The Bureau* (NY: Norton 1979) pp.179–80; G. Gordon Liddy, *Will* (NY: St Martin's Press 1996) pp.78–9; Anthony Marro, 'FBI Break-In Policy', in Athan Theoharis (ed.) *Beyond the Hiss Case: The FBI, Congress and the Cold War* (Philadelphia: Temple UP 1982) p.103; Andrew, *For The President's Eyes Only* (note 37) pp.217–18.
84. George Blake, *No Other Choice* (NY: Simon & Schuster 1990) pp.6, 10–12; Michael Smith, *New Cloak, Old Dagger* (London: Gollancz 1996) p.116.
85. Wright, *Spycatcher* (note 82) pp.79–80.
86. Donald P. Steury (ed.) *On the Front Lines of the Cold War: Documents on the Intelligence War in Berlin, 1946 to 1961* (Washington DC: Center for the Study of Intelligence 1999) pp.401–5; Murphy, Kondrashev and Bailey, *Battleground Berlin* (note 82) pp.423–8.
87. For the 95,000 personnel figure, see US House of Representatives, Committee on Appropriations, *Department of Defense Appropriations for 1975*, 93rd Congress, 2nd Session, 1974, Part 1, p.598. For NSA personnel at Ft Meade, see US House of Representatives, Committee on Appropriations, *Department of Defense Appropriations for 1972*, 92nd Congress, 1st Session, 1971, Part 3, p.536; US House of Representatives, Committee on Appropriations, *Department of Defense Appropriations for 1975*, 93rd Congress, 2nd Session, 1974, Part 3, pp.340, 663.
88. Tom Mangold, *Cold Warrior* (NY: Simon & Schuster 1991) p.264.
89. *Report to the President's Foreign Intelligence Advisory Board on Intelligence Community Activities Relating to the Cuban Arms Build-Up: 14 April through 14 October 1962*, undated, p.27, National Security Files: Countries: Cuba, Box 61, Kennedy Library, Boston, Massachusetts.
90. Nigel West, *Games of Intelligence* (NY: Crown Publishers 1989) p.136.
91. Walter Laqueur, *A World of Secrets* (NY: Basic Books 1985) p.24.
92. *The Joint Study Group Report on Foreign Intelligence Activities of the United States Government*, 15 Dec. 1960, p.48, Dwight D. Eisenhower Library, Abilene, Kansas; US Senate, Committee on Armed Services, *Department of Defense Authorization for Appropriations for Fiscal Year 1994 and the Future Years Defense Program*, 103rd Congress, 1st Session, 1993, Part 7, pp.452, 462; John L. McLucas, 'The US Space Program Since 1961: A Personal Assessment', in R. Cargill Hall and Jacob Neufeld (eds.) *The US Air Force in Space: 1945 to the Twenty-first Century* (Washington DC: USAF History and Museums Program 1998) pp.85–6.
93. *The Joint Study Group Report on Foreign Intelligence Activities of the United States Government*, 15 Dec. 1960, p.47, Dwight D. Eisenhower Library, Abilene, Kansas.
94. For Sigint detections of the unusual surge in Soviet shipping traffic to Cuba, see Msg, *Unusual Number of Soviet Passenger Ships En Route to Cuba*, 24 July 1962, Msg, *Possible Reflections of Soviet/Cuban Trade Adjustments Noted in Merchant Shipping*, 31 July 1962,

Msg, *Further Unusual Soviet/Cuban Trade Relations Recently Noted*, 7 Aug. 1962, Msg, *Status of Soviet Merchant Shipping to Cuba*, 23 Aug. 1962, and Msg, *Further Information on Soviet/Cuban Trade*, 31 Aug. 1962, all located at www.nsa.gov/.
95. US Senate, Committee on Armed Services, *Authorization for Military Procurement, Research and Development, FY 1969*, 90th Congress, 2nd Session (Washington DC: GPO 1968) p.42; US House of Representatives, Committee on Armed Services, *Hearings Before the Special Subcommittee on the USS. Pueblo: Inquiry into the USS. Pueblo and EC-121 Incidents*, 91st Congress, 1st Session (Washington DC: GPO 1969) p.635; Laqueur (note 91) pp.165–7; Bouchard, Vol. II, p.519.
96. 'US Electronic Espionage: A Memoir', *Ramparts*, Aug. 1972, pp.43–4; Chet Flippo, 'Can the CIA Turn Students Into Spies?', *Rolling Stone*, 11 March 1976, p.30.
97. HQ USASA, *Historical Summary of the US Army Security Agency, FY 1968 – 1970*, pp.73–4, INSCOM FOIA; *CIA: The Pike Report* (London: Spokesman Books 1977) pp.139–40; James H. Polk, 'Reflections on the Czechoslovakian Invasion, 1968', *Strategic Review* 5/5, pp.31–2; 'US Electronic Espionage: A Memoir', *Ramparts*, Aug. 1972, p.43.
98. *History of the Collection Research Division, USAF ACSI: 1 January – 30 June 1969*, pp.50–1, USAF FOIA; Henry Kissinger, *The White House Years* (Boston: Little Brown 1979) pp.183–4; Patrick Tyler, *A Great Wall: Six Presidents and China* (NY: Century Foundation Books 1999) pp.65, 68.
99. Seymour Hersh, 'Was Castro Out of Control in 1962?' *The Washington Post*, 11 Oct. 1987, pp.H1–H2.
100. Seymour M. Hersh, *The Price of Power: Kissinger in the Nixon White House* (NY: Summit Books 1983) pp.257n, 526n.
101. Jerry Mooney Deposition, pp.238–9, RG-46, Records of the US Senate, Select Committee on POW/MIA Affairs, Depositions Box 14, NA, Washington DC.
102. George C. Herring (ed.) *The Secret Diplomacy of the Vietnam War: The Negotiating Volumes of the Pentagon Papers* (Austin: U. of Texas Press nd) pp.61–2.
103. 'US Electronic Espionage: A Memoir', *Ramparts*, Aug. 1972, pp.43–4; Chet Flippo, 'Can the CIA Turn Students Into Spies?', *Rolling Stone*, 11 March 1976, p.30.
104. See for example, Document 122, *Current Intelligence Weekly Report*, 5 Oct. 1961, in Department of State, *Foreign Relations of the United States 1961–1963: Volume 5, Soviet Union* (Washington DC: GPO 1998); and NSA Report, 25 Nov. 1963, JFK Assassination Records, NSA, Box 1, Record No. 144-10001-10056, NA, CP. For Teufelsberg Sigint successes, see Markus Wolf, *Man Without a Face* (NY: Times Books 1997) p.294; Klaus Eichner and Dr Andreas Dobbert, *Headquarters Germany* (Berlin: Edition Ost 1997) pp.225–7; Geraghty, *BRIXMIS* (note 69) pp.204–5.
105. 'HQ, US Army Security Agency, Europe (TAREX), Frankfurt am Main, West Germany: The End of My Army Career', undated, located at www.cdstrand.com/areas/usasaeur.htm.
106. Albert D. Wheelon and Sidney N. Graybeal, 'Intelligence for the Space Race', *Studies in Intelligence* 5/3 (Fall 1961) p.4; David S. Brandwein, 'Telemetry Analysis', ibid. 8/3 (Fall 1964) pp.28–9, both in RG-263, NA, CP.
107. Confidential interviews. See also US Senate, Committee on Armed Services, *Department of Defense Authorization for Appropriations for Fiscal Year 1994*, 103rd Congress, 1st Session, 1993, Part 7, pp.452, 456.
108. Frank Eliot, 'Moon Bounce Elint', *Studies in Intelligence* 11/1 (Spring 1967) pp.63–4, RG-263, Entry 27, NA, CP; Edward Tauss, 'Foretesting a Soviet ABM System', ibid. 12/4 (Winter 1968) pp.22–3, RG-263, Entry 27, NA, CP; David S. Brandwein, 'Interaction in Weapons R&D', ibid. (Winter 1968) pp.18–19, RG-263, Entry 27, NA, CP; Donald C. Brown, 'On The Trail of Hen House and Hen Roost', ibid. 13/1 (Spring 1969) pp.11–19, RG-263, Entry 27, Box 16, NA, CP; Gene Poteat, 'Stealth, Countermeasures, and Elint, 1960–1975', *Studies in Intelligence* 42/1 (1998) pp.53–4, 57–8, all located in RG-263, NA, CP.
109. OPNAVINST S3270.1, *Employment and Operating Policy for the US Navy HFDF Nets*, 18 May 1984, pp.2–4, Navy FOIA; OPNAVINST 02501.5E, *Cryptologic Tasks Assigned to Fleet Commanders in Chief*, 24 June 1969, p.3, Navy FOIA; NSGINST C3270.2,

Bullseye System Concept of Operations, 30 June 1989, p.3, via Dr Jeffrey T. Richelson; NWP-5, *Naval Cryptologic Operations*, pp.3-3-3-4, Navy FOIA; *1984 Annual History Report for the Headquarters Naval Security Group Command*, 5 June 1985, Section 10, Item 10.2.1, COMNAVSECGRU FOIA; Desmond Ball, 'The US Naval Ocean Surveillance Information System – Australia's Role', *Pacific Defence Reporter*, June 1982, pp.45–6.
110. Patrick E. Tyler, 'Laxness Cited At U.K. Spy Agency', *Manchester Guardian Weekly*, 10 Nov. 1985, p.15.
111. 'US Electronic Espionage: A Memoir', *Ramparts*, Aug. 1972, p.41.
112. SNIE 11-13-68, *US Intelligence Capabilities to Monitor Certain Limitations on Soviet Strategic Weapons Programs*, 18 July 1968, p.5, RG-263, NA, CP.
113. HQ USASA, *Historical Summary of the US Army Security Agency, Fiscal Years 1968 – 1970*, p.61, INSCOM FOIA; US Senate, *Final Report of the Select Committee to Study Government Operations with Respect to Intelligence Activities*, 1975, Book I, p.27; General Bruce Palmer Jr, *The 25-Year War: America's Military Role in Vietnam*, (Lexington:U. Press of Kentucky 1984) pp.63, 167.
114. According to one source, in 1980 NSA consisted of 40,000 personnel, for which see Bob Woodward, *Veil* (NY: Simon and Schuster 1987) p.71. Another report indicated that in 1980 NSA had a total manpower strength of 30,000, for which see Commission on the Roles and Capabilities of the United States Intelligence Community, *Preparing for the 21st Century: An Appraisal of US Intelligence* (Washington DC: GPO 1996) pp.96, 132. For personnel at Ft Meade, see US House of Representatives, Committee on Appropriations, *Military Construction Appropriations for 1981*, Part 1, p.474.
115. 'CRYPTOLOG Interviews NSA Employee Gene Becker', *Cryptolog*, Spring 1996, p.19.
116. In Nov. 1982, Prime pleaded guilty to charges of spying for the Russians and was sentenced to 35 years in prison. Mark Urban, *UK Eyes Alpha* (London: Faber 1996) p.6; Philip Taubman, 'US Aides Say British Spy Gave Soviet Key Data', *New York Times*, 24 Oct. 1982, pp.1, 15.
117. *Report of the Commission on the Organization of the Government for the Conduct of Foreign Policy* (Washington DC: GPO, June 1975), Vol. 7, p.26; Marchetti and Marks, *The CIA and the Cult of Intelligence* (note 82) p.168; 'Eavesdropping on the World's Secrets', *US News & World Report*, 26 June 1978, p.45.
118. *Intelligence Community Experiment in Competitive Analysis: Soviet Strategic Objectives: An Alternative View: Report of Team 'B'*, Dec. 1976, p.9, RG-263, NA, CP.
119. *Report of the Defense Panel on Intelligence*, 1975, p.8, declassified and on file at the National Security Archives, Washington DC; *Report of the Commission on the Organization of the Government for the Conduct of Foreign Policy* (Washington DC: GPO, June 1975), Vol. 7, p.26.
120. Loch K. Johnson, *Secret Agencies: US Intelligence in a Hostile World* (New Haven, CT: Yale UP 1996) p.178.
121. William Drozdiak, 'A Suspicious Eye on US "Big Ears": Europeans Fear Listening Posts Eavesdrop on Their Businesses', *Washington Post*, 24 July 2000, p.A1.
122. Anthony Kenden, 'US Military Satellites', *Journal of the British Interplanetary Society*, Feb. 1985, p.64; Christopher Anson Pike, 'Canyon, Rhyolite and Aquacade: US Signals Intelligence Satellites in the 1970s', *Spaceflight*, #37, 11 Nov. 1995, p.381; Jonathan McDowell, 'US Reconnaissance Satellite Programs, Part 2: Beyond Imaging', *Quest* 4/4, pp.41–2.
123. Confidential interview.
124. The Sigint intercept in question was a transcript of a radio-telephone conversation between General Secretary Leonid Brezhnev and Defense Minister Marshal Andrei Grechko on the last day of a summit meeting with President Nixon before the signing of the SALT I Treaty. During the conversation, Grechko assured Brezhnev that the huge SS-19 ICBM, which had begun test firings at Tyuratam in March 1972, could be placed inside the existing SS-11 ICBM silos, thus bypassing the provision of Article II of the SALT I Treaty, which limited increases in silo dimensions to 15 per cent. According to publicly

available information, American negotiators 'maneuvered with it [the Sigint intercepts] so effectively that they came home with the agreement not to build an antiballistic missile defense system'. Hersh, *The Price of Power* (note 100) p.547; David Kahn, 'Big Ear or Big Brother?', *New York Times Magazine*, 16 May 1976, p.62; 'Eavesdropping on the World's Secrets', *US News & World Report*, 26 June 1978, p.47; Walter Andrews, 'Kissinger Allegedly Withheld Soviet Plan to Violate SALT I', *Washington Times*, 6 April 1984, p.1; Bill Gertz, 'CIA Upset Because Perle Detailed Eavesdropping', *Washington Times*, 19 April 1987, p.2A.
125. US House of Representatives, Select Committee on Intelligence, *US Intelligence Agencies and Activities: The Performance of the Intelligence Community*, 94th Congress, 1st Session, 1975, Part 2, pp.658–9, 680–1; US Senate, *Final Report of the Select Committee to Study Governmental Operations With Respect to Intelligence Activities*, 94th Congress, 2nd Session, 1976, Book I, p.85; *CIA: The Pike Report* (London: Spokesman Books 1977) pp.141, 143, 147; 'Eavesdropping on the World's Secrets', *US News & World Report*, 26 June 1978, p.47.
126. CINCPACFLTINST S3251.1B, *CLASSIC WIZARD Reporting System*, 16 Dec. 1983, CINCPACFLT FOIA; Hugh Lanning and Richard Norton-Taylor, *A Conflict of Loyalties: GCHQ 1984–1991* (Cheltenham: New Clarion Press 1991) p.51; 'U.S. Electronic Espionage: A Memoir', *Ramparts*, Aug. 1972, p.41; Nicholas Daniloff, 'How We Spy on the Russians', *Washington Post Magazine*, 9 Dec. 1979, p.24; Duncan Campbell, 'How We Spy on Argentina', *New Statesman*, 30 Apri 1982, p.5; George C. Wilson, 'Soviet Nuclear Sub Reported Sunk', *Washington Post*, 11 Aug. 1983, p.A9; Tyler, 'Laxness Cited At U.K. Spy Agency' (note 110); '3 Soviet Submarines Said to Patrol Atlantic Box', *New York Times*, 6 Oct. 1986, p.A6.
127. Lanning and Norton-Taylor, *A Conflict of Loyalties* (note 126) pp.63–4; Willis C. Armstrong *et al.*, 'The Hazards of Single-Outcome Forecasting', in H. Bradford Westerfield (ed.) *Inside the CIA's Private World* (New Haven, CT: Yale UP 1995) p.254; Robert M. Gates, *From the Shadows* (NY: Simon & Schuster 1996) p.133.
128. Jack Anderson, 'Project Aquarium: Tapping the Tappers', *Washington Post*, 2 Dec. 1980, p.B15.
129. For the 75,000 NSA personnel figure, see Declaration of Dr Richard W. Gronet, Director of Policy, National Security Agency, 14 June 1989, p.5, in CIV. No. HM87-1564, *Ray Lindsey v. National Security Agency/Central Security Service*, US District Court for the District of Maryland, Baltimore, Maryland.
130. Memo, Vice Admiral Bobby R. Inman, USN to Special Assistant, Office of the Secretary of Defense, *Transition Coordination*, 9 Dec. 1980, Section VIII, Modernization Objectives, NSA FOIA; Angelo Codevilla, *Informing Statecraft: Intelligence for a New Century* (NY: The Free Press 1992) p.124.
131. Johnson, *Secret Agencies* (note 120) p.21.
132. Bob Drogin, 'NSA Blackout Reveals Downside of Secrecy', *Los Angeles Times*, 13 March 2000, p.A1.
133. Col. Lori M. Sadler, USMC, 'Improving National Intelligence Support to Marine Corps Expeditionary Forces', *American Intelligence Journal*, Summer 1992, p.51.
134. Confidential interviews.
135. Among the successes enjoyed by NSA during the 1980s outside of the Soviet Union was the role Sigint played in the following crises: Aug. 1981 Gulf of Sidra incident with Libya; the covert war fought against Nicaragua and El Salvadoran insurgents during the early to mid-1980s; the bloody intervention in Lebanon from 1982 to 1984; the Sept. 1983 shootdown of Korean Airlines Flight 007 off Sakhalin Island; the American invasion of Grenada in Oct.–Nov. 1983; the *Achille Lauro* incident of Oct. 1985; the bombing of the La Belle Disco in Berlin and the resulting American air strikes against Libya in March–April 1986; and the American invasion of Panama in Dec. 1989.
136. Mark Kramer, 'In Case Military Assistance Is Provided to Poland: Soviet Preparations for Military Contingencies, August 1980', *Cold War International History Project Bulletin*, No. 11, Winter 1998, p.104; Confidential interviews.

THE NSA AND THE COLD WAR 65

137. Jeffrey T. Richelson, *A Century of Spies* (NY: Oxford UP 1995) p.386; Andrew, *For the President's Eyes Only* (note 37) p.475.
138. 'The World According to Oliver North', *Washington Post*, 21 Dec. 1986, p.D1.
139. Major A. Andronov, 'American Geosynchrenous Sigint Satellites', *Zarubezhnoye Voyennoye Obozreniye*, No. 12, 1993, pp.37–43.
140. Jeffrey T. Richelson, *The US Intelligence Community – Third Edition* (Boulder, CO: Westview Press 1995) pp.172, 179; Jeffrey T. Richelson, *America's Space Sentinels: DSP Satellites and National Security* (Lawrence: UP of Kansas 1999) p.153; 'Soviet Missile-Motor Plant Shut By Explosion, Pentagon Says', *Washington Post*, 18 May 1988, p.A29; Peter Almond and Paul Bedard, 'Explosion Deals Serious Setback to New Soviet ICBMs', *Washington Times*, 18 May 1988, p.A1; Andronov, 'American Geosynchrenous Sigint Satellites' (note 140) pp.37–43.
141. Count de Marenches and Christine Ockrent, *The Evil Empire: The Third World War Now* (London: Sidgwick & Jackson 1988) pp.99, 101; Tim Weiner, *Blank Check: The Pentagon's Black Budget* (NY: Warner Books 1990) p.153; Mohammad Yousaf and Mark Adkin, *The Bear Trap* (London: Leo Cooper 1992) p.94; John K. Cooley, *Unholy Wars* (London: Pluto Press 1999) pp.96–7, 172; David B. Ottaway, 'US Concerned About Soviets' Use of Bombers in Afghanistan', *Washington Post*, 1 Nov. 1988, p.A27; Scott Shane and Tom Bowman, 'Espionage from the Frontlines', *Baltimore Sun*, 8 Dec. 1995, pp.20A–21A.
142. Alexander Dallin, *Black Box: KAL 007 and the Superpowers* (Berkeley: U. of California Press 1985) pp.58–61; Richard Halloran, 'Soviet's Defenses Called Inflexible', *New York Times*, 18 Sept. 1983, p.A1; Walter Pincus, 'The Soviets Had The Wrong Stuff', *Washington Post*, 18 Sept. 1983, p.C5; Dusko Doder, 'Soviets Said to Remove Air Officers', *Washington Post*, 5 Oct. 1983, p.1; Bill Gertz, 'Soviet 007 Tape Revealing', *Washington Times*, 15 Aug. 1992, p.A1; William L. Norton, *Briefing on The Re-Organization of Soviet Air and Air Defense Forces* (Falls Church, VA: E-Systems Melpar Division 1984) pp.29–33, paper presented at the Strategy 84 Conference, Washington DC, 12 March 1984.
143. Desmond Wettern, 'Soviet Pilot Wrongly Identified Intruding Cessna', *Jane's Defence Weekly*, 20 June 1987, p.1289.
144. Angelo Codevilla, *Informing Statecraft: Intelligence for a New Century* (NY: The Free Press 1992) p.124.
145. For its work on 'Ivy Bells', the *Halibut* earned Navy Unit Citations in 1971, 1974 and 1975, and a Presidential Unit Citation in 1972. Sherry Sontag and Christopher Drew, *Blind Man's Bluff* (NY: Public Affairs 1998) pp.158–83; Michael Dobbs, 'KGB Chief Details US Spy Operation', *Washington Post*, 3 Sept. 1988, p.A1; Norman Polmar, 'How Many Spy Subs', *Naval Institute Proceedings*, Dec. 1996, p.87. For the Russian perspective, see Nikolai Brusnitsin, *Openness and Espionage* (Moscow 1990) pp.13–14; and N. Burbiga, 'A Fishy Day at the CIA', *Izvestia*, 1 March 1994.
146. Previously, US counterintelligence officials had believed that the operation had been compromised by CIA officer Edwin Lee Howard, who before being fired from the CIA in 1983 had been trained to retrieve the TAW tape recordings from the communications tunnel south of Moscow. Pete Earley, *Confessions of a Spy: The Real Story of Aldrich Ames* (NY: Putnam's 1997) pp.117, 193; James Risen, 'In Defense of the CIA's Daring-Do', *Los Angeles Times*, 4 Jan. 1996, p.A1. Photographs of the TAW tap and recording equipment is shown in Brusnitsin, *Openness and Espionage* (note 146) pp.14–15.
147. Jay Peterzell, 'Spying and Sabotage by Computer', *Time*, 20 March 1989, p.25.
148. Richard Harwood, 'Spies Overshadowed by the Bureaucracy', *Washington Post*, 8 Dec. 1985, p.A24.
149. Bob Drogin, 'NSA Blackout Reveals Downside of Secrecy', *Los Angeles Times*, 13 March 2000, p.A1.
150. Ronald Reagan, 'Remarks at Dedication Ceremonies for New Facilities', 26 Sept. 1986, *Weekly Compilation of Presidential Documents*, No. 39, 29 Sept. 1986, p.1278.
151. *Remarks at a Presentation Ceremony for the National Security Agency Worldwide Awards*

in Fort Meade, Maryland, May 1, 1991, http://csdl.tamu.edu/bushlibrary/papers/1991/91050101.html.
152. Harry Rowe Ransom, *The Intelligence Establishment* (Cambridge: Harvard UP 1970) p.127.
153. Ferris, 'Coming in from the Cold War (note 4) p.92.
154. Neil Munro, 'The Puzzle Palace in Post-Cold War Pieces', *Washington Technology*, 11 Aug. 1994, p.14.
155. McConnell, 'The Future of SIGINT' (note 7).

3
GCHQ and Sigint in Early Cold War, 1945–70

RICHARD J. ALDRICH

Two immediate observations can be made about the place of the Government Communications Headquarters (GCHQ) in during the early Cold War. First, far less has been written about this organisation than either of the other two British secret services. A recent bibliography on British secret service catalogues hundreds of items. Only half a dozen relate to GCHQ after 1945. By contrast the literature on post-war SIS and MI5 is vast.[1] Second, by most forms of measurement, whether it we take the volume of product, the size of budget, or numbers of personnel, GCHQ was the most important service. Personnel is the easiest indicator to track. In 1966 GCHQ and its attendant supporting collection organisations commanded about 11,500 staff. This was not only more than SIS and MI5 combined. It was also larger than the entire Diplomatic Service, including the Foreign Office in London and all its overseas embassies and consulates.[2]

How can this curious disparity be accounted for? Those who prefer a conspiratorial approach will identify a purposive element in the improbably low profile of GCHQ. In the summer of 1945 the British Joint Intelligence Committee (JIC) deemed signals intelligence, together with deception, to be the two areas that were absolutely beyond the pale in terms of the writing of the history of the Second World War. Both the JIC and the London Signals Intelligence Board (LSIB) devoted some time to the issue of how these matters might be effectively hidden. A programme of official histories, in which these matters would be airbrushed out, assisted by historians who were aware of these matters, was viewed as a key part of the process. The JIC did not expect these secrets to remain intact for long. But to their surprise it was nearly 30 years before Bletchley Park became a familiar part the wider historical vocabulary, beginning with Group Captain Frederick Winterbotham's *Ultra Secret* in 1974. By contrast, books about the role of SOE in the Second World War began to appear in the late 1940s.[3]

68 SECRETS OF SIGNALS INTELLIGENCE DURING THE COLD WAR

Yet historians needed no help from the authorities to go badly off track. The arcane matters of modern cryptanalysis are not an immediately attractive subject for historical writing. The work of special operations or secret agents, dependent on human beings for their progress, has seemed more accessible and more comprehensible. Market forces have also played their part. Since the 1960s British popular culture has developed a strong appetite for revelations about Soviet agents in government, especially 'molemania', with its rich tapestry of revelation, governmental embarrassment and tales of nefarious doings in high places. In 1963 the Profumo Scandal confirmed the growing extent of public fascination with such subjects. Thereafter, in both East and West, secret services themselves capitalised on public tastes, sponsoring more semi-official accounts of defecting agents. Kim Philby's *My Silent War* (1968) must surely count as one of the most successful. The resulting literature offers a misleading sense of the broad contours of British secret service and GCHQ figures very little in this landscape.

The 1990s saw some marginal change. The end of the Cold War triggered the release of a small, but significant proportion, of the vast archives of GCHQ and NSA for the early Cold War period. Perhaps more importantly, the increasingly central place of communications and information, together with forecasts of the 'knowledge-based economy' has required GCHQ to adopt a more public profile as its deals with issues of information security. Yet the majority of GCHQ's activities during the last half-century still remain unknown. This short essay can do no more than identify the extent to which some recent archival releases allow us to peer a little further into a largely unexplored territory.

LONG-RANGE PLANNING

The critical change at GCHQ towards the end of the war was not targets, but attitude. Led by an aggressive managerial figure, Sir Edward Travis, and having brought in highly intelligent university-educated staff who enjoyed free-thinking and free-speaking, an atmosphere of constructive self-criticism developed. In 1945 all of Britain's established secret services were still somewhat antiquated in their approach to operations, reflecting the interwar years of moribund leadership and under-funding. But GCHQ had the keenest appreciation of this fact and was the most active in seeking to transform itself.

Wartime Government Code and Cipher School (GC&CS), the predecessor to GCHQ had not been a real intelligence service, but instead a code and cipher breaking centre. It had also been something of an underdog, working in the shadow of SIS, and without even the limited organisational intelligence structures of SIS and MI5. Wartime expansion and contact with the Americans, had opened the eyes of GCHQ to what was possible. Now a core of determined individuals in GCHQ were eager to promote change. In 1944 they began the long-range planning that would turn wartime Bletchley Park – with its chess players and crossword puzzlers – into GCHQ, Britain's premier post-war secret service, with a strong sense of identity, a large budget and predatory designs on other bodies. Three key figures were instrumental in this change. First, Gordon Welchman, the man behind Bletchley Park's Hut Six. Second, Harry Hinsley, who would serve as the 'sherpa' for the Anglo-American Commonwealth Sigint summits of 1945–48. Third, Edward Crankshaw, who had handled wartime Sigint discussions with the Soviets.

On 15 September 1944, only weeks after the liberation of Paris, GCHQ set up a committee to study its post-war future. These three influential GCHQ officers set out their future vision for Travis. They surveyed the entire British intelligence scene, calling for a more centralised 'Foreign Intelligence Office' as part of a coherent British national intelligence organisation. Led by Gordon Welchman, they pressed for a comprehensive body dealing with all forms of Sigint, and also with a modern signals security organisation with the latest communications engineering. This would become a truly modern 'Intelligence Centre' governing all types of interception activities.

Before the war with Germany, they conceded, GC&CS had been 'little more than a cryptographic centre' with no ability to sift, collect or interpret material. There was little appreciation of the coming importance of electronic engineering, and little conception of the coming need for large-scale planning to cope with the exploitation of intelligence produced on an industrial scale. University-based recruiting had saved them. This had produced some natural leaders who, together with the few pre-war figures of wide outlook on the permanent staff, had made 'a passable show in this war'.

The Japanese war, with its need for major organisations overseas, would be ongoing for some time – probably into 1946 – or so they thought. They concluded that Japanese traffic presented a more complicated problem to which they could not make a major contributions. The Americans were ahead on Japanese systems and they should be left to it. Sigint for figures such as Admiral Lord Mountbatten and General Sir William Slim fighting in South East Asia was not considered a high managerial priority.

Instead Welchman's group made a hard-nosed proposal. There were few people in GCHQ with real ability in general planning and strategic co-ordination. Indeed, they said, 'it would be difficult to count as many as a dozen'. This should not be wasted on the Japanese war. So, they insisted: 'as soon as the German war is over, as many as possible of the few potential planners should be set to work in the direction of our three immediate objectives, instead of devoting more of their time to Japanese problems'. GCHQ, should not lose touch with developments in the field of Japanese Sigint problems, since there were issues worth learning from this sphere. But the approach was to extract technical benefits from the Japanese war, not to expend resources upon it. For British commanders such as Slim in Burma the tag of the 'Forgotten Army' was wholly appropriate.[4]

GCHQ moved quickly. They felt time was against them and so it was 'imperative to make an approach to the present Prime Minster at the earliest possible moment'. Any successor to Churchill, they reasoned, however sympathetic, could not have a real appreciation of 'the fruits of intelligence in this war' or Churchill's keen appreciation of the importance of tight security. In Churchill they had a heavyweight advocate and so they wished to strike while the iron was hot. They feared a return to the pre-war situation of under-recognition of what Sigint could achieve, noting that even now, the true scale of their wartime output was know to a 'very few' in 'high places'. Moreover, the really talented Sigint planners were newcomers, and soon they would return to their pre-war occupations, unless some positive action was taken to retain them. Quite simply this came down to cash. The postwar organisation had to have sufficiently high status to secure 'a sufficiently liberal supply of money to enable it to attract men of first rate ability'. They were thinking particularly of engineers and electronics experts, even now they had to subsist with 'amateurish engineering groups'. They were also sensitive to the shift to peacetime intelligence arguing that, in the post-war period, they would have to give equal wait to 'all types of intelligence about foreign countries, including scientific, commercial and economic matters', a tacit reference to the targeting of friendly states.[5]

These ambitions shaped the progress of GCHQ as it moved from its wartime site at Bletchley to new accommodation at Eastcote at Uxbridge on the suburban fringes of North-West London. By 1946 GCHQ had escaped the formal control of Major-General Sir Stewart Menzies, the Chief of SIS, to become more of a separate intelligence service in its own right. It quickly achieved supremacy in the new field of electronic intelligence, the monitoring of non-communication electronic signals from radars and missiles, known as Elint, and hitherto dominated by the three services. GCHQ's move into this field began in 1948 and was completed in 1952. But there were further battles ahead. It was 1969 before its wishes to control all aspects of signals work, including communications security (Comsec), were realised.[6]

In 1945 GCHQ continued to advocate a centralised Foreign Intelligence Office, tied closely to the Prime Minister and the Cabinet Office. William F. Clarke, who had served continuously from 1916, now applied his long-term experience to the issue of GCHQ in the post-war world. He warned that the 'enormous power wielded by the Treasury' might be brought to bear. As in 1919, work on military ciphers might cease in favour of concentration on diplomatic material only. This, he insisted 'may be disastrous' and the resulting damage to ongoing cryptographic research might mean that in a future conflict, enemy military traffic might prove inaccessible. It was also essential, he counselled, to build up the prestige of GCHQ. Its very secrecy was its worst enemy, ensuring that many in high places did not know of its true value.

There was also the 'potential danger' of a Labour government coming to power, recalling the interwar Labour government and its aversion to things secretive. Clarke also paused to consider Roosevelt's emerging United Nations, observing that if the new international organisation took the step of abolishing all code and cipher communications this action 'would contribute more to a permanent peace than any other'. 'This however' he conceded 'is probably the counsel of perfection' and seemed to him highly improbable. Instead codemaking and codebreaking seemed likely to be major activity in the post-war world.[7]

FROM WAR TO PEACE

It is all but impossible to draw a distinction between GCHQ's work on wartime Germany and her growing work on the Soviet Union in the 1940s. Knowledge of Germany required the tracking of events on the Eastern Front and involved learning as much as possible about the Soviet effort. British intelligence began to value the Germans for their knowledge of the Soviet Union as soon as Ultra came on stream. German messages used to send their own Sigint summaries back to Berlin were, in turn, intercepted. This 'second-hand' Sigint proved to be London's best source on the performance of the Soviet forces. As early as 1943 the JIC were able to produce detailed and accurate reports on the capabilities of the Soviet Air Force based on Luftwaffe Sigint material.[8]

In July 1944 the Combined Intelligence Priorities Committee began consulting at Bletchley Park about what material they wished to scoop from an occupied Germany. Suitably briefed, by early 1945, Intelligence Assault Units were moving into Germany with the forward elements of Allied formations, looking for all kinds of German documents, experimental weapons and atomic plant. Combined Anglo-American Target Intelligence Committee (TICOMs) teams were despatched from Bletchley Park to Germany to seek out cryptographic equipment and Sigint personnel. They were not disappointed. Stopping at various German headquarters along the way they ended up at Hitler's Berchtesgaden retreat in the Bavarian Alps, where they found a Luftwaffe communications centre and a large amount of communications equipment. Eventually German POWs were persuaded to lead them to a vast haul of materials buried nearby and four large German lorries were loaded to capacity with contents that were then unearthed. The teams returned to Bletchley Park with their haul on 6 June 1945.[9]

By 22 July 1945, the US Army European Theatre Interrogation Centre had completed a dossier on the 'German G-2 Service in the Russian campaign' running to over 220 pages. This gave considerable attention to the role of the Wehrmacht's Signal Intelligence Liaison Officers, 'the most important man in the circle of the G-2's sources', delivering the fruits of Germans Signal Reconnaissance Regiments tasked with wireless interception on the battlefront. Soviet radio discipline, they concluded, was very good, and much depended upon the interpretation of radio silence or knowing the transmission habits of particular operators.[10] The United States was soon seeking to reconstruct the German service if only to ensure the security of the communications of the nascent German administration. In 1947, Dr Erich Hutenhain laid the foundation of a new German crypto service based at Camp King, Oberursel, co-located with the early Gehlen Organisation. Inevitably, this unit had to be treated to surprise briefing on the inadequacy of wartime German Enigma machines.[11]

GCHQ's corporate takeover of the Axis Sigint effort was not limited to Germany. There were also dividends in post-war Italy. This flowed partly from the fortuitous coincidence that the deputy chief of the SIS station in Rome from 1944, Sheridan Russell, had previously worked at Bletchley Park and was sensitive to the fact that the Italians were talented cryptanalysts. By 1944 he was second in command of the SIS station in Rome under his namesake, Brian Ashford Russell.[12] Safe regular employment and a steady income was offered to many elements of Italian intelligence after 1943.[13]

But there were others bidding in the same market and a substantial remnant of 80 Sigint staff under Major Barbieri continued to work for the Germans at a station near Brescia until April 1945. When this latter group were finally interrogated in Rome in mid-1945 they proved to have a large quantity of material, including photostatic copies of the codebooks of Turkey, Rumania, Ecuador and Bolivia. They had reconstructed codebooks of France, Switzerland and the Vatican. They also had smaller amounts of British and American traffic. By 1945 Barbieri's unit was concentrating on French diplomatic traffic, 'a large number being messages to Paris either from BONNET [Ambassador] in NEW YORK or from CATROUX [Ambassador] in Moscow'. This traffic offered insights into subjects as diverse as Soviet-Yugoslav relations, Soviet policy in Germany, French economic negotiations with the United States and French plans for exploiting the Saar coal mines in Germany.[14]

Under British control, this Italian unit worked continuously into the post-war period without deviating from its French target. Major Barbieri's Sigint unit was an SID (Italian Army Intelligence) element within 808 Communication Service Battalion. Barbieri was proud of his efforts, but pressed for more staff. So many of the best cryptographers, he complained, had been captured by the French in Africa, adding 'the FRENCH are now employing them in their own service!'. Nevertheless, the British concluded that the Italians were 'doing remarkably well with the limited reserves at their disposal'.[15]

By mid-1946 the British were giving their Italians new tasks, including Soviet 'Taper' five-figure traffic. British liaison officers with the Italians were working closely with GCHQ in Britain on the identification of new 'Taper groups'. Remarkably, some of the Italian operators did 'not know that they are intercepting Taper traffic'. Occasionally an operator, after intercepting several typical Taper messages, would note that 'the procedure signal ... is often used by the Russians'. It was obvious that senior Italian Sigint officers knew that Taper traffic 'which had been taken with so much depth and continuity for the past month' was Soviet intelligence traffic.[16] The process was productive yet precarious. British Sigint officers handled Barbieri's organisation carefully lest they do something that might 'lead to them asking what is done with traffic they are passing' and then refusing to co-operate further.[17]

GCHQ AFTER 1945

GCHQ shrank at the end of 1945. The pressure to demobilise, combined with the end of the need for operational Sigint, affected even the privileged ranks of signals intelligence. British Army Sigint collection units shrank from 4,000 personnel in December 1945 to about 1,000 by March 1946.[18] Re-organisation was facilitated by relocation. Some of its equipment was constructed at the laboratories of the Post Office Research Department at Dollis Hill in North London and it was no coincidence that Travis chose to move the organisation to Eastcote near Uxbridge in North-West London, only a few miles from Dollis Hill. Here it remained until 1952 when his successor, chose to relocate to Cheltenham.[19]

In the late 1940s the key target for GCHQ was the Soviet atomic bomb. The British Chiefs of Staff were fascinated by the problem of Britain's relative vulnerability to attack by weapons of mass destruction and wanted forecasts on this crucial issue. The JIC ordered Britain's codebreakers to focus their efforts upon this, together with other strategic weapons systems such as chemical and biological programmes, ballistic rockets and air defence. Although the JIC placed these subjects in a special high priority category of priority it was to no avail. The Soviet bomb, took the Western allies by surprise in late August 1949.

Other Soviet activities, including espionage and diplomatic initiatives constituted second and third priorities, but here too there were thin pickings. Many Soviet messages employed onetime pads which, if correctly used, could not be broken, although some machine-based military ciphers seem to have been read. The extent to which Britain was surprised by the Tito–Stalin split in 1948 underlines limited success enjoyed against its diplomatic targets. Secure Soviet communications were only part of the problem. Moscow and its satellites used landlines which could not be easily intercepted, instead of wireless transmissions. It was these problems that prompted the British to follow the Soviets down the path of more extensive physical bugging in the mid-1950s.[20]

High priority targets aside, GCHQ was nevertheless providing Whitehall with large quantities of material in the late 1940s, albeit of a secondary and tertiary order. They continued to attack the communications of many states with vulnerable cipher systems. Some were persuaded to adopt Hagelin-type machines previously used by the Axis, in the belief that these machines provided a secure means of communication. This was a belief that GCHQ did nothing to undermine. The JIC had requested material on subjects such as Arab nationalism and the relations of Arab states with the UK and USA, the attitude to the Soviet Union, France, Italy and the Arab states towards the future of the ex-Italian colonies, especially Libya. GCHQ was also urged to focus on the Zionist movement including its intelligence services. These subjects proved more accessible. In 1946 Alan Stripp, a codebreaker who had spent the war in India working on Japanese codes, suddenly found himself redeployed to the Iranian border. During the Azerbaijan crisis of 1946 he worked on Iranian and Afghan communications for the duration of the crisis.[21]

Although GCHQ was always the predominant post-war British secret service much of its activity was hidden by the use of the signals units of the armed services for interception. Each of the three services operated half a dozen sites in Britain. GCHQ also had civilian outstations including a Sigint processing centre at 10 Chesterfield Street in London, a listening post covering London at Ivy Farm, Knockholt in Kent and a Post Office listening post at Gilnahirk in Northern Ireland.[22]

GCHQ had overseas stations hidden with embassies and high commissions in countries such as Canada and Turkey. There were also service outposts. In the Middle East, the base of Ayios Nikolaos, just outside Famagusta on Cyprus became a critical intelligence centre, receiving Army and RAF Sigint units as they gradually departed from Palestine, Iraq and Egypt.[23]

Further east, the Royal Navy maintained its shore intercept site at HMS Anderson near Colombo on Ceylon, and the Army began reconstruction of its pre-war Sigint site at Singapore. But the main British Sigint centre in Asia after 1945 was Hong Kong, initially staffed by RAF personnel. Here, together with help from Australia's Defence Signals Directorate, they captured Chinese and Soviet radio traffic.[24]

Despite London's decision to give GCHQ the lion's share of British intelligence resources and the tendency to bury some of the costs in other budgets, resources were tight. On 22 January 1952, the Chiefs of Staff had met together with the Permanent Under Secretary of the British Foreign Office to review plans for improving British intelligence. GCHQ came out on top in this exercise. Its cutting edge programmes, mostly in the area of computers and 'high speed analytical equipment' for communications intelligence were given 'highest priority', and government research and supply elements were instructed accordingly. The Admiralty were beginning a new programme to build better receivers for ground-based and seaborne 'Technical Search Operations'.

Elint was no less critical and so new airborne radio search receivers and direction-finding (D/F) projects run out of the Central Signals Establishment were also to be given 'all possible priority'. The Chiefs of Staff continually reiterated the 'very great importance' to speeding up development and construction in these 'very sensitive' areas.[25]

By November 1952 a major review of British intelligence was underway. The process was prolonged by the primitive nature of available managerial instruments. Patrick Reilly, who liased between SIS and the Foreign Office, confessed that no one knew what Britain spent on intelligence. Now, 'for the first time', Sir Edward Bridges from the Treasury and a committee of permanent secretaries was assembling some figures so they could review intelligence costs in the context of the overall defence budget. The Chiefs of Staff wanted 'increased expenditure on intelligence' within the general programme of Korean War re-armament, but were unsure of the figures or how much detail to give to ministers. All were clear that in the short term the emphasis should be 'for Sigint'. Eric Jones, the Director of GCHQ, reported that he was busy filling the 300 extra staff posts recently authorised. GCHQ had proposed a further increment for an extra 366 staff as to follow. GCHQ was moving from strength to strength.[26]

GCHQ AND ITS ALLIES

As early as 1945, most English-speaking countries had committed themselves to post-war signals intelligence cooperation. Policy-makers at the highest level had come to expect a world in which a global Sigint alliance rendered enemy intentions almost transparent. They were not about to relinquish that privilege willingly. In the autumn of 1945, when President Truman was winding up OSS, he was also giving permission for American Sigint activity to continue and approved negotiations on continued Allied cooperation.[27] All desired the maximum option. On 19 November 1945, Admiral of the Fleet Sir Andrew Cunningham, the Chief of the Naval Staff attended a critical meeting of the British Chiefs of staff. There was 'much discussion about 100 per cent cooperation with the USA about Sigint'. he recorded, adding 'decided that less than 100 per cent was not worth having.' In Ottawa, George Glazebrook recommended to the Canadian JIC that Canada enhance her independent Sigint effort in order to stake claim in this secretive and emerging cooperative system. 'It is paramount' he insisted 'that Canada should make an adequate contribution to the general pool.'[28]

Yet the way ahead was strewn with obstacles and the package of agreements, letters, and memoranda of understanding, often referred to as the UKUSA treaty, that sealed this alliance, was not completed until 1948. As this agreement emerged, Britain derived considerable benefit from her dominance over Empire-Commonwealth partners. The semi-feudal relationship which London enjoyed is no better illustrated than in Australia where Sigint operations were controlled by London. Only in 1940 did Australia established her own separate organisation. When this became the Australian DSD, formed at Albert Park Barracks in Melbourne on 12 November 1947, it remained in the shadow of GCHQ. Four Australian applicants for the directorship were rejected in favour of Britain's Commander Teddy Poulden, who filled the senior DSD posts with 20 GCHQ staff and communicated with GCHQ in his own special cipher. During the winter of 1946–47, a Commonwealth Sigint conference was held in London, chaired by Travis, during which each country received designated spheres of activity.[29]

Canada's Sigint organisation under the long-serving Lt. Colonel Ed Drake suffered similar treatment. On 13 April 1946 the Canadian Prime Minster, Mackenzie King authorized the consolidation of several wartime organisations into a small post-war unit of about 100 staff known as the Communications Branch of the National Research Council (CBNRC). Again, the senior post were filled by staff seconded by GCHQ, prompting them to say that CBNRC stood for 'Communications Branch – No Room for Canadians', and by the late 1940s Drake had resolved to offset this by developing better relations with the US Army Security Agency.[30]

However, the Americans were also inclined to give Canada second-class treatment. During the 1948 discussions of the UKUSA Agreement it became clear that the US Communications Intelligence Board was equally anxious to prevent an information free-for-all among all its signatories.

They preferred to hand material to the Canadians 'on a "need to know" basis' and were anxious to prevent a proliferation of liaison officers.[31]

Although GCHQ representatives were often over-awed by the scale of American Sigint resources, matters looked quite different from Washington. With the war over and an economising Republican Congress controlling the federal purse-strings, resources for American Comint interception activities were remarkably tight before 1950. As Matthew Aid has conclusively shown, this led to a state of parlous under-preparedness prior to the Korean War. It also prevented the European expansion that American Sigint had hoped for. In 1949, Army Security Agency interception units in Europe were still passing their product to GCHQ rather than back to Washington, for analysis. GCHQ retained primary responsibility for areas such as Eastern Europe, the Near East and Africa.[32]

The exchange of communications intelligence between these allies was of several types. A narrow range of Comint-producing agencies exchanged all manner of data, both raw and processed. A much wider range of bodies circulated the finished product. The key instrument was the 'Comintsum', a digest of latest 'hot' material which made its way around Comint-cleared centres. London would send 20 copies of this document to Washington on a regular basis, with two copies going to air force intelligence, two to army intelligence and so forth.[33]

UKUSA only codified and smoothed out what was clearly a pre-existing practice. Thus, as early as 28 April 1948, General Charles Cabell, Head of USAF Intelligence, reviewed the intelligence arrangements in support of the atomic strike plan 'Halfmoon'. 'At the present time' he noted with satisfaction 'there is complete interchange of communications intelligence information between the cognizant United States and British agencies. It is not believed that the present arrangements ... could be improved.'[34]

The 1950s saw the development of spheres of influence. For example relations with Norway were an American responsibility, while relations with the Swedes belonged to GCHQ, although this demarcation was not strictly adhered to.[35] GCHQ enjoyed the benefits of a panoply of bases provided by Britain's imperial and post-imperial presence. Although the empire was shrinking, the very process of retreat often rendered the new successor states more willing to grant limited base facilities to the departing British. These facilities seemed innocuous, and often termed 'communications relay facilities', however the reality was often different. Many countries were unwitting hosts to GCHQ collection sites.

Ceylon, which became independent in 1948, is a good example. Britain was allowed to retain a communications relay centre at HMS Anderson, close to the city of Colombo, on Ceylon, which was in practice a large Sigint collection site. In 1949 the Ceylonese government decided to develop the area where the aerial farm was located and so asked the British to move sites. London was amenable for officials were convinced that again, at the new site, 'the real purpose could be easily disguised'.³⁶

By October 1951 the new station that would replace Anderson was being planned and Dr John Burrough, a senior GCHQ official, was attending regular meetings at the Admiralty to discuss the technical problems. It proved difficult to identify a site which was not too remote and yet did not suffer from interference either from town or from naval transitions. Even the ignition systems of cars on a busy highway up to 500 yards away could cause unacceptable interference. Moreover this station was upgraded to monitor signals traffic from 'all bearings' and needed a facility that covered more than 400 acres. By 1952 they had decided on a site at Perkar, about two miles from the old station at Anderson.³⁷

Ceylon revealed how the remnants of empire helped to offset the imbalance between British and American capabilities. In January 1951, the US National Security Council noted that the Pentagon and the CIA had been keen to construct 'elaborate radio facilities to be operated by US personnel' in Ceylon. But Ceylon had resisted the idea of even a 'modest' US Navy radio station. They had accepted they were every unlikely to establish a foothold here and that any collection from this site would be conducted by the British.³⁸

Island locations, including Britain, were intrinsically attractive because they would be slower to be overrun in any wartime military operations. In September 1951, the 14th Radio Squadron Mobile, a unit of the US Air Force Security Service, was deployed to Cyprus and tasked with 'conducting communication intercept activities' across the Middle East. A further US Sigint site on Crete had been authorised and other in Greece and Iran were being investigated. The product, while shared with the British, was not disseminated to other NATO partners in Europe.³⁹

In time of war, the emergency evacuation plans for Cyprus called for US Sigint personnel to be relocated in British bases in the Suez Canal zone under a scheme codenamed 'Applesauce'. In Asia the reverse was true. Sigint units from Hong Kong would be relocated to US bases on Okinawa. British Sigint units in Hong Kong were joined by the Americans, anxious to expand capability following the outbreak of the Korean War. On 26 November 1951 the Senior US Liaison Officer in Britain informed the London Signals Intelligence Board of Washington's 'urgent need' for a US Air Force Sigint unit to be deployed to Hong Kong. Other sites were developed on Taiwan in an attempt to remedy the yawning intelligence gap in East Asia revealed by the outbreak of the conflict in Korea.⁴⁰

In the 1950s, Anglo-American relations were made easier in the Comint field by the arrival of the National Security Agency (NSA) which imposed some order upon the squabbling of the US armed services. In 1952 the Brownell report had recommended to Truman the creation of a strong centralising force. The three separate American armed services fought a desperate rearguard action against the creation of the NSA. General Samford of US Air Force intelligence denounced 'strong central control of the national COMINT effort' as a 'major error'. He also warned darkly about Comint slipping away towards civilian control under the office of the Secretary of Defense. But Truman's mind was made up and the NSA began to reshape American Comint.[41] However, the efforts of NSA to extend its control over Elint and to 'fuse' it with Comint processes would be more troubled and stretched on into the 1960s. American officials often envied the more centralised British model.[42]

Resources enforced increasing British dependency upon the United States during the early 1950s. Dependency could be disguised in the field of Sigint gathering, even in some aspects of processing. But in other areas, such as communications security or Comsec, the threadbare nature of the British effort was all too obvious. Many British machines used for enciphering communications were still of Second World War vintage and no longer offered the desired level of security. Britain and indeed other NATO countries could not afford to replace these machines rapidly. The United States offered assistance, allowing some of its cryptographic principles to be adopted by Britain and paying for new machines operating on combined circuits carrying communications between the US, UK and NATO.

The Combined Cipher Machine (CCM) widely used in the early 1950s and superior to the British Typex, was nevertheless a major subject of concern. American assistance allowed NATO to aim to dispose of it by 1955. While making available American cipher principles and machines for NATO use, it reserved a higher grade system, the CSP 2900 cipher machine 'for exclusive U.S. use'.[43]

A shared problem for London and Washington was the need to supply Comint to multi-national centres such as NATO and Supreme Headquarters Allied Powers Europe (SHAPE). NATO was considered to be insanitary from a security point of view. Even General Dwight Eisenhower's SHAPE HQ, where British and American officers only mingled with the French, was considered a major security problem. Yet SHAPE's efficacy in war depended on strong flow of Comint to support the direction of operations. In early December 1952 London hosted an Anglo-American conference to work out a solution with the Intelligence Chief at SHAPE. The US Army was to place a liaison officer at GCHQ to co-ordinate a flow of Comint to SHAPE and the NSA was to supply advanced cipher machines for the purpose. Remarkably, the whole system was to remain 'informal' to allow the material to stay in the hands of British and American officers only.[44] This was complemented by an arrangement where the senior intelligence officer dealing with Sigint at SHAPE was always British or American and never French.[45]

SPECIAL PARTNERS – AIR FORCES AND ELINT SHARING

The closest Anglo-American intelligence relationship during the immediate post-war period as probably that developed between RAF intelligence and the US Air Force. General Charles Cabell (later Deputy Director of the CIA) was head of US Air Force Intelligence as the USAF became fully independent of the US Army in 1947–48. While establishing an expanded intelligence organisation and getting to grips with being a fully independent service, the American found RAF intelligence to be an ideal partner. RAF intelligence was headed by the convivial Air Vice-Marshal Lawrence Pendred, who was anxious to cement the Anglo-American relationship. This growing friendship also reflected the fact that GCHQ had identified air power as a critical area for Sigint, especially those arcane forms of Sigint associated with strategic bombing. Sigint in the air was one of the major growth areas of the early intelligence Cold War.

Air intelligence was keen to develop 'Elint' or electronic intelligence. This involved the interception of electronic signals that did not carry messages, but instead offered information about subjects such as radar sites and air defences. Such information was invaluable for the operational planning for air attack against the Soviet Union. It was equally invaluable to anyone planning peacetime 'spy-flights' over Soviet airspace and looking for gaps in Soviet radar cover. Thus, in this area, air intelligence collectors were also consumers, not least to protect the security of their own missions. Elint was first developed by the Allies in the face of radio-guided German air raids during the Second World War and was later sited at the Central Signals Establishment at RAF Watton.[46]

Towards the end of the war it continued to be refined against Japan. An elaborate Elint unit was set up under Mountbatten's South East Asia Command under the improbable cover name of the 'Noise Investigation Bureau'. Elint equipped 'Ferret Aircraft' patrolled the night skies over Rangoon and then Singapore listening to Japanese radars in the spring and summer of 1945.[47]

Exchange in this area began early. General Curtis LeMay had been given permission to begin exchanging Elint with the British on an informal basis at the end of 1947, but it is likely that *ad hoc* cooperation began earlier. In 1948 Elint sharing was being brought within the growing body of Western Sigint agreements. At this time GCHQ was attempting to extend its control over service-based Elint activity. GCHQ approached Washington with a proposal to 'extend the present British-U.S. Comint collaboration to include countermeasures, intercept activities and intelligence' in the field of Elint. This was put forward by Colonel Marr-Johnston, the GCHQ liaison officer in Washington, who then held discussions with Captain J. Wenger, a senior naval cryptanalyst. He suggested coordinated patterns of 'Ferret' flights with the resulting intelligence being swapped 'via Comint channels'.[48] By 1952 GCHQ had achieved complete control over the Elint field in the UK and was managing relations with the Americans in this field.

Initially, the RAF was ahead in this new field. By 1947 a fleet of specially equipped as fleet of Lancaster and Lincoln aircraft patrolled the East German border, monitoring Soviet air activity. This was complemented by a programme of monitoring of basic low-level Soviet voice traffic by ground stations at locations such as RAF Gatow in Berlin. British 'Ferrets' began their first forays into the Baltic in June 1948 and the Black Sea in September 1948.[49] In 1948 they began to supplemented by American prototype B-29 'Ferrets' flying missions from Scotland to the Spitzbergen area. B-29 'Ferrets' were also supplied to the RAF under the Mutual Assistance Act from 1950.[50]

By 1948 much of the perimeter of the Soviet Union was covered. A British undercover team was operating in northern Iran monitoring Soviet radar in Caucasus as well as Soviet missile tests at Kasputin Yar on the edge of Caspian Sea. The team conducting this work were posing as archaeologists, a favourite British cover for all sorts of intelligence work, including atomic intelligence work ongoing in India at the same time. This information was useful for RAF crews flying aerial reconnaissance of this area from bases in Crete from 1948.[51]

The Royal Navy conducted operations around the Soviet northern periphery. In October and November 1949, the cruiser HMS *Superb* undertook a month-long Elint investigation of the Kola Peninsula and the naval base of Murmansk. The Royal Navy also maintained a chain of fixed stations in the UK and a forward listening station at Kiel on the Baltic.[52] The destruction of US Navy Elint aircraft off the coast of Latvia in April 1950 while trying to identify new Soviet missile bases seemed to indicate that aerial collection in these areas was more hazardous than ship-based or submarine-based collection. Further missions were postponed, but the outbreak of the Korean War resulted in enhanced demand for intelligence and operations resumed. From 1952 onwards much of this work was carried out by RB-50Gs operating out of RAF Lakenheath in East Anglia.[53]

Elint in northern areas was a multinational activity. During the war Bletchley Park had worked with the Norwegians and in 1946 the RAF were assisting the Swedish Air Force in investigating what were thought to be

Soviet rocket tests that intruded into Swedish air space. Washington took responsibility for cooperation with the Norwegians and encouraged reconnaissance in the area of Murmansk and Novaya Zemlya. By January 1949, detailed material on Soviet radars from Swedish intelligence was making its way to Washington via British representatives who had taken responsibility for cooperation with Sweden.[54] There was particular interest in the possibility that the Soviets might be attempting the further development of German stealth technology, such as radar absorbent coverings for submarine periscopes and snorkels.[55]

The rearmament that followed the Korean War prompted a major expansion of command communications and navigation which in return demanded a greater Elint input. Some Elint could be monitored from sites in Britain and these were expanded. For example in 1952 the 47th Radio Squadron of the US AFSS opened a station at Kirknewton airbase in Scotland listening to activity around the Kola Peninsula. But a great deal of traffic was short-range requiring collection by ships and aircraft.[56] By October 1952 Elint had become so large that liaison arrangements had to be expanded. London proposed the appointment of an additional officer, Squadron Leader J.R. Mitchell as 'liaison officer for GCHQ' specialising on Elint. Washington agreed and appointed William Trites and Forrest G. Hogg to equivalent roles in Britain.[57]

The Comint and Elint effort against the Soviet Air Force and associated strategic systems was one of GCHQ's key areas of achievement in the first post-war decade. The arrival of the first Soviet atomic bomb may have eluded it, but its subsequent operational deployment certainly did not. During the early 1950s the Joint Intelligence Bureau in London and the USAF target intelligence staffs had been busy exchanging sensitive data on 'the mission of blunting the Soviet atomic offensive'. This involved the early counter-force targeting of Soviet nuclear forces in the hope of destroying them on the ground before they could be used. This was a politically sensitive issue because it raised the issue of the use of nuclear weapons at an early stage in any future conflict. Nevertheless, senior officers in London had given particular attention to this matter because of the vulnerability of the UK. The Americans were impressed by the 'considerable progress that London had made on the counter atomic problem'. GCHQ and the RAF had amassed 'a significant amount of evaluated intelligence, particularly in the special intelligence field, which would be of the greatest value'. Most airfields and the operational procedures for Soviet strategic air forces in the European theatre had been mapped by 1952.[58]

The full Anglo-American intelligence exchange in this field was somewhat ironic given the different views held in London and Washington on nuclear strategic issues at this time. However, full intelligence exchange on targets continued regardless.

FLAPS AND SHOOT-DOWNS

Between 1956 and 1960 several 'incidents' reverberated upon intelligence-gathering from seaborne and airborne platforms. In each case ministers in London reacted more strongly than their counterparts in Washington, constraining the nature and frequency of subsequent operations. For the practitioners this underlined the value of working with allies. In the late 1940s and early 1950s the British had been more relaxed about forward operations, such as over-flights, and had passed their dividends to Washington. After 1956 the situation was reversed. London's hesitancy in the face of various flaps and shoot-downs accelerated the shift of momentum in the world of signals intelligence towards the United States.

The scale of political embarrassment that could be generated by bungled surveillance operations was first underlined by the infamous Commander 'Buster' Crabb incident. In April 1956 operations were mounted against the Soviet cruiser *Ordzhonikide* during the visit of Soviet leaders Bulganin and Khruschev to Britain. Despite some robust exchanges the visit went well and the Soviet delegation departed on 27 April 1956. But even as they left the press had begun to speculate about the mysterious disappearance of a British naval diver, Commander Lionel 'Buster' Crabb RNVR in the vicinity of the visiting Soviet warships. His headless body was recovered from the sea.[59] Prime Minister Eden intended to take 'disciplinary action' and told the ministers concerned to order their staff to cooperate fully with the inquiry.[60] Sir John Sinclair, the Head of SIS was replaced by Sir Dick White, Head of MI5.

Sir Edward Bridges, a somewhat nineteenth-century figure, conducted a thorough enquiry, employing the JIC mechanism to help him ferret out all aspects of the Crabb affair. Bridges rightly identified 'certain questions' of a broader nature arising out of the Crabb affair. On the one hand intrusive intelligence operations clearly had a capacity to cause international repercussions, but on the other hand the systems for their authorisation were unclear. Bridges recommended a new and broader enquiry reviewing all of Britain's strategic intelligence and surveillance activities. It would assess 'the balance between military intelligence on the one hand, and civil intelligence and political risks on the other'. Eden gave this job to Sir Norman Brook, the Cabinet Secretary, working with Patrick Dean, Chairman of the JIC.[61] This review had important consequences for intelligence. In April 1956, simultaneous with Khrushchev's visit to Britain, the first CIA U-2 jet reconnaissance aircraft had arrived at RAF Lakenheath and some U-2 work was Sigint-orientated. Eden now decided that this, and a host of other operations, had to go.[62]

Eden's review also impacted on naval Sigint. Even more secret than the U-2s were joint intelligence operations by British and American navies using submarines. But in the backwash from the Crabb affair, British submarine operations were cancelled and so the British half of the deal on Anglo-American submarine-derived Sigint could not be delivered. British officers in Washington spoke of their 'embarrassment' which would persist 'until we can make good our part of the bargain'. Their underlying concern was that Britain would be eclipsed by similar operations by American submarine commanders in the Atlantic which they were expanding 'so as not to be outdone by the Pacific submariners'. British Naval Intelligence wanted to keep their stake in the game and so urged not only that current operation be restored, but that it be followed by 'a bigger and better operation'.[63]

Rear-Admiral John Inglis, the British Director of Naval Intelligence in London, was agitated. The main scoop provided by this series of American operations had been a choice selection of short-range Comint and Elint: 'Considerable VHF voice, IFF and radar traffic' was recorded, mostly from airborne and coastal defences. The take was voluminous. Moreover, while the Soviets seemed prepared to repel 'unfriendly air intrusion' by contrast 'no difficulties were placed in way of submarine visitors' and Soviet anti-submarine capability seemed low. The Commander in Chief US Pacific Fleet was already pressing Washington to abandon the 12-mile restriction on operations near the Soviet coast. But the question now was, were there to be any further British operations?[64]

By the end of 1956 the Royal Navy felt things slipping away from it. Admiral Robert Elkins, the senior naval officer at the British Joint Staff Mission in Washington wrote to Mountbatten to voice his concern. As predicted, the US Navy was beginning its own independent operations off Murmansk. Initially, the American Office of Naval Intelligence had decided that the British Naval Intelligence Division was not to be informed. But Admiral Warder from the secretive American Op31 section tasked with this mission decided that it would be foolhardy not to draw on extensive British experience of similar operations in these waters. So the British Commander John Coote, who had been on the Murmansk run several times, was called in to brief the first American crew. But this was only on the understanding that he told no other British naval officers in Washington.

These new American submarine intelligence operations off Murmansk had been triggered by two factors. First, the cancellation of British operations. Elkins lamented 'we are no longer providing sufficient cover in an area where we have hitherto been a reliable and productive source'.

Second, the US Navy had used the reports of previous British intelligence operations off Murmansk to persuade the State Department that these activities were valuable while 'the risks of detection are negligible'. Elkins accepted that the British cancellations had been a high-level political decision. But he also warned that British prestige in the operational and intelligence fields, which was currently high, would soon suffer 'unless we resume these activities ourselves'.[65]

Later, under Harold Macmillan, intrusive operations using British aircraft, ships and submarines for photography and Sigint were gradually resumed. Moreover, between 1956 and 1960, 20 U-2 aircraft were involved in overflights. Some U-2 flights used British bases or pilots. Most of the deep penetration flights were launched from Adana in Turkey and six RAF pilots were based there. Because Turkey would not allow penetration directly into the Soviet Union, U-2s staged on to Peshawar in Pakistan before crossing the Soviet border. Along the southern border of the Soviet Union Soviet radar stations were more dispersed and a variety of attractive targets presented themselves including a range of missile testing centres in Kazakhstan, on the Caspian Sea and at Kapustin Yar on the Volga. Sites in Kazakhstan were of particular interest because of suspicions that the Soviets were working on anti-ballistic missiles there. Some of these flights substituted Sigint packages for cameras, however the Sigint package that the U-2 could carry was fairly light. Serious airborne Sigint activity of an intrusive variety was usually carried out by the American-modified version of the British Canberra jet, the RB-57D which, with improved engines and a vast wing-span, could reach nearly 60,000 feet, compared with the 70,000 ft available to the U-2.[66]

Even U-2 flights that were carrying out photography also served a useful purpose for Sigint collectors, triggering intense Soviet air defence activity. One U-2 flight launched from Bodo in Norway followed the Soviet-Polish border before heading for the Black Sea and landing in Turkey with the specific purpose of provoking Soviet air defence while its Elint package logged the characteristics of radars that locked on during the journey. Over 60 Soviet fighters had been launched against this U-2 flight and it was only a matter of time before the Soviets found a means to deal with the U-2s.[67] In 1959 a RB-57D was lost returning to Taiwan after it descended too early. The Norwegians, detecting which way the wind was blowing, resisted pressure from the American to use northern Norway as the host for 'Ferret' flights against Soviet missile tests.[68]

The loss of the Gary Powers U-2 aircraft occurred in May 1960. A month later an American RB-47 'Ferret' aircraft was lost over the Barents Sea engaged in maritime surveillance, very close to Soviet airspace. The latter aircraft had taken off from RAF Brize Norton in Britain. American and Norwegian Sigint stations had tracked the aircraft, but disputed its course, plotting it 30 miles and 23 miles respectively from the Soviet coast. The aircraft crew had received orders not to go closer than 50 miles. The Soviet coastal limit was 12 miles and the margin for error was small.[69]

The twin shoot-downs reverberated in Britain in the early summer of 1960. There was a public furore and questions in the House of Commons. Prime Minister Macmillan was bitter when the Gary Powers shoot-down contributed to the collapse of the East-West Summit in Paris, by which he had set much store. The Soviet Union exploited this to the full, threatening countries such as Britain and Japan, which hosted U-2 and RB-57D flights, with rocket attacks against the bases from which future flights were made over 'Socialist' countries. These threats were first made by the Soviet Minister of Defence, Marshal Rodion Malinovsky, on 30 May 1960 and were reiterated on 3 June to a packed press conference by Nikita Khrushchev himself. The JIC in London concluded that these threats were a bluff, nevertheless they induced a new climate of extreme caution on the part of Macmillan.[70]

The impact of these events in the summer of 1960 was similar to the Crabb affair in 1956. They served to crush a British plan for increased airborne surveillance of the Soviet fleet that had been emerging in the weeks and months immediately prior to the loss of the Gary Powers's U-2 aircraft. In early 1960 the First Sea Lord had held a meeting with the US Navy's Chief of Naval Operations (CNO) and agreed to an 'increased accent on surveillance'.[71] By March 1961, British plans for increased airborne surveillance of the Soviet fleet were put 'into cold storage indefinitely'.[72] Other long-established British programmes came to a close.

Macmillan now required the JIC to prepare a review of all aerial surveillance and submarine surveillance tasks so that he could assess provisionally the value of the intelligence gained from these sorts of activities.[73] These developments contained an element of irony. In the 1950s Britain and the United States had increasingly turned to technical means of examining the Soviet armed forces and Soviet scientific-technical developments, because human espionage inside the Soviet Union had proved increasingly hazardous and, with a few exceptions, notably unproductive. Forward technical surveillance was now proving to be less than risk free.

Safer alternatives certainly existed. As early as 1956 the British had begun cultivating an alternative form of seaborne surveillance: the possibility of gathering intelligence on the Soviet fleet from the relative safety of British trawlers operating in northern waters. This was a direct copy of the Soviet Sigint trawler that became ubiquitous by the 1960s. This sort of activity was less provocative, though not without risk.[74]

GCHQ AND SIGINT 87

Tensions between major powers often provide opportunities for minor powers. Turkey constitutes a good example of this, for throughout the 1950s, both British and American airborne and ground collection and operations had made considerable use of Turkey. In 1958 an American C-130 Hercules Elint aircraft had been destroyed by Soviet fighters on the border. Understandable nervousness at this had been heightened by the twin shoot-downs of 1960. On 27 March 1961 the Turkish General Staff placed new and severe restrictions on overflights, including a ban on approaching the border within 100km and a height ceiling of 40,000ft. The effect was to render Western perimeter flights for Elint purposes quite inoperable, and the Turks knew it.

The result was a high-level meeting attended by the Chief of the Turkish General Staff, General Sunay; the British Chief of Defence Staff; the American General Lemnitzer together with Captain W.D. Hodgkinson RN, a British Sigint specialist. The Turks were extremely evasive and it was over an hour and a half before they could bring the Turks to address the issue in hand. Instead the visiting party had to endure long diversions designed to avoid the matter of Sigint, while urging the assembled audience that Turkey needed more US equipment. Eventually they broached the key issue. British and American officers insisted that Sigint operations in Turkey were critical in supporting nuclear strike forces and were thus of 'vital importance to the maintenance of the deterrent'. Prolonged bargaining ensued. General Sunay agreed to remove the height restriction and allow flights within 50–70 km of the border. However, the local British air attaché, who was also present, expected the Turks to renege on the deal later warning that the Turks were trying to exploit the recent crises. What the Turks were really after was more American equipment for their own Sigint service and for their own jamming units. Further protracted bargaining marked the road to resumed operations.[75]

COMSEC AND GCHQ SECURITY

By the late 1950s Western signals intelligence was slowly entering a new era. High grade ciphers, for example onetime pads, used by the major powers, remained effectively impossible to break by the sweat of direct cryptanalysis, if they were employed correctly. As a result, increasing efforts were made to tap communications before they were enciphered. The era of large-scale bugging had arrived, accelerated by the development of transistors. Soviet efforts were revealed by the accidental location of a microphone in the office of British Naval Attaché in Britain's Moscow Embassy in July 1950. The British Air Attaché, who was testing a radio receiver, heard the voice of his colleague being broadcast loud and clear from another part of the building. An active search ensued but, alarmingly, the Soviets succeeded in removing the device before it could be found. In 1952 more bugs were found in the office of the American Ambassador, George Kennan, using 'a special British detector'. In 1956 conference rooms in the US EUCOM Building was found to be seriously compromised by listening devices.

Britain had not been slow to retaliate and in October 1952 Churchill ordered British defence scientists to begin a vigorous programme of developing British bugs for offensive use against the Soviets. By the late 1950s this was a busy field of activity.[76]

Bugging, direct tapping of landlines and the breaking of the communications traffic of minor states, ensured a stream of signals intelligence was routinely available to Whitehall. Little of this can be seen in the archives today due to the nature of security procedures attending it. Sigint material and ordinary working files never mixed. Before gaining access to Sigint, Foreign Office officials were required to attend a day course on Sigint security. Foreign Office staff could then go on the circulation list for BJs – Blue Jackets – the colour of the special folders in which Sigint material was circulated. This material was never to be referred to in ordinary Foreign Office paperwork and always remained its special blue jackets. BJs were circulated by special messenger, originating in the Permanent Under Secretary's department and always returned there after use. There, in a small office in this Department sat the Communications Security Officer, the workaday liaison with GCHQ. More humble files dealing with policy and correspondence lived in the Foreign Office registry. This hermetic separation has ensured the near invisibility of Sigint to post-war diplomatic historians.[77]

By the 1960s GCHQ found itself increasingly embattled. It needed more resources, for it was working in a field at the cutting edge of technology and engaged on 'many problems on the edge of what appears to be possible'. A large part of this strain was imposed by the growing threat to British communications.[78] London was forced to give attention to hostile Soviet Seaborne Comint-gathering activities. Shadowing Soviet naval vessels was unattractive, raising the possibility of an international incident, yet not to track them was risky. In September 1964, Peter Thorneycroft, the Minister of Defence, wrote to Prime Minster asking for permission to use naval vessels to shadow these sorts of craft during a forthcoming NATO exercise. This had previously been done with aircraft but during the last NATO exercise the number of flying hours expended on this problem had been enormous. Prime Minister Alex Douglas-Home was persuaded, partly because the JIC had now drawn up safer guidelines for the naval shadowing of Soviet Sigint collection vessels, known as the 'Sampan' rules.[79]

Prior to 1969, Comsec was not the responsibility of GCHQ, but of the London Communications Security Agency (LCSA) with its headquarters at 8 Palmer St, in London SW1. Inevitable turf fights developed between the well-resourced Cheltenham and its small London-based Cinderella sister.[80] LCSA nevertheless enjoyed a global network, consisting of regional Communications Security Committees, watching over British communications. They conducted their own monitoring of British radio, telephone and exchange links in exercises designed to simulate Soviet efforts and find the weak points in British security.[81]

Security within GCHQ itself also constituted a huge problem that was quite different to security problems in other areas of government. Elsewhere in government, anyone who was regarded as a possible security risk could be gradually transferred to less sensitive areas that acted as 'dumping grounds' for the unreliable. Such dumping grounds did not exist within GCHQ. There were also union problems. Although communism was not strong in the Britain, it was strong in certain unions, including the Electrical Trades Union (ETU), where it had extended its grip partly through a process of ballot-rigging. The ETU was important at GCHQ and the Americans were alert to this sensitive issue. In 1963 the CIA completed a long report on communism in Britain and identified Communist influence in the ETU as one of several awkward problems.[82] In the 1960s union disputes involving civilian wireless operators at the GCHQ/DSD site at Little Sai Wan in Hong Kong periodically halted intelligence output from that station.[83]

Even from the mid-1950s there was a gradual but deliberate drive to reduce some the civilian components of parts of the British Sigint programme and instead to replace them with service personnel. This was triggered by what the authorities termed 'an embarrassing domestic security case' involving the ETU at GCHQ in the spring of 1954. One result was a decision to bring together the monitoring elements of all the various services who contributed to GCHQ's collection effort. This meant welding elements of the Army, Navy, Air Force and Foreign Office radio services into something called the Composite Signals Organisation. This allowed better control, but suspicions about civilians remained. In the words of the Admiralty Director of Signals it 'clearly failed to solve the fundamental security problem of a "Y" Service ... which is civilian manned in a democratic country, and therefore with Trade Union affinities which no-one can guarantee cannot be communist (vide the E.T.U)'.[84]

The Navy's Sigint collection arm, the Admiralty Civilian Shore Wireless Service, was the only one which was entirely civilian. In the mid-1950s the Navy drew up proposals for complete 'navalisation'. The matter looked problematic for some grades would be lowered as a result of 'navalisation' and there was bound to be trade union hostility. Accordingly the Navy decided to move towards the objective gradually. Navalisation not only reflected a desire to de-unionise the collection process, but also its changing role. The Admiralty noted:

> Recent reports submitted to the London Signals Intelligence Board stress the increasing need to have "Y" stations close to the "Iron Curtain" and to increase the number of Special Operations. These requirements can only be met by uniformed personnel.

Some of these forward Sigint collection operations were deemed too sensitive for details to be committed to paper, and could only be discussed 'verbally'.[85] In the late 1950s the Royal Navy had six main collection activities. The largest was Perkar, the successor to HMS Anderson on Ceylon, which was earmarked for rapid expansion in war. There were also three 'forward area' stations that were wholly military and operated 'under "active service" conditions even in peace'. These were at Kiel in Germany, in Cyprus and in Turkey. Altogether somewhere over 1,000 persons were involved in this network.[86] In March 1955, the Cabinet Security Committee looked at the issue and decided that although there was definite security risk involved in trade union membership it 'should be accepted' and nothing radical should be done. The gradual drift to de-civilianisation continued, but it was clear that it 'would take many years to achieve'.[87]

The sensitive ETU issue distracted from the main security problem at GCHQ, which was the sheer scale of positive vetting required in such a large organisation. By the mid-1960s the problem had become worse. The scope and scale of security measures had increased with more personnel being subjected to positive vetting. Cogniscent of the damage inflicted on Macmillan by the Profumo affair, Harold Wilson gave the Paymaster General, George Wigg, the special brief of looking after government security matters. In 1966 Wigg was asked to conduct a review of security in the Diplomatic Service, GCHQ and the Ministry of Defence. GCHQ, he reported, was 'autonomous to a considerable extent' and vast in size. There were 11,500 people working on Sigint – 8,000 with GCHQ directly – and 3,500 as service personnel in the Combined Signals Organisation, working on collection. Of those working for GCHQ half were in Cheltenham and half 'scattered at listening stations at home and abroad'. With the exception of a very few ancillary staff, just over 600, everyone had to be positively vetted.

GCHQ had a large team of 21 investigating officers, many of whom were retired Cheltenham police officers. This was 'advantageous' as the many of GCHQ staff at Cheltenham were recruited locally and the ex-police had acquired 'considerable knowledge of the background and circumstances of the staff'. Despite increasing the investigating officers to 25, Wigg conceded that 'a backlog of positive vetting has built up'.

There was also the problem of document security. GCHQ's Comint product was tightly controlled under special procedures laid down by the UKUSA agreement. GCHQ policy and administrative documents were also closely controlled. But GCHQ's basic 'working material' was also highly secret and was not catalogued in registries. In other words it was easy for cryptanlysts to smuggle out papers that their own specialist branches were working on.[88]

Wigg had rightly identified security as the Achilles Heel of the vast Western post-war Sigint organisations. Their scale, together with the peculiar nature of their work and personnel, made effective and detailed positive vetting of their employees all but impossible. But small numbers of leaks or defections could do vast damage. This was endemic in a large-scale industrial information process that ensured that large numbers of individuals, often clerks and low-grade administrators, had access to a highly secret product. Moreover, security chiefs complained that the nature of the work attracted quirky individuals and attempting to spot potential security problems by looking for unconventional individuals was a hopeless approach. In the words of the Security Commission in 1983, 'because of the nature of GCHQ's work and their need for staff with esoteric specialisms, they attracted many odd and eccentric characters'.[89]

It was therefore almost inevitable that the Soviets would succeed in recruiting individuals such a Geoffrey Prime, perhaps the Soviets' most useful source within British signals intelligence since John Cairncross. Geoffrey Prime, who spent much of his espionage career as an RAF corporal doing interception work in Berlin, and never rose to a civilian rank much above that equivalent to sergeant, was nevertheless able to do intense damage. Moreover, despite an outspoken admiration for the Soviet Union and all things Communist and a long history of sexual crimes, he passed easily through five positive vettings.[90] Curiously, Prime's offer to work for the Soviets, probably made during his time in Berlin, coincided almost exactly with a similar offer from an American naval warrant officer, John Walker. Walker served the Soviets for an even longer period and their contemporaneous activities provided the Soviets with a remarkably detailed picture of Western Sigint operations.[91]

Both London and Washington had received a clear warning about the scale of damage that could be inflicted by single individuals. Espionage by William Weisband, a clerk in the US Army Sigint organisation called the Army Security Agency, had revealed Venona to the Soviets in 1948 and ended the flow of new traffic of this sort that could be broken. Weisband also did great damage to widespread Western interception of mid-level

Soviet and Eastern Bloc Sigint, such as divisional army traffic, which had been a profitable source until 1948.

George Blake had compromised the Berlin Tunnel before it began and although the material taken by that operation was probably unaffected, the duration of this operation was short.

Then in 1960 two officials from the NSA, Martin and Mitchell, had made an ideologically motivated defection to the Soviets. At least one of these individuals had already been identified as having displayed deviant behaviour, but had still been allowed to pass his vetting. After their defection the United States informed its British signals intelligence partners and the usual damage assessments were set in train. Although serious damage to Sigint was expected, the British reaction to the Martin and Mitchell affair was not over-anxious, perhaps reflecting a relief that the defectors to the East were for once not British-employed.[92]

The scale of what the Soviets discovered about Western signals intelligence efforts during 1945–70 was considerable. This sits awkwardly with assertions by Western Sigint organisations about the continued sensitivity of so many of their early Cold War operations, despite the wider context of the Open Government approach announced in the 1990s. After the accidental discovery of Geoffrey Prime it must have been abundantly clear to Cheltenham that some of their activities were thoroughly understood by the Soviets. Indeed, many 'vital' secrets of the West were uncovered, and in some cases widely advertised, by the East at an early stage of the Cold War. Some Western agencies continued to pretend that these matters remained very secret when clearly they were not. This approach is perhaps a symptom of the intense security-mindedness that characterises work in this field. Nevertheless, the practical result has been that British and American Sigint agencies have kept some of these things secret only from their own domestic populations in the West.

NOTES

I would like to acknowledge research leave supported by the British Academy, which helped to support the preparation of this paper. I am indebted to many individuals for their observations, including Matthew Aid, Ralph Erskine, David Stafford, Cees Wiebes and John W. Young. Errors remain the responsibility of the author.

1. Philip Davies, *The British Secret Services: A Bibliography* (Oxford: Clio 1996). Some of the literature of GCHQ after 1945 includes: Patrick Fitzgerald and Mark Leopold, *Stranger on the Line: The Secret History of Phone Tapping* (London: The Bodley Head 1987); Nigel West, *GCHQ: The Secret Wireless War, 1900–86* (London: Weidenfeld 1986); Richard Aldrich and Michael Coleman, 'The Cold War, the JIC and British Signals Intelligence, 1948', *Intelligence and National Security* 4/3 (July 1989) pp.535–49; Andy Thomas, 'British Signals Intelligence After the Second World War', *Intelligence and National Security* 3/4 (Oct. 1988) pp.103–10.

Much of the important writing about GCHQ is submerged within wider accounts of global networks. See in particular: James Bamford, *The Puzzle Palace: America's National Security Agency and Its Special Relationship with GCHQ* (London: Sidgwick 1983); Jeffrey Richelson and Desmond Ball, *Ties that Bind: Intelligence Cooperation Between the UKUSA Countries* (Boston: Allen and Unwin 1985); Olaf Riste, *The Norwegian Intelligence Service, 1945–70* (London and Portland, OR: Frank Cass 1999); Bradley F. Smith, *The Ultra Magic Deals and the Most Secret Special Relationship, 1940–1946* (Shrewsbury: Airlife 1993); Christopher Andrew, 'The Growth of the Australian Intelligence Community and the Anglo-American Connection', *Intelligence and National Security* 4/2 (April 1989) pp.213–57; Christopher Andrew, 'The Making of the Anglo-American SIGINT Alliance', in H.B. Peake and S. Halperin, *In the Name of Intelligence* (Washington DC: NIBC Press 1994); Desmond Ball, 'Over and Out: Signals Intelligence (Sigint) in Hong Kong', *Intelligence and National Security* 11/3 (July 1996) pp.474–96; W. Wark, 'Cryptographic Innocence: The Origins of Signals Intelligence in Canada in the Second World War', *Journal of Contemporary History* 22 (1997) pp.639–65.
2. Wigg to Wilson enclosing 'The Organisation of Security in the Diplomatic Service and Government Communications Headquarters', 17 Aug. 1966, PREM 13/1203, PRO.
3. LSIB Mtgs. Summary No.13, 18 July 1945, WO 203/5126, PRO.
4. Welchman, Hinsley and Crankshaw to Travis, 'A Note on the Future of G.C. & C.S.', 17 Sept. 1944, HW 3/169, PRO. On Slim, see Richard J. Aldrich, *Intelligence and the War Against Japan: Britain, America and the Politics of Secret Service* (Cambridge: Cambridge UP 2000) pp.316–17.
5. Part 1, 'The General Problem of Intelligence and Security in Peace', (personal for Director), Preliminary Draft, Sept. 1944, HW 3/169, PRO.
6. Bamford, *Puzzle Palace* (note 1) 317, 335.
7. Clarke, 'Post War Organisation', HW 3/30. I am indebted to Ralph Erskine for this reference.
8. Aldrich and Coleman, 'The Cold War, the JIC and British Signals Intelligence' (note 1) pp.538–40; F.H. Hinsley *et al.*, *British Intelligence in the Second World War* (London: HMSO 1980) II, pp.618–19.
9. P. Whitaker and L. Kruh, 'From Bletchley Park to the Berchtesgarten', *Cryptologia* 11/3 (1987) pp.129–20.
10. 'The German G-2 Service in the Russian Campaign', 22 July 1945, WO 208/4343, PRO.
11. Van der Meulen, 'Cryptologic Services of the Federal Republic', paper to the German Intelligence Studies Conference, Tutzing, 1999, and private information.
12. Sheridan Russell, *Sheridan's Story, 1900–1991* (privately published 1993).
13. L. Donini, 'The Cryptographic Services of the Royal (British) and Italian Navies', *Cryptologia* 14/3 (1990) pp.97–127.
14. AFHQ to MI8 London, 10 July 1945, 'SID Cryptographic Documents', WO 208/5073, PRO. Also AFHQ to MI8 London, 'Diplomatic Interception in Northern Italy', 16 May 1945, and Annex, 'SIM Success on British and US Diplo', ibid.
15. McKane to DDY, 'Italian Sigint Service', 15 May 1946, ibid..
16. McKane to Director LSIC (for Head of TA Group), 31 May 1946, enclosing 'Further notes on Italian Cover', ibid.
17. McKane to K.H. Sachse, (LSIC), 'Italian Intercept Organisation', 4 June 1946, ibid.
18. Y Services Summary, Dec. 1995 and March 1946, WO 212/228, PRO.
19. R. Lewin, *Ultra goes to War* (London: Hutchinson 1978) pp.129–33; R.V. Jones, *Reflections on Intelligence* (London: Heinemann 1989) p.15.
20. JIC (48)(0)(second revised draft), 'Sigint intelligence requirements' 1948, 11 May 1948, L/WS/1196, IOLR.
21. A. Stripp, *Codebreaker in the Far East* (London: Frank Cass 1988) pp.50–60; Jones, *Reflections* (note 19) pp.14–16.
22. Thomas, 'British Signals Intelligence' (note 1) pp.103–4.
23. Ibid. pp.106, John Sawatsky, *For Services Rendered* (Markham, Canada: Penguin 1983) pp.23–4.
24. Ball, 'Over and Out' (note 1) pp.32–44; Thomas, 'British Signals Intelligence' (note 1) p.107; Sawatsky, *For Services Rendered* (note 23) pp.25–6.

25. Eubank (COS) to Rowlands (MoS), 31 Jan. 1952, DEFE 11/350; Eubank to DRPC, 31 Jan. 1952, ibid.
26. COS (52) 152nd mtg. (1) Confidential Annex, 4 Nov. 1952, DEFE 11/350, PRO.
27. Christopher. Andrew, *For the President's Eyes Only: American Presidents and Intelligence From Washington to Bush* (London: HarperCollins 1999) pp.156–63.
28. Entry for 21 Nov. 1945, Cunningham and diary, MSS 52578, British Library; Wark, 'Cryptographic Innocence' (note 1) pp.558–9; Andrew, *President's Eyes Only* (note 27) p.161.
29. Andrew, 'Australian Intelligence Community' (note 1) pp.223–5; Bamford, *Puzzle Palace* (note 1) pp.314–17; Richelson and Ball, *Ties that Bind* (note 1) pp.141–5.
30. Wark, Cryptographic Innocence' (note 1) pp.639–5; Sawatsky, *For Services Rendered* (note 23) p.29. I am much indebted to the guidance of Matthew Aid on these matters.
31. Brigadier General USAF, Acting Director of Intelligence, Walter R. Agee, to US Coordinator of Joint Operations, 7 June 1948, 'Proposed U.S.-Canadian Agreement', USAF D of I records, File 2 1200/2-1299, Box 40, RG 341, USNA.
32. Matthew Aid, 'US Humint and Comint in the Korean War: From the Approach of War to the Chinese Intervention', *Intelligence and National Security* 14/4 (Winter 1999) pp.15–50.
33. A-2 to Naval Communications Annex, 'Request for British Comintsum Publications, 19 March 1948, 21450, Box 41, USAF D of I records, RG 341, USNA.
34. Cabell to Air Police Division, 28 April 1948, 21200, Box 40, USAF D of I records, RG 341, USNA.
35. Riste, *Norwegian Intelligence Service* (note 1) pp.95–7; R. Tamnes, *The US and the Cold War in the High North* (Aldershot: Dartmouth 1991) pp.76–7.
36. UKHC Ceylon to Defence Dept., 21 April 1950, DO 35/2418, PRO.
37. Mtg at the Admiralty, 26 Oct. 1951, ibid.
38. Mtg No.80, 17 Jan. 1951, PSF – NSC meetings, Box 211, Harry S. Truman Library.
39. Ackerman to D of I, 21 Sept. 1951, 219900, Box 58, USAF D of I records, RG 341, USNA.
40. Young (USAF) to Coordinator USCIB, 'Site Requirement', 15 July 1952, 224100-, Box 66, USAF D of I records, RG 341, USAF.
41. Samford (D of I USAF) to Twining, 6 Aug. 1952, 224400, Box 66, USAF D of I records, RG 341, USNA.
42. Brief on Fifth Report to the President by PBCFIA (Recommendation concerning Fusion of Comint-Elint Activities), 11 March 1960, File: 1960 Meetings with the President Vol. 1 (5), Box 4, Presidential Subseries, Special Assistant Series, OSANA, WHO, Dwight D. Eisenhower Library, Abilene, Kansas.
43. US JCS decision on J.C.S. 2074/14, A Report by the Chairman, US Armed Forces Security Agency Council on 'Security of the Combined Cipher Machine', 19 May 1952, Note by the Secretaries, Papers of the Joint Chiefs of Staff, 1951-3 CCS 311 (1-10-42) Sec.14, RG 218, USNA.
44. 'Results of US-UK Conference in London on Comint Service to SHAPE and its Subordinate Commands', 23 Dec. 1952, 3-1, Box 70, USAF D of I records, RG 341, USNA.
45. 'Memorandum from Colonel Fergusson to General Gruenther', B. Fergusson, *Hubble-Bubble* (London: Collins 1978) pp.54–6.
46. R.V. Jones, *The Wizard War: British Scientific Intelligence 1939–1945* (NY: Coward, McCann and Geoghegan 1978) p.92.
47. SEAC Noise Investigation Bureau report for May 1945, WO 203/4089, PRO.
48. Captain Wenger, US Navy Coordinator of Joint Operations, to Colonel R.P. Klocko, USAF, CJO 0001922, 12 March 1948, Memo: 'British proposal for liaison on "noise investigation"', USAF D of I records, File 2-1100/2-1199, Box 40, RG 341, USNA.
49. McMurtie (JSM), to Moore (Pentagon), 20 Nov. 1948, File 2-8300 – 2-8399, USAF D of I records, RG 341, USNA.
50. Partridge Memorandum, 'Northern European Ferret Flights', 20 Aug. 1947, File 2-800 – 2-899, USAF D of I records, RG 341, USNA.
51. I am indebted to Matthew Aid's forthcoming study of US Sigint for this point.
52. Air Technical Center, Air Technical Intelligence Study No.102-EL-23/51-54: Radio Frequency Transmissions, July–Sept. 1950, File 2-20944, USAF D of I records, RG 341, USNA.

53. Tamnes, *High North* (note 35) p.77.
54. Ibid. pp.50–2.
55. Air Technical Intelligence Study, 'Soviet Electronic Countermeasures', 10 June 1951, 20034, Box 149, USAF D of I records, RG 341, USNA; Air Technical Intelligence Study, 'Soviet Air Communications', 12 July 1951, 20032, ibid.
56. Tamnes, *High North* (note 35) pp.116–17.
57. D of I USAF to US Air Attaché London, 'Liaison with GCHQ', 26 Oct. 1952, 235700, Box 68, USAF D of I records, RG 341, USNA.
58. AFOIN-T to D of I USAF, 16 April 1952, 223200, Box 64, USAF D of I records, RG 341, USAF.
59. R.R. James, *Anthony Eden* (London: Weidenfeld 1986) pp.436–7.
60. Eden to Bridges, M.104.56, 9 May 1956, AP20/32/78, Avon Papers, Birmingham University Library (BUL).
61. Prime Minister, Anthony Eden, to Minster of Defence, Antony Head, 22 Dec. 1956, AP20/21/228, Avon Papers, BUL.
62. Richard M. Bissell Jr, *Reflections of a Cold Warrior: From Yalta to the Bay of Pigs* (New Haven, CT: Yale UP 1996) pp.115–16.
63. Elkins (BJSM) to Mountbatten, 16 Oct. 1956, ADM 205/110, PRO.
64. Inglis (DNI) to Flag Officer, Submarines, 19 Oct. 1956, ibid; Coote, USS *Stickleback* report, ibid.
65. Elkins (BJSM) to Mountbatten, 31 Dec. 1956, ibid.
66. Mikesh, *Canberra, B-57* (Shepperton: Ian Allan 1980) XX; G. Pedlow and D. Welzenbach, *The CIA and the U-2 Program, 1954–74* (Washington DC: CIA 1998) pp.9–51; Tamnes, *High North* (note 35) pp.128–9.
67. Ibid. pp.128–9; Bissell, *Reflections* (note 62) p.121.
68. Tamnes, *High North* (note 35) pp.123–5.
69. S. Ambrose, *Eisenhower the President, Vol. 2, 1952–59* (London: Allen and Unwin 1984) p.584.
70. The JIC paper was JIC (60) 43 (Final), 'Soviet Threats Against Reconnaissance Flight Bases Following the U-2 Incident', and is summarised in DEFE 13/342, PRO.
71. Minutes, 'Surveillance Meeting', 26 April 1960, 16/W/160, ADM 1/27680, PRO.
72. Minute by Head of Military Branch II, 10 March 1961, ibid.
73. Memo from PS to VCAS to PS to S. of S., 'Aircraft Approach Restrictions – Operation TIARA/GARNET', Oct. 1960, AIR 20/12222, PRO. The JIC paper prepared for Macmillan was JIC (60) 62 (Revised), 1 Sept. 1960.
74. On continuing British submarine intelligence operations in the 1960s see Riste, *Norwegian Intelligence Service* (note 1) pp.228.
75. Mtg between Chief of Turkish Gen Staff, 25 April 1961 recorded by Capt. Hodgkinson, DEFE 25/11, PRO.
76. Richard S. Aldrich, *Espionage, Security and Intelligence in Britain 1945–1970: Documents in Contemporary History* (Manchester UP 1998) pp.147–9. See also Radford (JCS) memo, 'Clandestine Listening Devices', 6 April 1956, file: Presidential papers 1956 (8), Box 3, Presidential subseries, Special Assistant Series, OSANA, WHO, Dwight D Eisenhower Memorial Library.
77. Private information.
78. Memo to the RSB, 'Government Communications Headquarters', from GCHQ (GCHQ Ref.M/8087/100/1) 11 Feb. 1960, WO 195/14887, PRO. I am indebted to Rob Evans for drawing this document to my attention.
79. Thorneycroft to Douglas-Home, 12 Sept. 1964, DEFE 13/403, PRO; Jellicoe to Butler, 11 Sept. 1964, ibid.
80. Penney (LCSA) to Howes, 16 Oct. 1957, I – 1516, Mountbatten Papers, Hartley Library Special Collections, Univ. of Southampton.
81. COS (61) 466, Annex II, 'Communications Security Committee Middle East', AIR 19/1101, PRO.
82. CIA, Special Report – Office of Current Intelligence, 'The British Communist Party', OCI No. 275638, File UK General, 1963, Box 171, NSF Files, John F. Kennedy Memorial Library, Boston.

83. Ball, 'Over and Out' (note 1) p.482.
84. Lenox-Conygham (DSD), minute, 19 Nov. 1954, ADM 1/26478, PRO.
85. Head of C.E. Branch IV, 15 Feb. 1955, ibid.; Lenox-Conygham (DSD), 25 Feb. 1955, ibid.
86. YWP (P) 12, July 1955, ibid..
87. Head of C.E. Branch IV minute, 2 April 1955, ibid.
88. Paymaster General, George Wigg, to Prime Minster Harold Wilson enclosing 'The Organisation of Security in the Diplomatic Service and Government Communications Headquarters', 17 Aug. 1966, PREM 13/1203, PRO. This document is reproduced and discussed in J. Young, 'George Wigg, The Wilson Government and the 1966 Report into Security in the Diplomatic Service and GCHQ', *Intelligence and National Security* 14/3 (Autumn 1999) pp.198–209.
89. Quoted in D.J. Cole, *Geoffrey Prime* (London: Robert Hale 1995) p.64.
90. Cole, *Prime*, pp.54–76.
91. The most incisive assessment is offered in C. Andrew and O. Gordievsky, *KGB: The Inside Story* (London: Hodder 1990) pp.438–43.
92. Strong (JIB) to Minster of Defence, Harold Watkinson, 29 Aug. 1960, DEFE 13/9, PRO. Also private information.

4

Canada's Communications Security Establishment from Cold War to Globalization

MARTIN RUDNER

The Communications Security Establishment (CSE) is Canada's largest and costliest intelligence organization and the main provider of foreign intelligence to the Canadian government.[1] It is, arguably, also the most secretive component of the Government of Canada. For decades the very existence of CSE was unconfirmed, it has no statutory mandate, and virtually all details of its resources, objectives and operations are still shrouded in official secrecy.[2] What is known is that CSE collects, analyses and reports on signals intelligence (referred to as Sigint) derived from interceptions of foreign electronic communications, radio, radar, telemetry, and other electromagnetic emissions. In fulfilment of these foreign intelligence functions, CSE participates in international collaboration and exchanges as part of a special Sigint sharing arrangement with the United States, United Kingdom, Australia and New Zealand. CSE is also responsible for providing technical advice and guidance for protecting Canadian government communications and electronic data security.

CSE is a civilian agency of Canada's Department of National Defence (DND). Ministerial responsibility for CSE is vested in the Minister of National Defence, however in a unique bifurcation of executive authority, administrative and operational controls are divided between DND and the Privy Council Office (PCO), the federal government's central agency, headed by the Prime Minister. Administrative and financial matters are under the control of DND, through the Deputy Minister of National Defence, its most senior official, whereas policy and operational controls over CSE are exercised by the Deputy Secretary, Security and Intelligence in PCO. At the policy level, the direction and co-ordination of Canada's intelligence effort involves a complex web of PCO secretariats and inter-departmental committees.[3]

At the operational level, the actual staffing of Canada's Sigint interception land sites is undertaken not by CSE as such, but by specialized military detachments of the Canadian Forces Information Operations Group (CFIOG), working under the overall direction of CSE. CFIOG deploys about 1,000 personnel, mainly military Communications Research Operators (known colloquially as '291ers'), at Canadian Forces Base Leitrim (near Ottawa), also service the remote stations at Alert, Gander and Masset. An exchange arrangement with the United States has some 25 291ers posted to US Navy stations in California, Hawaii and Texas, while a similar number of American personnel are attached to the Leitrim facility.[4]

During the Cold War the Canadian signals intelligence effort was directed primarily at the Soviet Union and its Warsaw Pact allies. That lent Canada's foreign intelligence requirements a certain stability and predictability.[5] Following the collapse of Communism in Europe and the end of the Cold War, however, CSE found itself impelled to alter the scope and direction of its activities in response to shifting perceptions of the threat environment confronting Canada. A more variegated and volatile security situation had a far-reaching impact on Canadian foreign intelligence

FIGURE 4.1

Canadian Communications Security Establishment HQ buildings in Ottawa.
Communications Security Establishment, Canada

requirements. Thus, in 1991, for the first time ever, the federal Cabinet issued a directive on foreign intelligence priorities.[6]

The study that follows traces the historical evolution of CSE in performing its signals intelligence functions from the Cold War to this more diverse and globalized security agenda. Given the sensitivity of Sigint issues, this study relies on open sources.

THE BEGINNINGS OF CANADIAN SIGINT

Canada has never had a consolidated, dedicated foreign intelligence service, unlike most of its allies. Historically, Canadian requirements for foreign intelligence have been addressed through an array of functionally differentiated agencies, most of which were linked to international intelligence sharing arrangements. Canada's involvement in Sigint began prior to the Second World War, when the Royal Canadian Navy put in place a monitoring station on the West Coast to supply raw intercepts to the British Admiralty. During the war the Army, Navy and Air Force set up their own respective signals intelligence units in collaboration with their British counterparts.[7] These separate Sigint units were later combined into a so-called 'Joint Discrimination Unit'.

Meanwhile a civilian entity, styled the 'Examination Unit', had been established in 1941 to provide communications intelligence and cryptanalysis, primarily of diplomatic traffic, for the Department of External Affairs (as it was then). In April 1946, Prime Minister MacKenzie King approved the creation of a peacetime communications intelligence organization, and in September of that year the existing military and civilian units were merged to become the Communications Branch of the National Research Council (CBNRC).[8] In 1975 the functions of CBNRC were relocated in their entirety to DND, and reconstituted as the Communications Security Establishment.

No statutory framework for CSE (or its predecessor) was ever put in place. In fact, for virtually all this period the very existence of a Canadian signals intelligence capability was itself an official secret.

While the decision to create a peacetime Canadian Sigint capability preceded the onset of the Cold War, the looming confrontation with an expansionist Soviet Union gave a powerful impetus to this incipient foreign intelligence initiative. As it happened, a coincidence of events around the pivotal years 1945–49 underscored the strategic value of signals intelligence in the Cold War context. In 1945, a cipher clerk in the USSR embassy in Ottawa, Igor Gouzenko, defected, bringing with him documentary evidence of a Soviet espionage network.[9] Although there is nothing to indicate that the Gouzenko defection impacted directly on

Enigma cipher machine.

Communications Security Establishment, Canada

Canadian Sigint operations, the accompanying cipher material itself underscored the potential role for signals intelligence in the defence of Canadian and allied security.[10]

Meanwhile, in 1946, US codebreakers succeeded in deciphering previously intercepted Soviet KGB signals. This operation, code named 'Venona', paved the way for future Sigint attacks on Soviet diplomatic, military, and intelligence communications.[11] In so far as just knowing the capabilities of communications intelligence can suffice to give warning of target vulnerability, these Sigint organizations, technologies and operations were generally treated as matters of utmost secrecy.

By then, the senior echelons of the Canadian foreign policy and defence establishment would have become aware of the wartime contributions of 'Ultra' and 'Magic', the British and American Sigint breakthroughs against German and Japanese diplomatic and military communications, respectively.[12] They certainly knew of the ongoing British and American initiatives to develop new modalities for post-war cooperation in communications intelligence. Early on, in October 1945, the British Sigint organization, then styled as the Government Code and Cipher School (GC&CS), predecessor of what became in 1946 the Government Communications Headquarters (GCHQ), approached the Canadian authorities to solicit their participation in a combined Anglo-American communications intelligence initiative that would involve a complete sharing of intercepts. Aware that they could not achieve global Sigint coverage by themselves, the British sought to divide the world into tripartite spheres of cooperation, but asked that Canada permit Britain to represent its interests in negotiations with the United States. It is noteworthy that, at the time, GC&CS conceived of the tripartite agreement as involving just military and clandestine radio traffic but not diplomatic interceptions. The Canadian Joint Intelligence Committee (CJIC) agreed to cooperate and mandated Britain to negotiate with the Americans on Canada's behalf.[13] In March 1946, an British-US Agreement (BRUSA) was concluded on communications intelligence sharing, which also embraced Canada.[14]

Prior to the 1960s, most international (and long-distance domestic) telecommunications traffic everywhere in the world was carried by high frequency (HF) radio networks. This HF infrastructure served for telephones and telegraph, and diplomatic and military messaging. Since HF radio signals achieve their long range by bouncing between the ionosphere and earth's surface, they are vulnerable to interception as well as reception. HF radio signals can be readily intercepted with specialized antennae which can simultaneously monitor as many frequencies from as many bearings as may be desired, requiring only a suitable parcel of land in, ideally, a 'quiet' radio environment. Canada's geographic location provided particularly advantageous situations for intercepting HF communications across the northern regions of the USSR and East Asia and the adjacent waters of the Atlantic, Pacific and Arctic oceans.

After the Second World War, Canada, like Britain and the United States, shut down most of the Sigint listening posts that had been set up in wartime. While the Leitrim site near Ottawa was kept operational, most other Canadian interception facilities and Royal Canadian Navy radio intercept and high-frequency direction-finding (HF-DF) stations were closed or returned to the Department of Transport.

Prompted by the BRUSA agreement, from 1946 a network of interception facilities was set up across Canada to cover gaps in the tripartite arrangement with Britain and the United States. Existing facilities at Leitrim, Coverdale (New Brunswick), and Prince Rupert (British Columbia) were expanded, and new intercept sites were established at Whitehorse in the Yukon, Churchill on the Hudson Bay in northern Manitoba, and Ladner, near Victoria (British Columbia). The Whitehorse facility, activated in 1948, intercepted Soviet and other Asian radio traffic; Churchill, opened the same year, copied Soviet radio traffic across the Arctic; and Ladner provided coverage of the Soviet Far East.[15]

By the 1950s CBNRC was monitoring Soviet air force and air defence communications across the northern USSR from ten small radio intercept stations operated by the Royal Canadian Navy (Aklavik, Churchill, Coverdale, Frobisher Bay on Baffin Island; Gander, Newfoundland; Masset, British Columbia), Army Corps of Signals (Alert, Northwest Territories; Ladner, Leitrim) and Royal Canadian Air Force (Whitehorse).[16]

In addition, a small network of HF-DF stations was created out of reactivated wartime posts and new naval installations at Aklavik (Northwest Territories), Masset (British Columbia) and Coverdale. These HF-DF stations were fully integrated into the Atlantic and Pacific HF-DF networks of the US Naval Security Group, while communications intelligence was channelled through CBNRC.

Building upon the tripartite arrangement under the 1946 BRUSA Agreement, a Canada-US Communications Intelligence Agreement (CANUSA) was concluded in May 1948, which, *inter alia,* established parameters for bilateral exchanges of communications intelligence albeit on a rather more limited basis than did the BRUSA arrangement. Be that as it may, this Agreement provided the impetus for Canada to further extend its involvement in alliance Sigint activities. By the late 1940s, Canada had

FIGURE 4.3

Canadian Army Corps of Signals Sigint intercept station at Alert, Northwest Territories.
Communications Security Establishment, Canada

emerged as a modest but important source of strategically valuable signals intelligence on the Soviet Union and East Asia. Canada's collaboration with allied Sigint efforts, which was subsequently expanded into the wider ranging UKUSA alliance (see below), was valued not so much for this country's inherent capabilities in Sigint or its contributions to intelligence production generally, as for its geographic advantage in providing communications intelligence coverage of the Soviet Union, especially its Arctic and Far Eastern regions. Indeed, Canada's Sigint allies would have cause to lament Canada's meagre capacity to offer exchanges of intelligence product.[17] Nevertheless, Canadian geography made up for the otherwise lamentable 'terms of trade'. By November 1957, CBNRC had given up its attempts at machine cryptanalysis, reducing Canada's role to that of a mere supplier of raw intercepts to its more highly capable, better equipped Sigint allies.

CANADA'S SIGINT COLLECTION EFFORT

Up to the present most of the foreign intelligence provided to the Canadian government by virtue of Canada's own intelligence collection capabilities derives from signals intelligence provided by CSE. Canadian Sigint operations collect intelligence by means of sophisticated, covert interception technologies designed to intercept terrestrial, microwave, radio, and satellite communications along with other electromagnetic emissions. These intercepts are then processed through technologically advanced computer systems programmed to search for specific telephone numbers, voice recognition patterns, or key words, and to decrypt text.

Canada also has access to Sigint collected by its allies in the UKUSA signals intelligence alliance (see below). This unique alliance links Canada's CSE to the United States, through its National Security Agency (NSA); the United Kingdom, through GCHQ; Australia, through its Defence Signals Directorate (DSD); and New Zealand, through the Government Communications Security Bureau (GCSB). The UKUSA alliance provides CSE with a shared global capacity to collect and deliver real-time Sigint intercepts on targeted objectives to selected clients within the Government of Canada.

The clandestine and broadly intrusive function of Sigint has had important implications for political control and accountability, oversight and legal compliance relating to the privacy of Canadians. Ultimate political control over intelligence in the Canadian parliamentary system is vested in the Prime Minister. As head of government, the Prime Minister bears overall responsibility for Canada's national security and the safeguarding of the country's territorial integrity.

Parliament has traditionally played a very limited role in regard to foreign intelligence generally since most detailed information on budgets, operations and the performance of the organizations concerned, including CSE, must necessarily remain classified. However, along with all other Canadian government departments and agencies, CSE and other components of the intelligence community are subject to scrutiny and review by the Auditor-General of Canada, the Canadian Human Rights Commission, the Privacy Commissioner, and the Information Commissioner, as well as the courts.

In 1996, the government took a step towards creating a more public accountability framework for CSE by appointing a CSE Commissioner with a mandate to review and report upon its activities in order to determine their compliance with the law. Assurances have been given repeatedly in ministerial pronouncements and in reports of review agencies like the Privacy Commissioner and CSE Commissioner to the effect that Canadian Sigint operations respect the laws of privacy and do not intentionally target Canadians or monitor their domestic private communications. Nevertheless, there is some deliberate ambiguity as to the extent to which interceptions of foreign targets may incidentally capture communications to or from Canadians.

The methods utilized to intercept targeted local communications are obviously highly sensitive. There are several ways in which local in-country interception operations could have been mounted. What is noteworthy is that CSE shared some of the technologies of its UKUSA partner organizations that enabled them to surreptitiously intercept telephonic or digital communications, sift them for messages to or from targeted individuals or organizations, and decrypt the enciphered content.

Cryptanalysis represented a vital part of Canada s early Sigint collection effort. At the outset, CBNRC provided the mathematical and cryptological skills to decipher intercepted Soviet bloc communications. However by the late 1950s this cryptanalytical effort had to be mostly abandoned.[18] Historians claim that no Soviet diplomatic communications were ever decrypted after 'Venona', because of KGB penetrations of NSA and GCHQ thwarted subsequent codebreaking efforts.[19]

Over the next two decades Canadian signals intelligence was but minimally involved in serious cryptanalysis. What was done was mostly undertaken manually, as few computer resources were deployed in Canada's Sigint effort. It was only in the early 1980s that one of CSE's IBM 370 mainframes was made available for cryptanalysis, even though NSA was reportedly doubtful whether this computer could generate results.

Nevertheless, CSE was now able to break into certain cipher keys that yielded up intelligence to Canadian requirements. Yet, by the time this system achieved a minimal capacity for codebreaking, around spring 1981, CSE cryptanalysts were already acknowledging that more powerful computational technologies would be required for operational effectiveness.[20]

COLD WAR SIGINT OPERATIONS

Canadian signals intelligence operations during and after the Cold War may be considered in terms of four types of interception, in accordance with the location and technologies deployed. Local in-country interception operations were mounted within Canada, targeting communications to or from this country. External interception operations targeted communications in foreign countries from Canadian diplomatic posts. Long-range operations targeted communications and electromagnetic emissions abroad from interception facilities in Canada. Later, specialized facilities were installed to also monitor satellite communications links. The primary targets for each of these types of interception during the Cold War were the diplomatic, military and espionage communications of Soviet Bloc countries. Other countries' communications were also sometimes targeted.

Local Sigint operations mounted within Canada during the Cold War targeted mainly the Soviet Bloc diplomatic and consular missions, trade and commercial offices, and organizations and individuals suspected of involvement in espionage or subversion.[21] Canadians were also intercepting the radio transmissions from Soviet research stations in the Arctic, allowing intelligence analysts to monitor their scientific experiments.[22] No official confirmation of these sensitive operations was ever forthcoming. A 1956 operation ('Dew Worm') to secrete listening devices in the Soviet embassy in Ottawa was a failure, as was another attempt to penetrate the Polish consulate in Montreal (Operation 'Satyr').[23]

The Soviets returned the compliment by way of surreptitiously installing radio-intercept posts in their KGB residencies in Ottawa and Montreal to monitor Canadian communications. Moreover, the KGB radio-intercept post in New York succeeded in intercepting communications traffic between the Canadian permanent mission to the United Nations and Department of External Affairs.[24]

In parallel with these local and external operations Canadian signals intelligence also undertook long-distance Sigint intercepts from interception stations in Canada. Long-distance HF radio intercepts enabled Canada and its allies to eavesdrop on internal Soviet (and other Warsaw Pact) military, naval, rocket force and air force communications networks across the Arctic. These Soviet Bloc armed forces HF radio networks were generally less well protected than political-level and diplomatic communications, and could be intercepted and processed with contemporary technologies. Sigint interceptions of HF communications played a key role in that strategically vital polar theatre by way of providing distant early warning of the Soviet order of battle and potential first strike capability, intelligence of primary significance during the Cold War for the defence of Canada and North America.

By the mid-1970s, however, the USSR seemed better able to effectively protect its high-level communications against interception.[25] By then Canadian Sigint operations were also targeting other perceived threats to Canada's national security and territorial integrity. Among the countries now targeted were those whose foreign policy behaviour was considered inimical to Canada and its allies, and those whose embassies or representatives were suspected of engaging in illegitimate political activities, inappropriate dealings with Canadian residents, support for subversive or terrorist groups, or illicit arms procurements. With the election of a separatist government in the Province of Quebec, CSE allegedly began monitoring communications traffic between the governments of Quebec and France, according to disclosures by a disaffected former employee.[26] Such operations were ostensibly mounted by CSE itself, some say with support from Sigint allies in Norway and the United States.

Canadian and allied Sigint interceptions of in-country Soviet communications helped to fill the information void in these otherwise closed, secretive, unfriendly regimes. Information needed simply to manage bilateral relations, or to assess international behaviour and risks, which in other societies would have been open source, could only be acquired in the context of Soviet secretiveness by intelligence means. Sigint interceptions of local communications was one of the most effective, least risky, means of penetrating the Iron Curtain of secrecy. Soviet countermeasures were deployed in the 1970s to frustrate Sigint operations run from the US Embassy. It was suspected that the electromagnetic radiation may have caused the American ambassador to become ill with leukaemia, but this fear was later allayed.[27] There is no indication that any Canadians were affected by countermeasures against the listening post in the Canadian Embassy.

During the 1970s CSE, acting at behest of NSA, began mounting external interception operations from Canadian diplomatic posts abroad in an operation codenamed 'Pilgrim'. Microwave systems in most countries converge on their capital cities, rendering some of their most sensitive communications traffic vulnerable to embassy-based interception operations. Embassy-based Sigint stations were also effective for intercepting official car phone communications transmitted by short-range radio. External communications interceptions provided, at the time, a unique aperture into in-country telecommunications.[28] State-of-the art communications monitoring and processing equipment was supplied by NSA, which also trained Canadian personnel and guided the targeting.

This equipment, and the personnel, were surreptitiously located in certain Canadian embassies and consulates. The first such interception operation, 'Stephanie', was mounted from the Canadian embassy in Moscow beginning in the autumn of 1972, and ran for about three years.[29] A subsequent operation, 'Sphinx', was run in the late 1980s. The first permanent intercept site was reportedly established in 1983 at the Canadian High Commission in New Delhi, as Operation 'Daisy'.[30]

Among the other capital cities where Canada is said to have run external Sigint collection operations from diplomatic or consular posts were Abidjan ('Jasmine'), Beijing ('Badger'), Bucharest ('Hollyhock'), Rabat ('Iris'), Kingston, Jamaica ('Egret'), Mexico City ('Cornflower'), Rome, San José (Costa Rica), Warsaw and possibly Tokyo. All the intelligence collected by Canadian embassy-based interceptions was actually remitted to NSA for deciphering and analysis, since at the time Canada lacked a capacity to do this. It was ironical that for the want of cryptanalytical capability Canada was unable to process the take from its own external Sigint collection efforts, but had to rely on partners for this intelligence product.

In the early 1980s Canadian Sigint was even targeting non-security related economic targets of opportunity as part of Operation 'Aquarian' aimed at foreign embassies and consulates, even those of friendly or indeed allied countries. CSE intercepts were said to have been instrumental in enabling Canada to out-compete the United States in a US$5 billion wheat sale to China in 1981.[31]

Although the NSA partially funded the modernization of Canadian communications interception facilities in the 1960s, the number of stations was reduced to just six by the early 1970s. Frobisher Bay, Whitehorse, Churchill, Coverdale and Ladner were all closed down. A new station was activated in Inuvik (to replace Aklavik, closed in 1961), Northwest Territories, and a naval HF-DF station was opened in Bermuda in 1963. Following the transfer of Sigint responsibilities to CSE in 1975, a complex of specialized Sigint antennae and processing stations was constructed at Leitrim, Alert, Gander, Whitehorse (now closed) and Masset, staffed with military personnel from what is today the Canadian Forces Information

Operations Group.³² By the late 1990s the interception stations at Alert, Gander and Masset were fully automated and would henceforward be remotely controlled from the central CSE collection facility at Leitrim.

CANADA AND THE UKUSA AGREEMENT

Canadian involvement in an international Sigint alliance structure began in stages between 1946 and 1948 with separate arrangements with Great Britain and the United States, and culminated in an expanded five-power globally-capable architecture for the sharing of technological capabilities and intelligence product. After Canada was included in the 1946 BRUSA Agreement, albeit as an affiliate of the Britain, GCHQ sought to achieve further synergy and intelligence connectivity by mobilizing the Sigint efforts of the self-governing Dominions (as they were then), under its own leadership, of course. During the winter of 1946–47 the British convened a conference of the Dominions' signals intelligence services with the aim of creating a Commonwealth Sigint organization headed by GCHQ and having a global interception capability. Although this objective was too ambitious for the time, the conference did succeed in nurturing the development of close, even intimate working relationships among the Sigint organizations of the UK, Australia and Canada, in particular.

This British attempt to mobilize dominion support for an 'Old' Commonwealth Sigint network coincided with an acute crisis in Anglo-American intelligence cooperation, prompted by the post-war Labour government's controversial sale of jet engines (and, as alleged at the time, jet aircraft) to the Soviet Union. Furious at what they saw to be a betrayal of Western interests, the Americans reacted by placing intelligence cooperation 'under review' and stopping any further disclosures of intelligence 'sources', 'methods of acquisition', and 'information pertaining to cryptography and cryptographic devices' – all the essentials of communications intelligence sharing.³³ The resulting freeze no doubt reinforced Britain's desire to create an alternative, Dominions-based arrangement for Sigint cooperation.

Around the same time, in early 1948, the United States moved swiftly, to avoid being outflanked, to negotiate separate bilateral communications intelligence cooperation agreements with Canada and Australia. Thus Canada found itself entangled by circumstances in competing Sigint alliances with contending allies: the British-inspired Commonwealth Sigint Organization (CSO) Agreement of 1947 and the Canadian-US Communications Intelligence Agreement (CANUSA) of 1948. In any event, faced with a deteriorating Cold War situation in Europe, the British and Americans resolved their differences by April 1948, paving the way to the signing in June of the UK-USA Security Agreement (UKUSA) on communications intelligence cooperation, the UKUSA alliance. Informed sources maintain that UKUSA is not a single treaty document but rather a set of Anglo-American agreements, memoranda of understanding and exchanges of letters which have been acceded to also by Canada, Australia and New Zealand.[34] Details of these agreements remain highly classified. This framework agreement created a tight, resilient collaborative arrangement between the First Party, the American NSA, and the Second Parties, the Sigint agencies of Great Britain, Canada, Australia and New Zealand, for cooperation in the sharing of Sigint technologies, in targeting and operational matters, and in exchanges of foreign intelligence collection.[35]

It became an underlying principle of UKUSA that the partner countries did not target one another or their respective nationals.[36] As an expression of the intimacy of their cooperation, CSE (and its Australian and New Zealand counterparts) exchange liaison officers with the otherwise highly secretive Sigint organizations of the United States (NSA) and Great Britain (GCHQ). This pattern of liaison exemplifies the hub-and-spokes configuration of the UKUSA relationship. Be that as it may, it is clear that most of Canada's foreign intelligence collection activities have taken place within the collaborative Sigint framework of UKUSA.

The UKUSA connection has had implications for Canada's intelligence role in other international security contexts. Thus, as a partner in UKUSA, Canada was likewise involved in a so-called CANUKUS intelligence grouping within NATO. This tripartite Canada-UK-US intelligence grouping was said to have contributed the bulk of the input into the annual NATO Military Committee assessments of Soviet military power.[37] Other than German intelligence, which provided particular knowledge of Eastern Germany, CANUKAS furnished a preponderant share of NATO's intelligence requirements. Most of this joint intelligence input was derived from Sigint, including CSE product.

SATELLITE COMMUNICATIONS AND ECHELON

The inauguration of the space age in communications after the 1960s was to greatly expand global telecommunications traffic while engendering an enhanced role for UKUSA in signals intelligence. The development of space based technologies has served both to facilitate global telecommunications and, conversely, to intercept these communications from space itself and on land. Space-based Sigint satellites and their processing facilities are exceptionally costly; the latest renditions cost to the order of US$1 billion apiece. Since 1968 at least three classes of Sigint satellites ('Canyon'; 'Rhyolite'/'Aquacade'/'Magnum'/'Orion'; 'Jumpseat' /'Trumpet') as well as several classes of dedicated Comint satellites ('Chalet/Vortex/Mercury') have been launched by the United States, the only country to have deployed space technologies for the interception of communications.

While particulars about American Sigint satellites launched after 1990 remain classified, the apparent expansion of the relevant ground centres associated with these satellites seems to indicate that space-based collection systems have grown in significance. Canada did not possess Sigint satellite technologies of its own, however the UKUSA arrangement allowed CSE to share in satellite-based Sigint collection and also to task – within certain parameters – US satellites to respond to specific Canadian foreign intelligence requirements.

Ever since the 1970s a rapidly increasing share of international telecommunications traffic has been relayed by Intelsat (International Telecommunications Satellite Organization) satellites and other regional communications satellites. At first just two specialized ground interception stations, one British and the other American, were sufficient to achieve UKUSA monitoring of all Intelsat traffic across the world. However, subsequent refinements to Intelsat satellite design impelled the UKUSA

FIGURE 4.4

Canadian satellite interception dish, Leitrim, Ontario.

Communications Security Establishment, Canada

alliance to build a chain of six intercept stations over the years in order to maintain global coverage, and to link these in a functional network. The launching of Soviet and other regional communications satellites spurred the building of other suitably-situated Sigint interception facilities to augment this UKUSA network. One of these operated under CSE aegis at Leitrim, Ontario, ostensibly targeted on Latin American satellite communications.

American Sigint satellites yielded a prodigious flow of intercepted telecommunications traffic requiring powerful computers to process, search, filter, and identify material of intelligence interest. CSE involvement in the UKUSA network of ground-based installations for satellite Sigint collection demanded a substantial upgrading of its technological base. The first satellite interception dish was installed at Leitrim in late 1984; another medium-size dish was erected in 1986. Staffing likewise had to be augmented and trained to analyze and disseminate the ensuing intelligence product.

To deal with this surge in Sigint collection after 1984 CSE undertook a revitalization and enlargement of its intelligence processing capacity and cryptanalytic capabilities. Early in 1985 CSE acquired its first supercomputer for cryptanalysis, a Cray X- MP/11. CSE staffing grew from around 600 personnel in the late 1970s to some 720 in the mid-1980s, and to about 900 by the end of the decade. By the late 1990s there were four satellite dishes operating at Leitrim.

By the 1990s, extensive refinements to UKUSA satellite interception technologies had made possible a virtually seamless global intelligence collection capability for the various modalities of signals intelligence collection: local in-country, external, HF long distance and space based. This quantum leap forward towards a convergence and meshing of Sigint technologies reached its zenith in the tightly integrated and networked interception and processing system known as 'Echelon'.[38] Highly secret still, 'Echelon' had its origins in the computerized processing and networking technologies which evolved since the 1970s and were greatly enhanced in the 1990s. Compared to earlier Sigint systems deployed during the Cold War, which were designed primarily to intercept diplomatic, espionage and military communications, 'Echelon' had a broad banded capacity to monitor virtually all types of electronic communications among public and private sector organizations and individuals in almost every country.

The 'Echelon' system links together an array of large-scale computer processing capabilities so as to enable the various UKUSA intercept stations to function as parts of an integrated, virtually seamless Sigint network. These interception and processing technologies are able to sort through vast flows of telecommunications traffic to identify specifically targeted messaging. At the operational heart of this integrated Sigint processing and networking system is the so-called 'Echelon' Dictionary computer. These Dictionary computers, which can store a comprehensive database on designated targets, including names, topics of interest, addresses, telephone numbers and other criteria for target identification, are emplaced only in certain 'Echelon'-linked Sigint interception facilities, though not in Canada. Given the tight networking achieved under 'Echelon', every participating interception facility's Dictionary computer contains not only its parent organization's designated keywords but also a list for each of the other partner Sigint agencies.

While CSE may not have its own 'Echelon' production capability, this networking arrangement enables Canada to post its search lists with the 'Echelon' Dictionaries at other partners' facilities. Intercepted communications would be processed through these inter-connected Dictionaries, with targeted intercepts being forwarded automatically to the listing organization. The reciprocity arrangement under UKUSA gives partner Sigint organizations virtually automatic access to Canadian interception modalities – local in country, external, HF long distance, or satellite downlinked – without Canada necessarily being aware of their targets – while in return CSE gets to share and participate in the global capabilities of the 'Echelon' system.

SIGINT TECHNOLOGY ACCESS AND SHARING

'Echelon' was designed to be a shared, collaborative Sigint collection network. The technologies behind 'Echelon' and other high capacity Sigint modalities were for the most part American in origin. The technologies developed for signals intelligence purposes were so specialized and of such advanced complexity that only experienced US defence contractors and niche suppliers could design and manufacture this purpose-built equipment for NSA, and then only with government technical and financial backing.[39] Some of this equipment was made available to other partner Sigint organizations. Among the American technologies reportedly procured by CSE were Cray supercomputers, 'Echelon' systems and their miniaturized versions ('Oratory') for outstations, miniaturized interception and processing equipment for embassy-based interceptions, high-capacity/ high-speed information retrieval technologies, and high-speed traffic/topic analysis search engines, *inter alia*. CSE and other Sigint partner organizations also relied on NSA training facilities for their cryptanalytical and other technical specialists.

This networking system demonstrated its robustness, and the burden-sharing capabilities of the UKUSA arrangement, when the NSA main computer system crashed calamitously for four days in January 2000. What was described as a 'system overload' shut down the computers used to process collected Sigint intelligence from 24 to 28 January, causing an unprecedented breakdown in the processing and analysis of raw intercepts.[40] Nevertheless, Sigint interceptions continued uninterrupted, and the processing of incoming intelligence was shunted to other components of the 'Echelon' system for the duration of the NSA outage. CSE was likely to have been involved, underscoring the high degree of systems integration among UKUSA partners and the particular value of this capability to the senior partner, the United States.

Whereas 'Echelon' was conceived as a shared network, there are suggestions that its actual workings are asymmetric. According to New Zealander Nick Hagerty's disclosures about Government Communication Security Board (GCSB) involvement in 'Echelon', each participating Sigint organization can only access that system for its own stipulated targets, and does not necessarily share any of the intelligence generated for other partners.[41] Participating organizations may request intelligence product from other partners' 'Echelon' Dictionaries, but actual access is effectively controlled by that country. If that is the case, Canada might not be able to receive output of the whole 'Echelon' network even though a considerable portion of CSE's own intelligence collection probably goes to serve other UKUSA partners requirements. It seems likely that only the NSA colossus, by virtue of its size and leadership role within 'Echelon' can access the full global potential of the system. For lesser players like CSE these controls on 'Echelon' access render the reciprocal sharing of signals intelligence under UKUSA in effect asymmetrical.

This asymmetry is also manifest in the targeting of Sigint satellites. All the Sigint satellites available to UKUSA are proprietary US craft embodying American technologies, though the uplinking and downlinking networks can also involve other partners' facilities. Under the 'Echelon' system these US satellites could be tasked in effect through the Dictionary mechanism. However, notwithstanding the sharing principle underlying UKUSA, the orbital positioning and targeting of these satellites remain exclusively under the control of the United States. While the US has sometimes been willing to reposition satellites so as to hover and zero in on targets requested by UKUSA allies, such requests were not without their difficulties and the response was entirely at American discretion.[42]

It should be noted, parenthetically, that the NSA also transferred certain of its Sigint technologies to the American private sector. Once these technologies had become operationally obsolete, they were spun off to commercially successful civilian applications. However, the tables were turning by the late 1990s, when it became apparent that private sector-inspired developments in certain areas of information and communications technology, like for example encryption, were beginning to run ahead of governmental Sigint capabilities. Indeed, NSA has come under increasingly sharp criticism from congressional intelligence committees for not keeping pace with advances in communications technology.[43]

CSE sought to promote the local development of Sigint technologies in niches where Canada enjoyed some particular competitive advantage and where Canadian solutions might also possibly spin off to commercial applications. Over the years Canada's high-tech industry achieved demonstrated strengths in information and communications technology. Since an integrated market for Canadian and American defence industries already existed, it was considered possible that a Canadian Sigint technology could be readily marketable to NSA and other partner organizations.

Two particularly relevant areas of niche technology where Canada seemed to enjoy competitive advantages were continuous speech recognition, software that translates verbal into digital text, and speaker/voice recognition, software than can identify individual talkers. In 1990 CSE awarded the first of a series of contracts to the Centre de recherche informatique de Montreal (CRIM) to design and build a word-spotting technology for Comint applications that could function reliably even in poor conditions.[44] After encountering insurmountable difficulties, CRIM proposed instead in 1993 to concentrate on developing a voice/topic identification module in collaboration with some American defence contractors. Further contracts were let, but progress towards an operational topic spotter system was still only in the experimental phase seven years later. Also in 1993, CSE commissioned CRIM to produce a workable speaker identification system. There are indications that this was achieved, and in 1995 NSA reportedly procured a Voice Activity Detector and Analyser which may have incorporated Canadian technology.[45]

CANADA'S POST-COLD WAR SIGINT AGENDA

These advances in Sigint technology and capabilities coincided with the ending of the Cold War and the adoption of new, more globalized priorities for Canada's foreign intelligence. In 1991, for the first time, the Government of Canada adopted an intelligence directive setting out its priority requirements for foreign intelligence collection. These priority requirements have updated almost annually since then. Among the current priorities are international terrorism, ethnic and religious conflict, proliferation of weapons of mass destruction, illegal migration, transnational organized crime, economic (counter-)espionage, and trade intelligence.[46] These emergent objectives were given operational expression in Sigint targeting, utilizing the enhanced technological capabilities that were now available.

Although CSE does not disclose its operational targets, the various annual reports of government agencies, occasional media reportage and other disclosures give indication of the persistent security challenges and new priorities shaping Canada's foreign intelligence agenda.

While the expanded foreign intelligence requirements identified certain new objectives, this in no way implied a relegation of *traditional Canadian security concerns*. Indeed, Canadian intelligence assessments perceive an ongoing espionage threat from Russia and other former Cold War adversaries.[47] They also assess security risks arising from newly assertive powers like China or India with hegemonic ambitions in regions of strategic significance to Canada; countries trying to evade internationally mandated sanctions or Canadian embargoes; warring states attempting to interfere with peacekeeping or preventive-diplomacy initiatives; rogue states like Iran, Iraq or Libya seeking to exploit a presence in Canada for nefarious purposes; or even nominally friendly countries whose perspectives on certain key issues relating to national security may conflict with those of Canada. Thus, clandestine French activities in support of Quebec separatism were closely monitored and countered by Canadian intelligence services.[48]

The security concerns of Canadian intelligence extended as well to the inappropriate activities of foreign governments trying to exercise improper influence on Canadian decision-making or public opinion, as when China attempted surreptitiously to buy control of local Chinese-language print and broadcast media outlets in order to manipulate sentiment in the aftermath of the 1989 Tiananmen Square massacre.[49]

CSE plays a part in helping to defend Canadian sovereignty and strategic interests by collecting operational intelligence on international security threats for Canadian government departments; providing counter-intelligence support by monitoring clandestine activities; and protecting Canada's communications systems against foreign intrusion.

International terrorism figures prominently among the security concerns for Canadian foreign and security intelligence.[50] Many of the world's terrorist groups have established a presence in Canada, virtually all of them relating to ethnic, religious or nationalist conflicts elsewhere in the world.[51] Among the international terrorist organizations or fronts active in Canada are

Hizballah and other Shiite Islamic terrorist organizations from the Middle East, the Palestinian Hamas, the Provisional Irish Republican Army (PIRA), the Liberation Tigers of Tamil Eelam from Sri Lanka, the Kurdistan Workers Party (PKK) from Turkey, and every significant Sikh terrorist group from India. These organizations established Canadian sanctuaries in order to raise and transfer funds, procure weaponry and material, set up operational bases, and to cover infiltration across the border to the United States or overseas.

Operational responsibility for security intelligence against terrorist threats to public safety or national security is vested mainly in the Canadian Security Intelligence Service (CSIS), working together with other government departments (e.g. Citizenship and Immigration, Department of Justice), Royal Canadian Mounted Police (RCMP) and local police services. It may be presumed that CSE monitors the international communications of suspected terrorist elements based in Canada as well as the activities of complicit foreign groups trying to operate through Canada to attack friendly countries. In a recent instance, Sigint interceptions helped foil an alleged conspiracy by a Montreal-based cell of the Algerian 'Armed Islamic Group' (GIA) to commit a terrorist bombing attack in the US during the New Year's 2000 celebrations.[52]

Transnational crime was recognized by the Group of 7 (G-7) Summit in Halifax in 1995 as a national security threat to many facets of public order – political, economic, social and environmental.[53] Since then international criminality has emerged as a priority concern for foreign and security intelligence in Canada and other UKUSA countries.[54] Organized criminal enterprises originating in Eastern Europe, Asia, North and South America, and Africa span the world, moving money, people and goods across borders, including Canada's.

Even more threatening than the traditional transnational crimes like trafficking in drugs and arms, money-laundering, and tax evasion, are the larger-scale, potentially more devastating instances of major international fraud, corruption and the manipulation of political and financial systems, which can destabilize democratic governments, subvert legitimate institutions, undermine social order, and distort economic activities.

UKUSA operations against international crime extended to the creation of a dialogue forum involving the five partners with other European countries, the International Law Enforcement Telecommunications Seminar (ILETS), which aimed to coordinate design standards for telecommunications equipment and software so that they remain accessible to legal surveillance. This of course implied that global telecommunications would remain vulnerable to covert interception, which some in the

European Union have come to regard as a significant threat to their commercial interests and privacy rights.[55]

Canada has not been immune to these types of transnational criminality. In 1995 CSIS indicated that Canada's intelligence community would take on a role in combating transnational crime, primarily through the provision of international criminal intelligence and strategic analyses to law enforcement agencies.[56] As part of this combined effort it may be expected that CSE would target the international communications of criminal personalities or organizations.

International commercial crime is especially vulnerable to Sigint interceptions, given its inescapable dependence on electronic means of voice and data communications. Sigint interceptions could offer a unique aperture into illicit transactions and criminal activities that threaten the integrity of Canadian financial and commercial institutions. As well, Sigint could contribute timely information on the bona fides of certain large commercial entities operating out of turbulent regions of the world and seeking to do business in Canada.[57] The intelligence collected can serve to aid law enforcement and the effectiveness of financial and commercial regulatory agencies. Furthermore, it could help inform Canadian foreign policy decision-making regarding the countries concerned.

Canadian foreign and security intelligence concerns are also directed at the connection between transnational criminality, on the one hand, and terrorist racketeering and criminal collaboration with insurgency movements elsewhere, on the other. In one of the more notorious instances, the Liberation Tigers of Tamil Eelam established an underground network among Tamil sympathizers across Canada and also became extensively involved in racketeering to generate financing for their insurgency war in Sri Lanka.[58] Their criminal activities are alleged to have included drug trafficking partnerships with Pakistani heroin producers, immigrant smuggling, commercial fraud, and extortion from Tamils residing in this country and elsewhere. Sigint operations can provide law enforcement agencies and foreign policy-makers with timely intelligence about attempts by transnational criminal elements to undermine the integrity of other countries and influence our own in ways detrimental to the laws and interests of Canada.

Canada participates in virtually the entire array of global and regional initiatives to counter the *proliferation of weapons of mass destruction* and their delivery systems. Canadian nuclear capabilities are devoted exclusively to peaceful purposes. Its non-proliferation foreign policy is aimed at ensuring that Canada's nuclear exports are utilized solely for intended, non-military purposes, and to promote the evolution of a comprehensive and effective non-proliferation regime. By way of supporting this non-proliferation policy, CSE operations aim at identifying attempts by countries of proliferation concern to acquire Canadian weapons-related technology and expertise. Intelligence produced by Sigint helps keep the Government of Canada and its allies alert to proliferation threats.[59]

THE ECONOMIC INTELLIGENCE CONUNDRUM

Many countries, from major powers to other smaller trade-dependent nations, have made the collection of economic intelligence an increasingly significant function of their respective foreign intelligence services. Economic intelligence is expected to identify opportunities and warn of threats to national economic and commercial interests. As early as 1970 the former Executive Director of the US Foreign Intelligence Advisory Board assigned economic intelligence a priority equivalent to diplomatic, military, technological intelligence.[60] Canada's post-Cold War intelligence directives identified economic espionage and competitiveness among its priorities for targeting.[61]

The implications of economic intelligence collection inject a competitive impulse, not to say conflicts of interest, into the otherwise cooperative ethos of UKUSA. To deal with this, a consensus seems to have emerged among the UKUSA partner organizations to the effect that commercial firms are not allowed to actually task Sigint operations for their own commercial purposes. Doing so could have posed operational risks, and is in fact unnecessary. Rather, the practice seems to have been for each UKUSA country to mandate its own national intelligence assessment organization and relevant government departments to task and receive economic intelligence from Sigint sources. Decisions on whether to disseminate this economic intelligence to the private companies were typically taken by these other governmental instrumentalities and not by the Sigint organizations themselves. Thus, for example, it is reported that Australia's Defence Signals Directorate regularly remitted commercially relevant Sigint to the Office of National Assessments, which in turn disseminated pertinent information to interested government departments and also private firms.[62]

Until recently Canadian efforts in economic intelligence seem to have been primarily defensive in orientation. According to intelligence sources, Canada's chief concern in this domain has been to counter economic espionage, defined as 'clandestine, deceptive, coercive or illegal activity carried out or facilitated by a foreign government aimed at obtaining access to Canadian proprietary information and/or technology for reasons of economic advantage'.[63] CSIS carried the main responsibility for countering economic espionage in the context of its security intelligence mandate, however Sigint doubtless made a contribution. One indication of growing CSE involvement in this domain was its 1995 effort to recruit additional staff with qualifications in economics, commerce and international business, in order to build up its own analytical capacity in economic intelligence.[64]

CSE operations in economic intelligence have gone rather beyond the strictly defensive to also help promote Canadian economic competitiveness and commercial objectives in world markets. Accounts published by reliable journalists claim that CSE provided Canadian policy-makers and negotiators with economic intelligence pertaining to international trade negotiations, including the plurilateral negotiations with Mexico on the North Atlantic Free Trade Agreement (NAFTA) of 1994; the 1995 multilateral ('Uruguay Round') trade negotiations; the Asia Pacific Economic Cooperation (APEC) Ministerial and Leaders' meetings in Vancouver in 1997; and bilateral negotiations with South Korea on their procurement of Candu nuclear reactors and with China on wheat sales.[65] The targeting of international economic and business affairs remains, of course, a highly delicate matter, all the more so in view of Canada's overwhelming trade dependence on the United States.

CSE efforts in economic intelligence do not appear to have provided Canadian commercial firms with access to Sigint products, at least not directly. The Canadian government has no identifiably dedicated unit either in the Privy Council Office, which coordinates Canada's intelligence effort, or in the intelligence agencies, or in the Department of Foreign Affairs and International Trade or Industry Canada, which could handle the interface between commercially-relevant intelligence and the private sector. Indeed, the peculiar structure of Canadian industry would greatly complicate any provision of government-sourced commercial intelligence to the private sector. Much of Canada's large-scale industry consists of subsidiaries of foreign firms which would make the dissemination of commercial intelligence highly problematic. To be sure, there are important Canadian industrial enterprises in the telecommunications, aircraft, power generation and civil engineering sectors, industries that are generally dependent on politically determined markets, but there is no evidence that the Canadian government supplies these firms with commercial intelligence in support of their marketing ventures.

Of course, government officials may sometimes provide advice and counsel by way of helping to promote Canadian trade, without necessarily revealing their sources in economic intelligence. Canada's crown corporations present a somewhat different challenge for economic intelligence: these enterprises, established by the federal and provincial governments, control important sectors of the export economy, including grain exports, energy exports, and export insurance and finance, where commercial intelligence can yield competitive advantages in government-to- government negotiations. However, it is questionable whether any intelligence so garnered was actually shared with crown corporations like the Canadian Wheat Board or Atomic Energy Canada Limited, or whether government negotiators themselves used this information to shape their bargaining positions on such public sector transactions as wheat sales to China or Candu sales to South Korea.

CSE is also responsible for Canadian information technology security (ITS). Canada has state-of-the-art industrial capabilities in various sectors of information technology, and Canadian companies have been targeted by foreign governments for economic or industrial espionage.[66] Some of the foreign governments engaging in technological espionage are recent adversaries while others are erstwhile friends and allies. Moreover, certain of these information technologies can have dual use, and may be vulnerable to redeployment by weapons proliferators or even terrorists. As the lead federal agency for ITS, CSE provided technical information, tools and expert services to government departments and private industry in areas of Network Security, Internet Security, Cryptography and Public Key Infrastructure. The aim of CSE industrial programmes is to collaborate with Canadian industry to develop advanced ITS products and services.[67]

It is inherently difficult to assess the operational performance of intelligence agencies. According to the 1996 Auditor-General's report, CSE has made a significant effort to cost its operations and products and identify gaps in its collection of signals intelligence in relation to national priorities and the specific requirements of client departments.[68] The government's own assessment of the performance and value of Sigint is indicated in its resource commitments to CSE, both funding and staffing. In the early post-Cold War period, government budgetary appropriations for CSE were estimated at C$113 million for fiscal year 1995–96, a reduction of about 10 per cent in real terms from 1990–91 (i.e. Cold War) levels. This compared favourably with the sharp cutbacks that took place in federal spending generally, including (indeed especially) national defence. While a declining trend continued for virtually all government departments and agencies, the nominal CSE budget for 1999/2000 of C$109 million suggests that Sigint continued to fare better than most other government services.[69]

Staffing has remained stable at approximately 900 (exclusive of Canadian Forces Information Operations Group personnel), with the proportion of analysts probably expanding. DND defence planning

CANADA'S COMMUNICATIONS SECURITY ESTABLISHMENT 121

guidelines project a 6 per cent increase in CSE's budget over the next five years.

FUTURE CHALLENGES

Signals intelligence collection provides Canada's policy-makers and security establishment with a capacity to cope with risk and threats to Canadian interests in an otherwise uncertain and volatile global security environment. Canada has a comparatively small population, yet it is a member of the G-7 and is extensively engaged in international relations in security, trade and finance, social affairs, environment, development, peacekeeping, and global governance. These international involvements entail a requirement for foreign intelligence in support of policy-making and the conduct of bilateral and multilateral relations. CSE has been able to provide this intelligence in part by dint of its own Sigint capabilities, but more significantly through the extended capabilities available to Canada under the UKUSA arrangement.

Current trends in Sigint imply two major challenges for CSE's future capability to perform its signals intelligence collection and processing functions.

The first of these challenges stems from ongoing trends in communications technology which tend to favour communications security over penetration and protection over interception.

A second set of challenges arises from prospective changes in the dynamics of UKUSA once competition outstrips cooperation in the emergent globalized agenda for intelligence collection, and in particular economic intelligence. It is ironic that these challenges derive from existing arrangements that have served CSE well, but are now developing in directions that can jeopardize the future capacity of CSE to respond to Canada's foreign intelligence requirements.

The technological lead in computers and information technology once enjoyed by Sigint organizations has now been very largely dissipated.[70] Widely available technologies today offer others, including potential adversaries, the same technical advantages to protect their communications as Sigint hitherto had to monitor this traffic. As a result, access to global communications networks is likely to become increasingly problematic for signals intelligence. This will become even more challenging as international telecommunications shifts over to high capacity optical fibre networks which reportedly cannot be intercepted by current Sigint technologies.[71] Intrusive access would be necessary for interceptions. Clandestine operations of this type would be risky, and could become politically unacceptable.[72]

Sigint advantages in cryptanalysis are likewise dissipating in face of rapid advances in civil and commercial cryptography along with the development of more effective cryptographic security systems.[73] Indeed encryption is becoming widespread very rapidly as electronic commerce expands, an increasingly problematic trend for communications intelligence collection, in particular. It is clear that CSE and its Sigint partner organizations were unsuccessful in their bid to constrain private sector cryptography by arguing for 'public key escrow' and similar systems ostensibly to support law enforcement (as distinct from signals intelligence) requirements. Innovative and costlier technologies will have to be deployed in future in order to stretch cryptanalytical capabilities sufficiently to extract the intelligence required.

The transition from the Cold War to a new, more globalized Sigint agenda poses certain other challenges for the future of UKUSA operational solidarity and intelligence sharing. It has been a principle of UKUSA cooperation that its Sigint activities do not target one another or their respective nationals (including corporations). Whenever Sigint intercepts incidentally implicate nationals of the partner countries, steps are taken to protect the anonymity of the individual(s), or enterprise(s) in the handling and sharing of the intelligence. This principled understanding was necessary in order to ensure compliance with national law and self-interest in the partner countries while facilitating inter-group collaboration and sharing of signals intelligence collection; it also served to mitigate conflicts of interest.

Once the Cold War was over, however, the adoption of a more broadly globalized agenda for foreign intelligence collection by each of the UKUSA partner countries, including Canada, had far reaching implications for the shared Sigint enterprise. Unlike the focused Sigint effort of the recent past, the more broadly targeted post-Cold War intelligence directives adopted by the UKUSA governments were not entirely congruent one with the other. Differences and asymmetries in priorities created a potential for conflicts of interest over Sigint targeting and intelligence collection. Although UKUSA partners remain committed to the principle of refraining from targeting each other or their respective nationals, nevertheless a former US National Security Council official Howard Teicher made a point of commenting:

> I would never say never in this business because, at the end of the day, national interests are national interests ... sometimes our interests diverge. So never say never – especially in this business.[74]

Arguably, the risks of conflicts of interest within UKUSA are greatest in the increasingly important Sigint domain of economic intelligence. It is here that the UKUSA ethos of cooperation may be most vulnerable to protectionist impulses and dysfunctional competition. In as much as UKUSA countries are major trading partners between and among themselves, they are often engaged in trade negotiations or dispute settlement procedures at the bilateral, regional (e.g. APEC, NAFTA) and multilateral (e.g. World Trade Organization) levels. Since these economies are also competitors in many world markets, they are frequently keen commercial rivals. In the circumstances, Sigint economic intelligence operations that *never* targeted other partners' commercial interests or negotiating stances would probably be deemed irrelevant by domestic policy-makers, and yet any effort to systematically target allies' proprietary commercial, technological or policy secrets would compromise UKUSA collaboration and render the 'Echelon' alliance highly problematic. Nonetheless, press accounts describe the activities of friendly and even allied countries in eavesdropping on one another in order to gain negotiating advantages at bilateral and multilateral meetings on international trade.[75]

Hence the paradox of cooperation/competition that confronts Sigint in the domain of economic intelligence. Economic intelligence collection which is timely and informative for competitive advantage can be inherently undermining and destructive of operational cooperation and technology sharing. Yet any turn of events that would tend to constrain collaboration in UKUSA would substantially weaken CSE's capacity to achieve near global access to Sigint facilities to meet Canada's foreign intelligence requirements. Foreign intelligence is an essentially competitive enterprise in which countries seek their own advantage, and in which all gains are differential, asymmetric gains.

Another area of potential conflict of interest among UKUSA partners concerns the use of Sigint interceptions for law enforcement purposes.[76] In targeting transnational crime, Sigint operations must take account of the mandatory legal and technical prerequisites governing interceptions for law enforcement purposes, as distinct from interceptions of communications intelligence. Not only must this distinction be recognized and observed, it must be observed operationally and reciprocally across the multiple legal jurisdictions of UKUSA so as not to compromise the bona fides of law enforcement. Any blurring of this distinction would risk dangerous illegalities and human rights transgressions and the gathering of inadmissible evidence. It is pertinent to acknowledge in this regard that Canadian jurisprudence is more protective of privacy rights than many other legal systems, including that of the United States.[77]

It is questionable whether Canadian law or Charter of Rights and Freedoms can be applicable to Sigint operations that task CSE facilities to target alleged transnational criminality at the behest of UKUSA partners. The legal issues implicit in Sigint-derived evidence have never been tested before Canada's courts. Whenever questions have been raised, mere reference to Canada's 'international' obligations has sufficed to defer detailed inquiries. Up until now Canadians have been generally (albeit tacitly) willing to countenance Sigint interceptions for 'security' purposes, however broadly defined. Were there to be perceived violations of law and human rights, however, these are unlikely to be politically acceptable to government and public. Yet for Canada (or another partner country) to impose national legal or human rights standards unilaterally onto Sigint interceptions might well jeopardize future UKUSA collaboration against transnational crime and other sensitive targets.

The more Canada's foreign intelligence requirements become globalized in future, the greater will be CSE's reliance on UKUSA sharing arrangements and the more its operational activities will become exposed to the underlying risks. The prospect of any lessening of these cooperative Sigint capabilities, whether due to technological trends, differential interests of partners, or legal dilemmas, could severely circumscribe Canada's capacity for foreign intelligence collection. Canada depends on CSE to manage its own resources and international linkages in a way that safeguards its future capacity to respond to Canadian foreign intelligence requirements in an increasingly predatory international environment.

NOTES

1. Federal government expenditure on signals intelligence and information technology security involves the combined budgets of CSE itself and the Canadian Forces Information Operations Group, which provides operational personnel for its interception facilities, and which together exceed spending on the domestic security intelligence organization, the Canadian Security Intelligence Service.
2. CSE (like its predecessor, the Communications Branch of the National Research Council) was established by Order-in-Council, that is by cabinet decree, rather than on the basis of formal enabling legislation. There is very little information in the public domain regarding CSE. Some carefully crafted official information is available in the *Report of the Auditor General of Canada, 1996, The Canadian Intelligence Community – Control and Accountability* (Ottawa: Nov. 1996) Chapter 27; in annual reports of the Office of CSE Commissioner; in infrequent officials' testimony before Parliamentary committees; and in snippets of other periodic reports (e.g. DND budgetary documents, The Privacy Commissioner's *1995–96 Annual Report*). The CSE's own website (URL: www.cse.dnd.gc.ca) concentrates on its public information technology security mission. There have been occasional newspaper articles on CSE activities and references to it in studies of other Canadian intelligence organizations or allied Sigint organizations. An unofficial website prepared by Bill Robinson on *The Communications Security Establishment: An Unofficial Look Inside Canada's Signals Intelligence Agency* is accessible at the URL: http://watserv1.uwaterloo.ca/~brobinso/cse.html.
3. For a synopsis of the structure of government control and accountability over the Canadian intelligence community, see the *Report of the Auditor General of Canada, 1996*, Chapter 27:

The *Canadian Intelligence Community. Control and Assessment*, paras. 27.66-27-94 (URL: www.oag-bvg.gc.ca/domino/reports.nfs/html9627ce.html).
4. Peter Hum, 'I Spy', *Ottawa Citizen*, 10 May 1997.
5. Auditor-General, *The Canadian Intelligence Community* (note 2) para. 27.30
6. Ibid. para. 27.82.
7. On the wartime history of Canadian signals intelligence see John Bryden, *Best-Kept Secret: Canadian Secret Intelligence in the Second World War* (Toronto: Lester 1993); Wesley Wark, 'Cryptographic Innocence: The Origins of Signals Intelligence in Canada in the Second World War', *Journal of Contemporary History* 22 (1987) pp.639–65.
8. Kevin O Neill, *History of CBNRC* (1987)[Classified]. Parts of this internal history have been released in abridged form under the Access to Information Act.
9. Vladislav Zubok and Constantine Pleshakov, *Inside the Kremlin's Cold War. From Stalin to Khrushchev* (Cambridge, MA: Harvard UP 1996) p.146.
10. For an article placing Gouzenko's defection into the larger foreign policy context relating to Canadian involvement in the Cold War, see Robert Bothwell, 'The Cold War and the Curate's Egg: When did Canada's Cold War Really Begin?', *International Journal* 53/3 (Summer 1998).
11. Nigel West, *Venona: The Greatest Secret of the Cold War* (Toronto: HarperCollins 1999).
12. While most CBNRC personnel were Canadian, for several years senior staff came from Britain's GCHQ, giving indication of the early and close working relationship established between the British Sigint organization and its emergent Canadian counterpart.
13. Bryden, *Best Kept Secret* (note 7) pp.280–1; Christopher Andrew, 'The Making of the Anglo-American Sigint Alliance', in Hayden Peake and Samuel Halperin (eds.) *In the Name of Intelligence: Essays in Honor of Walter Pforzheimer* (Washington DC: NIBC Press 1994) p.105.
14. Stephen Dorril, *MI6. Inside the Covert World of Her Majesty's Secret Intelligence Service* (NY: Free Press 2000) pp.54–5. Australia likewise consented to being represented in the alliance by Great Britain.
15. Bryden, *Best Kept Secret* (note 7) pp.291–2; Wark, 'Cryptologic Innocence' (note 7) p.659. I am indebted to Matthew Aid for making available his impressive historical records on Canadian Sigint.
16. Robinson, *The Communications Security Establishment* (note 2) Sigint sites; O'Neill, *History of CBNRC* (note 8) Chap.2; Matthew Aid, communication to author.
17. Bryden, *Best Kept Secret* (note 7) p.296; Memo, Agee to Coordinator of Joint Operations, *Proposed US-Canadian Agreement*, 7 June 1948, RG-341, cited in communication from Matthew Aid.
18. Bryden, *Best-Kept Secret* (note 7) p.326. For a summary of Canadian cryptanalysis in the service of signals intelligence, see Bill Robinson, 'The Fall and Rise of Cryptanalysis in Canada', *Cryptologia* (Jan. 1992) and 'Cryptanalysis at CSE', *The Communications Security Establishment* (note 2).
19. Mark Urban, *UK Eyes Alpha* (London: Faber 1996) p.6.
20. Robinson, 'Cryptanalysis at CSE' and 'The Fall and Rise of Cryptanalysis in Canada' (note 18).
21. Richard Cleroux, *Official Secrets* (Toronto: McLelland & Stewart 1991) p.266.
22. N.C. Gerson, 'Collaboration in Sigint: Canada-US', *NCVA Cryptolog* (Spring 1999); Matthew Aid personal communication.
23. Peter Wright, *Spycatcher* (New York: Viking 1987).
24. Christopher Andrew and Vasili Mitrokhin, *The Mitrokhin Archive. The KGB in Europe and the West* (London: Allen Lane, The Penguin Press 1999) pp.451, 448, 850, footnote 63.
25. Urban, *UK Eyes Alpha* (note 19) p.6.
26. Eldon Black, *Direct Intervention. Canada-France Relations 1967–1974* (Ottawa: Carleton UP 1996) refers to Canadian 'security authorities' providing intelligence on French communications with 'dubious contacts in Quebec' (pp.50–1). For reports on CSE monitoring Quebec separatist communications with France, see Doug Gilmour, 'WCC Members Likely Targets for Defence Monitors – Ex-Spy', *Edmonton Journal*, 15 Oct. 1982;

Peter Moon, 'Canadian Agency Safeguards its Role in World Spy Game', *Globe and Mail*, 30 March 1987; Gerry Arnold, 'Officials Deny Report of Canada-France Spy Feud', *Ottawa Citizen*, 22 May 1992; Mike Frost and Michel Gratton, *Spyworld: Inside the Canadian and American Intelligence Establishments* (Toronto: Doubleday 1994); cited in Robinson, 'Eavesdropping on the Quebec separatist movement', *The Communications Security Establishment* (note 2).

27. Andrew and Mitrokhin, *The Mitrokhin Archive* (note 24) p.453.
28. On one occasion, at least, an external interception operation was reportedly mounted in an allied country at the invitation of that government. Thus, on the eve of the British general election of 1983, GCHQ was alleged to have conveyed a personal request from Prime Minister Margaret Thatcher for Canadian Sigint assistance in monitoring communications of two of her cabinet ministers ostensibly 'to find out not what they were saying, but what they were thinking'. CSE involvement was sought due to the extraordinary political sensitivity of the operation, which made it inappropriate for GCHQ itself to undertake. An interception facility was set up at the Canadian High Commission in London and the 'take' was delivered to GCHQ. This alleged episode was revealed by former CSE employee, Mike Frost, in a CBS '60 Minutes' program and reported in 'Spy Agencies List in on Diana', *Sunday Times* (27 Feb. 2000).
29. Frost and Gratton, *Spyworld* (note 26) pp.19, 72, 76; Bruce Livesey, 'Trolling for Secrets: Economic Espionage is the New Niche for Government Spies', *Financial Post*, 28 Feb. 1998.
30. Frost and Gratton, *Spyworld* (note 26) pp.183, 191
31. Canadians reportedly underbid the United States on this wheat deal after having intercepted a car phone conversation between the US Ambassador and Ottawa Embassy discussing the American negotiating position. Cf. 'The Murky Side of Trade', Livesey, 'Trolling for Secrets'.
32. Jeffrey Richelson and Desmond Ball, *The Ties that Bind: Intelligence Cooperation Between the UKUSA Countries* (London: Allen & Unwin 1985) p.144. All the outlying receiver stations are now remote controlled from Leitrim.
33. Cited in Dorril, *MI6* (note 14) p.56.
34. On the unusual character of the UKUSA arrangement see Jeffrey Richelson, *The US Intelligence Community* (NY: Ballinger 1989) esp. Chap.12; Robinson, 'The UKUSA Community', in *The Communications Security Establishment* (note 2); Richelson and Ball, *The Ties that Bind* (note 32) pp.142–3 *et passim*; on the origins of UKUSA see Andrew, 'The Making of the Anglo-American Sigint Alliance' (note 13).
35. One of the rare explicit official references to the UKUSA agreements was made by the Deputy Clerk, Security and Intelligence, Privy Council Office, in testimony before the House of Commons Standing Committee on National Defence and Veterans Affairs, 2 May 1995.
36. Margaret Bloodworth, Deputy Clerk, Security and Intelligence, Privy Council Office, evidence presented to the House of Commons Committee on National Defence, 2 May 1995.
37. Urban, *UK Eyes Alpha* (note 19) pp.32–3.
38. Very little has been revealed officially about 'Echelon' by any of the UKUSA governments. Among the seemingly better informed sources are *Interception Capabilities 2000*, the Report to the Director-General for Research of the European Parliament prepared by Duncan Campbell (1999); and the disclosures about New Zealand involvement in Nick Hager, *Secret Power. New Zealand's Role in the International Spy Network* (Nelson, NZ: Craig Potton 1996), esp. Chap. 2 and 'Exposing the Global Surveillance System', *Covert Action Quarterly* (Winter 1997).
39. One of the rare descriptions of contemporary Sigint equipment is provided in the Technical Annexe to – *Interception Technologies 2000* (note 38).
40. 'NSA System Inoperative for Four Days', *Washington Post*, 30 Jan. 2000.
41. Hager, 'Exposing the Global Surveillance System' (note 38).
42. For an account of the British experience in persuading the US to reposition its Sigint satellite to provide intelligence coverage at the time of the Falklands War (1982) see Urban, *UK Eyes Alpha* (note 19) p.57.

43. 'NSA System Inoperative for Four Days'.
44. *Interception Technologies 2000* (note 38) Technical Annexe, paras. 33–6.
45. Ibid. para. 36.
46. Auditor-General, *The Canadian Intelligence Community* (note 3) para. 27.31.
47. Canadian Security Intelligence Service, *1997 Public Report*, Parts 1, 3 URL: www.csis-scrs.gc.ca/eng/publicrp/pub1997e.html.
48. Cf. Black, *Direct Intervention* (note 26) pp.50–1.
49. According to the media disclosure, telephone intercepts were part of this counterintelligence operation: 'CSIS warned Ottawa of Beijing Media Plot', *Globe and Mail*, 9 Feb. 2000.
50. CSIS *1997 Public Report*, Part 2.
51. CSIS, *Trends in Terrorism, Perspectives*, Report 2000/01, 18 Dec. 1999.
52. Newspaper accounts describe the role of Sigint interceptions in unravelling what appears to have been a complex Islamic terrorist conspiracy: 'US Probe Ties Bomb Plot to Bin Laden Group', *Washington Post*, 20 Feb. 2000; see also 'Calls Said to Link Woman to Man with Explosives', *New York Times*, 13 Jan. 2000, 'Canada Adds Details on Algerians' Suspected Bomb Plot', *New York Times*, 21 Jan. 2000, 'Algerian Charged in Bombing Plot Aids FBI Probe', *Washington Post*, 21 Jan. 2000.
53. Samuel Porteous, 'The Threat from Transnational Crime: An Intelligence Perspective', *CSIS Commentary #70* (Ottawa: CSIS, Winter 1996).
54. CSIS *1997 Public Report*, Part 3; 'Whose (sic) Being Spied On?' BBC, 13 Sept. 1999.
55. 'Comment les Etats-Unis espionnent l'Europe', *Le Monde*, 23 Feb. 2000.
56. CSIS *1995 Public Report and Outlook* (Ottawa: Canadian Security Intelligence Service 1995). The RCMP Economic Crimes Directorate is responsible for investigating commercial crime in Canada. Parliament is currently considering a bill to set up a dedicated Financial Transactions and Reporting Analysis Centre to monitor international money movements.
57. Porteous, *The Threat from Transnational Crime* (note 53).
58. Ibid.
59. CSIS *1997 Public Report*, Part 3.
60. *Interception Technologies 2000*, para. 97; Samuel Porteus, 'Economic/Commercial Interests and Intelligence Services', *CSIS Commentary #59* (Ottawa: CSIS, July 1995).
61. See Auditor-General, *The Canadian Intelligence Community*, para. 27.27.31; CSIS, *1997 Public Report*, Part 3.
62. *Interception Technologies 2000* (note 38) para. 100.
63. CSIS, *1997 Public Report*, Part 3.
64. Porteus, 'Economic/Commercial Interests and Intelligence Services' (note 60).
65. Livesey, 'Trolling for Secrets' (note 29).
66. CSIS *1997 Public* Report, Part 3.
67. On the Canadian Information Technology Security program operated by CSE see the CSE website at URL: www.cse.dnd.gc.ca.
68. Auditor-General, *The Canadian Intelligence Community* (note 3) Chap. 27: *The Canadian Intelligence Community. Control and Assessment*, paras. 107–9.
69. Robinson, *The Communications Security Establishment* (note 2).
70. According to Lt. Gen. Michael V. Hayden, Director of NSA, the Sigint agency is already lagging behind private sector technological developments in telecommunications (Address to Kennedy Political Union of American University, 17 Feb. 2000). See also *Interception Capabilities 2000* (note 38) paras. 106–8.
71. The possible existence of a new, top secret American technology to intercept fibre optic traffic is mentioned in *Le Nouvel Observateur* [Paris], (10–16 Dec. 1998) pp.10–22.
72. NSA has sought to overcome this growing cryptanalytical constraint by collaborating with CIA in a joint initiative, the Special Collection Service, which undertakes clandestine intrusive operations to intercept traffic from otherwise secure targeted systems: 'Une alliance secrète entre la NSA et la CIA', *Le Monde*, 22 Feb. 2000.
73. Simon Singh, *The Code Book* (London: Fourth Estate 1999) examines the evolution of codes up to current experimentation with quantum cryptography, which is said to be absolutely unbreakable; see also *Interception Capabilities 2000* (note 38) paras. 109–10.

74. Cited in *Interception Capabilities 2000* (note 38) para. 105.
75. Cf. 'The Murky Side of Trade Meetings'; Duncan Campbell and Paul Lasmar, 'The New Cold War: How America Spies on Us for its Oldest Friend – The Dollar', *The Independent*, 2 July 2000.
76. Cf. *Interception Capabilities* 2000, 'Policy Issues for the European Parliament' (note 38) para. 4.
77. Cf. Paul Palango, *The Last Guardians* (Toronto: McLelland & Stewart 1998) pp.165–7.

5
The Bundesnachrichtendienst, the Bundeswehr and Sigint in the Cold War and After

ERICH SCHMIDT-EENBOOM
(Translated by William Fairbanks)

Thanks to the films coming out of Hollywood and articles appearing in the mass media, the world's second oldest profession owes much of its aura to the misleading impression that most of its operational accomplishments have come from the activities of daredevil secret agents, who while operating deep behind the enemy's lines ransacked the safes of foreign powers, thereby gaining access to the most carefully guarded secrets of the enemy. But in fact, the everyday activities of these intelligence agencies have about as much in common with what is portrayed in spy movies as sports by amateurs have anything to do with attempts to break athletic world records.

The truth is that the clandestine spying activities of the world's intelligence agencies account for a very small percentage of the intelligence being collected. According to publicly available materials, modern intelligence services obtain about 75 per cent of their information from open sources, with only 25 per cent coming from secret intelligence. Of these secret intelligence reports, three-quarters come from Signals Intelligence (Sigint), which in German is known as *Fernmelde-und Elektronische Aufklärung* (Telecommunications and Electronic Reconnaissance, FmElo). And most of the information being collected is of little use to intelligence analysts. Of the about 100,000 intelligence reports that landed in 1989 on the desks of the analysts of the West Germany federal intelligence service, the *Bundesnachrichtendienst* (BND), only about 450 of these reports were considered to contain useful information.[1]

Not only the amount and diversity of the telecommunications transmission media increased enormously during the 1990s, but alongside them also the greed of the intelligence services to suck information from them by using increasingly more ingenious methods. Yet, the ether was not always the most worthwhile medium for intelligence hunting expeditions.

THE EARLY YEARS

Shortly before the end of World War II, on 10 April 1945, Hitler relieved Major General Reinhard Gehlen from his post as head of the *Wehrmachtsabteilung Fremde Heere Ost* (Foreign Armies East, FHO) of the General Staff. With 50 steel suitcases loaded with microfilms he took refuge in the 'Alpenfestung' and surrendered himself on 20 May 1945 in the vicinity of Miesbach to the American CIC (Counterintelligence Corps). The CIC however sent the former Nazi general to a special prison in Oberursel. Here in July 1945, Gehlen drew the attention of Brigadier General Edwin L. Sibert, head of the secret service in the US occupation zone. Because of the fact that the East–West conflict was already looming over the horizon and the US intelligence services hardly had any knowledge about the Soviet armed forces, Sibert sent Gehlen to Washington in August, where he stayed until 7 July 1946.

In the United States, in June 1946 the future chief of the German secret service successfully negotiated an agreement allowing him to resume his intelligence work mainly aimed at collecting military information from the Soviet Union and her satellite states. Gehlen could start with 50 personnel of the *Organisation Gehlen* (OG), which was financed by the United States, On 6 December 1947 the OG deployed to Pullach in the southern part of Munich. On 1 April 1956 the organization was taken over by the Federal Republic as the *Bundesnachrichtendiens*t (BND) and placed under the supervision of the Bundeskanzler.[2]

When the OG began in 1946 under the patronage of the OSS/CIA to primarily spy on the Soviet military, Human Intelligence (Humint) was the principal method by the Germans to collect intelligence behind the Iron Curtain. Two examples of OG Humint operations will suffice. During Operation 'Hermes', tens of thousands of German prisoners returning from Soviet POW camps were interrogated by Gehlen operatives for information about what was going on behind the Iron Curtain. And during Operation 'Fahrrad', the trash of Soviet military garrisons in occupied Austria was scoured by OG agents looking for useful intelligence information which the Russians had thrown away.[3]

Conceptual thinking about an independent West German Sigint service started late through the former Wehrmacht lieutenant general, Adolf Heusinger,[4] the General Staff chief of operations till he was arrested after the plot on 20 July 1944 and working for OG from March 1948 till 1952. Under the covername Dr Horn he was responsible for the analysis section in the OG. After 1950 he was also one of the most important advisers of Chancellor Konrad Adenauer in the field of the rearmament of West Germany.[5] On 30 December 1950, Heusinger wrote a lengthy memorandum entitled 'Thoughts about a future German radio-electronic reconnaissance', which recommended that OG devote more resources to Sigint collection.[6] He used two sources. First of all, thoughts about the 'Development of radio intelligence in the former German Wehrmacht'. Besides this source, he was also privy to detailed information about the, at that time, prevailing Sigint structures of the Italian and French armed forces. But whatever might happen – the later Inspector General of the Bundeswehr Heusinger wrote that the structures should be developed independently from the degree of their future integration within the relevant military arrangements of the European Defense Community.[7] The planned European defense community failed in 1954 due to the opposition of the French parliament.[8] Heusinger's main idea was to create a communication intelligence (Comint) organization with a secret department subordinated to a federal institute tasked to collect political, military and economic intelligence in peacetime, while the future army should be equipped with mobile communication intelligence units only.

Like the OG itself, the earliest West German Sigint effort was a creation of the US Army. In October 1945, some German experts guided by Gehlen's colleague and in 1947 chief of his Department I (Humint) Hermann Baun, performed with American equipment the first Sigint operations against Soviet troops in Germany from the so-called Bluehouse near Oberursel.[9] In 1946, the US Army CIC was instructed to assemble former German intelligence officers who had worked during World War II in the FHO and the *Reichssicherheitshauptamt* (Department of Federal Security, RSHA) at Camp King near the town of Oberursel, Germany. That same year, the CIC also invited experts of the former Sigint service of the Wehrmacht to the town of Bad Vilbel. The aim was to persuade these former German intelligence officers to create and operate Sigint intercept stations for the CIC. One of the invitees was Wilhelm Breede, who had served as the Wehrmacht's chief cryptanalyst at Smolensk on the Eastern Front, who had then had transferred in 1945 to General Gehlen's FHO.[10]

General Gehlen recruited a small number of radio intercept and cipher experts for his OG, such as Leo Hepp, who served as the Director of the OG Communications Reconnaissance Department at Pullach from 1948 until 1956. His successor was Albert Praun, who was also a former general who worked in the Wehrmacht radio intelligence organization during World War II. Hepp returned to BND headquarters at Pullach after a stint with the Bundeswehr from 1968 to 1970, and was named the Director of Department II, which managed the BND's Sigint effort.

In 1948, the first OG Sigint personnel moved into their new headquarters at Kransberg Castle near the Hessian town of Butzbach. The first chief of the Kransberg station was Friedrich Boetzel, a former general who ran the Wehrmacht's radio intelligence section during the latter part of World War II. From Butzbach, radio traffic originating from the Soviet Union was intercepted. It was also here that the decryption of all OG radio intercepts took place under the direction of a former Wehrmacht cryptanalyst named Erich Hüttenhain. In 1948, construction began on the first West German High Frequency Direction Finding (HFDF) network, which stretched from Bremen in the north to Butzbach, and southwards to Chiemsee in Bavaria.[11]

Under Hepp's leadership, the first OG listening posts were established.[12] For example, in 1952 a listening post was opened in the Bavarian town of Tutzing, which had the mission of monitoring radio traffic coming from East Germany. In order to conceal the station's activities, its operations building was disguised as a commercial company called *Südlabor GmbH*. This listening post was closed in 1991.[13]

In 1952 the US Army opened an intercept station at Lauf an der Pegnitz that was manned by former Wehrmacht Sigint personnel. The first chief of the Lauf station was Wilhelm Flicke, who ran the station during World War II. The Gehlen Organisation took control of the station in 1952, but the US Army agreed to subsidize the continued operation of the Lauf site until 1956.[14] A confidential document entitled 'Surveillance Areas of the Lauf Listening Post', dated November 1952, shows that the station's main target was monitoring Czechoslovakian border patrol traffic and the radio networks of the Czech internal security service. Using 14 radio intercept operators, about 2,000 Czech radio transmissions were intercepted monthly. Two radio intercept operators were assigned to listen to the radio traffic of the Polish armed forces and the Polish Internal Security Corps, but only 90 radio messages were intercepted monthly. Two more intercept operators monitored Hungarian radio traffic, but because of radio reception problems it was only possible to intercept about 100 Hungarian military messages a month. Finally, the Lauf station was only able to copy police and weather forecast radio transmissions coming from Romania.[15]

Before the *Bundesamt für Verfassungsschutz* (Agency for Constitutional Protection, BfV) was established in 1950, OG also had the responsibility for carrying out the counterintelligence mission. It was also involved in communications security, as well as intercepting and analyzing the electronic communications stemming from the Soviet and East European intelligence services and their clandestine sources in West Germany. Those functions were later to be taken over by the electronic communications surveillance service (*Funkbeobachtungsdienst*), which was subordinated to the telecommunications branch of the *Bundesgrenzschutz* (Federal Border Guard, BGS). The BGS operated four listening posts outside the towns of

Heimerzheim, Leer, Lübeck and Rosenheim. The radio traffic intercepted by these stations, some of which was obtained with computers, was used by the BfV, the BND, the *Zollkriminalamt* (customs criminal branch, ZKA) and for the BGS's own intelligence requirements.[16]

Despite the developments described above, technological intelligence gathering played a decidedly limited role during the early years of the BND. In 1957, after the OG became a federal agency, being the *Bundesnachrichtendienst*, Sigint was represented in the BND headquarters staff by only two small staff units: the Technology and Research Department, and the Telecommunications/Radio Surveillance Unit of Department III, which was the Directorate of the BND. However, one must remember that the BND was still a relatively small organization. As of 1957, the BND's total staff consisted of only 1,245 employees.[17]

Thirty years later in 1987, Department 2 (Technological Surveillance) of the BND managed the Sigint collection and processing efforts. This unit had during the largest extent of the BND, about 7,500 employees and a staff of 2,200 as well as several hundred more employees in Section 14 B for mail and telecommunications control, in central operations for ciphering, and in Sub-Department 63 for intelligence services technology and technological support with its three sections handling intelligence services' technology/ operational control, secret transactions, and technological physics. All were involved with Sigint. A measure of how greatly the size of the BND's Sigint effort increased can be deduced by looking at personnel growth between 1976 and 1997. The BND's personnel strength was equal in both these years about 6,000. But the number of BND employees directly engaged in Sigint-related duties increased from 475 employees in 1976 to 1,450 employees in 1998, an increase of more than 300 per cent.[18]

THE SIGINT EAVESDROPPERS OF THE BUNDESWEHR

Along with the rearmament of the Federal Republic of Germany, in 1956 the West German armed forces began creating military units for the interception of telecommunications traffic and electronic emissions. At the request of the BND, the Bundeswehr formed mobile Sigint collection units along the borders with East Germany and Czechoslovakia for monitoring radio traffic coming from military manoeuvres taking place in East Germany, which supplemented the larger permanent stations run by the BND. Several Bundeswehr Sigint officers and NCOs were trained in their new responsibilities at BND intercept and direction-finding establishments.[19]

In July 1957, the West German Army activated a temporary Sigint processing and analysis center in the town of Bergisch-Gladbach, and in

1959 the Bundeswehr's Telecommunications Battalion 51 began eavesdropping on the HF radio traffic of Soviet military forces stationed in East Germany. From 1960 until 1962, the first Bundeswehr HF direction-finding station operated out of the town of Cologne-Ostheim.

But the strengthening of the West German military's Sigint intercept and direction-finding units was not made a priority until after the Cuban Missile Crisis in October 1962 with the build-up phase lasting until the mid-1960s. During this period, nearly 20 HFDF sites and four new Sigint battalions being the Telecommunications Battalions 51, 110, 220, and 320 were activated. These HFDF stations were stretching from the town of Riepe in East Friesia in the north all the way to the Bavarian lake Chiemsee in the south.[20] But according to evaluations performed by the BND staff at Pullach, the Bundeswehr Sigint units were only able to begin performing limited military intercept missions beginning in 1963 due to personnel and equipment shortages.

With respect to the Sigint activities of the German Air Force, Telecommunications Regiment 71 in Osnabrück and Telecommunications Regiment 72 at Feuchtwangen were responsible for Sigint collection in the HF and VHF ranges against the high-level communications of the Soviet Air Force in East Germany, plus air-to-air and air-to-ground communications of Soviet and Warsaw Pact air defense units. During 1965–67, the German Air Force established Sigint intercept towers along the East German and Czech borders. These were Tower A at Grossenbrode, B at Thuraver Berg, C at Stöberhai, E at Schneeberg, and F at Hoher Bogen, followed in 1967 by an HFDF station at the village of Langenargen.[21]

The West German Navy also carried out HF eavesdropping operations from a listening post at the city of Flensburg (adjacent to the Danish border), as well as tactical Comint and Elint intercepts from a Sigint tower at Pelzerhaken. In 1958, a headquarters unit, designated *Marinefernmeldestab 70,* was created at Flensburg to manage all German Navy Sigint operations. From the early 1960s to the present, the German Navy has operated three Sigint ships, which mainly perform their mission in the Baltic Sea. When in 1971 the German Navy bought 16 Breguet Atlantic planes for use as maritime patrol aircraft, four of these planes were modified for Elint collection. These aircraft are still stationed at Nordholz Air Base in northern Germany. During their reconnaissance flights along the East German border and over the Baltic Sea during the 1980s, the German Sigint specialists on these aircraft were often accompanied by British and American intelligence officers. After the end of the Cold War, the Breguet Atlantic aircraft were used to monitor the United Nations embargo against Yugoslavia, and flew reconnaissance missions during the 1999 war in Kosovo.

However, even before the introduction of the Breguet Atlantic aircraft in 1971, the BND deployed aircraft for Sigint. The former defence minister (1958–62) and Bavarian prime minister, Franz Josef Strauss, recalls in his memoirs that the BND in May 1970 overheard from an airborne platform a telephone conversation between the DDR Prime Minister Willy Stoph and the DDR leader Walter Ulbricht about a recent visit by the West German Chancellor Willy Brandt.[22]

In 1957, the BND and Bundeswehr began negotiations to determine their respective Sigint collection and processing responsibilities. The Army's aim was to be assigned the area west of the line Riga–Grodno–Brest–Lvov, including all of East Germany, Poland, Czechoslovakia and Hungary. The BND would be tasked with Sigint coverage east of this line, meaning the western portion of the Soviet Union. During this first battle between the BND and the Bundeswehr, the BND claimed the overall responsibility for managing the German national Sigint effort, including a monopoly on sharing Sigint information with foreign intelligence partners. The BND was in the end successful, assuming a dominant position over the Bundeswehr in the Sigint field, which was laid out in a written agreement between the BND and the Bundeswehr signed in 1958.[23]

MODERNIZING THE BND

When Gerhard Wessel became the new director of the BND in May 1968, one of his first actions was to appoint a planning committee. In September 1968, this body proposed a new organizational structure for the agency, which would replace the insanely conspiratorial jungle that was the hallmark of General Gehlen's bureaucracy. As an example of the conspiratorial atmosphere that pervaded the BND headquarters at Pullach, when General Wessel became the new President of the BND, he discovered that he immediately became a target of a BND bugging operation. When Wessel became suspicious that his office and home in Pullach was bugged, he called in some former friends from the counter-espionage branch of the *Militärischer Abschirmdienst* (MAD), which was the Bundeswehr's security service. The specialists from Stuttgart checked his house, located on the grounds of the BND headquarters compound in Pullach, they found a large number of highly advanced eavesdropping bugs, which had been placed in all of the rooms of Wessel's villa, including the toilet.[24]

Starting in 1970, the BND's Sigint infrastructure underwent a massive overhaul under the Social Democratic governments of Chancellors Willy Brandt and Helmut Schmidt. One of the results of the planning committee's 1968 recommendations was the creation in 1970 of a new BND department dedicated almost entirely to Sigint called Department II, led by Robert

Burchardt. Department II was organized into three primary units: Sub-Departments IIA and IIB, which managed all BND Sigint collection, processing, analysis and reporting, Sub-Department IIC, which handled technological and operational support. The *Zentrale für das Chiffierwesen* (Code Coordination Bureau, ZfCh), the BND center for coding and decoding, also became a sub-department under the Burchardt's Administration and Legal Support Staff. Under General Hubertus Grossler, Burchardt's successor in 1974, Sub-Departments IID (Technological Development) and IIZ (Assignments/Operations) were added.

Also in the early 1970s, all of the BND listening posts scattered throughout Germany were given the uniform covername 'Federal Site for Telecommunications Statistics'.[25] Even today, the Sigint operations center of the BND in the Munich suburb of Stockdorf (internally this station is referred to as 'Object Stellwerk') is also publicly known as the 'Federal Site for Telecommunications Statistics, Main Site'. There is also at Stockdorf a 'Telecommunications Institute' with the codename 'Object Planet'. Today, at least four BND listening posts, such as those located in the towns of Heiligenhafen, Kassel, Bonn and Butzbach, still carry the public cover name 'Federal Site for Telecommunications Statistics', despite the fact that their real purpose has been publicly known for years.

As in any secret intelligence service, the BND assigned codenames to all of its technological intelligence collection programs in order to hide their activities. For example, Sigint collection systems were given codenames like 'Susanne' or 'Lerche'. Intercept sites were also given codenames. For example, one of the Sigint intercept towers situated on the East German border was called 'Gipfelkreuz', while the listening post located outside the town of Krailling was given the codename Object 'Dacapo'. This site was closed in 1994. Intelligence reports derived from specific Sigint collection operations or systems were also assigned their own separate codenames, such as 'Seelachs', 'Stimmgabel', 'Reingewinn', 'Hummer' or 'Tamburin'.

Among the few advantages that West Germany had as a frontline state was that it was perfectly positioned from a geographical standpoint to closely monitor the radio and electronic emissions emanating from the Warsaw Pact military forces in neighboring East Germany and Czechoslovakia. Its geographic position on the forward edge of the potential battlefield also meant that German Sigint could hear targets deeper in the Soviet hinterland than Western intercept sites situated in countries further back.

However, these opportunities were not fully exploited by the German government's intelligence services during the 25 years following the end of World War II. This is documented in a Top Secret memorandum written on 2 October 1970 by the BND's deputy director, Dieter Blötz. He stated, among other things, that:

The Sigint equipment employed in the FRG, especially at the BND's H sites (Horchposten) with its present staff of about 850 operatives and the Bundeswehr's Sigint units are far from sufficient in order to fully exploit the good opportunities offered in the FRG for the collection of information by electronic means. This is true for political as well as for military information. Due to special circumstances, the BND was barely able to maintain its Sigint coverage during the past two years at existing levels... The Sigint surveillance program needs long-term technological as well as long-term staffing development programs.[26]

This caused quite a few problems in those days because German Chancellery-Minister Horst Ehmke had demanded from the BND more political and less military intelligence information. Above all, he wanted to be better informed about developments in East Germany. But the BND's Sigint capabilities were not capable of responding to this request. On 27 January 1971 Deputy Director Blötz complained in a secret memorandum that the BND was producing virtually no intelligence information in the areas of politics, commerce or transportation, and that the sole existing intelligence expertise in these areas resided with the Bundeswehr.[27]

These military eavesdropping operations were known within the Bundeswehr as eavesdropping programs 'Kolchose' and 'Concordia'. It also comprised Operation 'Nachernte', which intercepted radio traffic passing along point-to-point radio links in Czechoslovakia and the 'Roman' and 'Spinne' programs, which intercepted East German military traffic for the Bundeswehr and East German political communications traffic for the BND that was carried on East German point-to-point command nets. This massive eavesdropping operation was called 'Laus'. The programs 'Kolchose' and 'Concordia' were combined in 1972 and funded in the 1973 budget with over 4 million DM.[28]

In September 1971 Blötz asked the BND's director of technology to determine how many new staff members would be required in order to conduct the 'Laus' Sigint operation around-the-clock. He also wanted to know which radio frequencies used on the East German point-to-point radio system could still be monitored.[29]

However, in order to intensify its Sigint collection operations, in April 1972 the BND requested from the Bundeswehr an extra 228 staff positions above the authorized manning level of 838 military personnel in order to raise the Sigint output of the BND.[30] At the same time, Blötz requested that the Bundeswehr ease the pressure on the Sigint program of the BND by assuming more responsibility in the area of Sigint on military targets. For this reason Blötz strongly protested against an internal proposal by the administrative director of his own intelligence service who planned in February 1975 to downsize by 20 per cent the number of military personnel assigned to the 'Concordia' Sigint program.[31]

From 1982 onward, under the Christian-Liberal government of Chancellor Helmut Kohl, the BND's Sigint reconnaissance programs grew dramatically in importance. Sigint was less dangerous than the work of clandestine agents overseas, and because high-quality Sigint could be traded to foreign intelligence services in return for a great deal of secret material. In 1988, the BND's Department 2 (Technological Surveillance) consisted not only of a headquarters staff unit, Sub-Department 20A (command support), but also three other sub-departments. These were:

UA 22 (Technological Surveillance 1), which consisted of three sections: 'Intelligence Acquisition, Military, Politics, Economy', 'Central Intelligence Processing', and 'Operational Procedures and Support' (planning and organization).

Then came UA 23 (Technological Surveillance 2, Radio Traffic Monitoring) with its three sections: 'Intelligence Acquisition Politics, Economy, Technology, Science', 'Central Intelligence Processing', and 'Operational Procedures and Support' (technological evaluation).

Finally, there was UA 24 (Support/Intelligence Technology), which consisted of five sections 'Control and Project Processing', 'Intelligence Technology', 'Technological Provisions', 'Technological Data Processing Support', and the 'Telecommunications Liaison Service', which handled Sigint liaison with German intelligence stations abroad.

Whenever Sigint traffic is analyzed, intelligence analysts typically comb through available computer databases using key word searches as they look for pertinent information. During the 1970s, the BND often used the computer facilities of its American partner services, such as the American Sigint computer database called 'Cherry Glove',[32] which the BND used in its in-house 'Austin' project, a system to filter telecommunication by key words. However, during the 1980s it carried out cost-intensive modernization programs of its complete technological realm. This took place in the framework of the so-called 'Nose Ahead' undertaking. This venture was proclaimed in 1993 by the executive officer of Department 2, BND Admiral Gerhard Güllich. During the 1990s, the German intelligence community became increasingly interested in monitoring cellular telephone traffic as well as the communications passing through the Internet as the number of network providers in the international marketplace increased.

Under Gerhard Wessel, President of the BND from 1968 to 1979, the BND was not the only body within the German intelligence community whose Sigint efforts took giant steps ahead. Wessel also intensified Sigint cooperation and liaison programs. Domestically, the BND improved its Sigint relationship with the Bundeswehr, despite the fact that the relationship among the two parties was oftentimes contentious and even sometimes somewhat disagreeable. Improved international cooperation was achieved by fine-tuning common Sigint interests between the BND and its foreign intelligence partners.

Pursuant to the terms of the secret 'Zugvogel' agreement, dated 10 October 1969, and a separate document entitled 'Guidelines for the Cooperation Between the Bundeswehr and the Bundesnachrichtendienst in the Fields of Telecommunications Surveillance and Electronic Surveillance', a division of labor between the military and the BND was worked out. According to the agreement, the BND's president was responsible for overall planning, the assignment of tasks, and the overall coordination of the German national Sigint effort.[33]

To help the German Navy with its staffing and equipment problems, BND Vice President Blötz requested in late 1974 from the Americans three mothballed US Navy vessels which the German Navy wanted to use to improve their Sigint 'take' in the Baltic Sea. But at that time the Ministry of Defense was hoping to build three of its own Sigint collection ships, which it hoped would be approved by 1975.[34] On 5 December 1974 Blötz once again tried to obtain the ships from the Americans, and also suggested increased coordination of Sigint activities among American, British and German naval vessels. All three nations wanted to include Denmark in the program. Blötz wrote that he thought that getting Denmark to participate in this multinational program was 'attainable'.[35] But it was only in October 1976 that the Danish government finally agreed to participate in the seaborne Baltic Sea Sigint surveillance operation.[36]

Information to be exchanged pursuant to this agreement was intelligence information obtained by German Navy land-based Sigint stations, such as the German Navy's Sigint and direction finding stations at Marienleuchte on Fehmarn, or the Navy's Sigint Tower M at Pelzerhaken near Neustadt. At Pelzerhaken, the BND worked together with the intercept operators of the German Navy, just as it did at the Navy's HFDF site at Husum. Since the mid-1960s, the German Navy's primary HF listening post had been located at Twedterfeld near the city of Flensburg.

In September 1975, the BND agreed to allow the Bundeswehr temporarily to use its HFDF site in Husum.[37] In August 1976, the Army and the BND signed a formal bilateral Sigint cooperation agreement.[38] The BND wanted to concentrate its Sigint effort on strategic targets well beyond East Germany, such as Soviet strategic communications links, Soviet military air transport units, and on Soviet airborne and air mobile forces. One such typical Sigint program was called Operation 'Orlog', where up until the mid-1970s the air-to-air and air-to-ground radio traffic of Soviet aircraft was intercepted. The BND wanted create a warning mechanism in case the Soviet military began airlifting large numbers of Soviet troops to East Germany.[39]

However, in March 1976 Blötz sent a letter to General Harald Wust, the Inspector General of the Bundeswehr, stating that the German Sigint Coordination Committee could only gradually transfer to the Bundeswehr the Sigint mission of monitoring the telecommunications networks of Soviet military forces in East Germany and Czechoslovakia. This included Sigint targeting against Soviet air and air defense units, as well as against the communications of the East German Army. In the letter it was also pointed out that the West German Air Force and Army were still a long way away from being able to carry out these missions on their own. Because the Bundeswehr could not assume these missions, this would force the BND to employ one-quarter of its Sigint staff on military-related Sigint missions in Eastern Europe. This meant that the BND would have to commit 100 out of its 412 Sigint intercept operators and 19 of its 63 Sigint staff at BND headquarters in Pullach.

It should be noted that since 1976, the Bundeswehr was supposed to be responsible for Sigint collection against the East German military, as well as Soviet forces in East Germany and Czechoslovakia. At the time the Bundeswehr employed 4,000 soldiers in its active Sigint units, which was considerably more than the number of Sigint personnel working for the BND. Sigint coverage of Poland, Hungary, and the western military districts of the USSR was the responsibility of the BND. However, there were severe disagreements between the BND and the Bundeswehr. The main issue was who was responsible for Sigint collection deep inside the Soviet Union, such as mainline command nets in the Moscow area, which the Bundeswehr would have loved to monitor, but could not get permission from the BND to do so.

GROWING COMPETITION

After Defense Minister Hans Apel of the Social Democratic Party (SPD) formed a new military Sigint Processing and Analysis Center, the *Amt für Nachrichtenwesen der Bundeswehr* (ANBw) in Bad Neuenahr-Ahrweiler in June 1978, he received strong criticism from the opposition party Christian Social Union (CSU). The CSU member of parliament Friedrich Zimmermann accused Apel of wanting to start another international intelligence service during his term of office. He warned Apel about the existence of a civilian as well as military foreign intelligence-gathering agencies, which would lead to a 'hopeless squandering of resources, but not to more efficiency'. With those words, Zimmermann carefully voiced the BND's unwillingness that soldiers presently seconded to Pullach, would be posted to the ANBw.[40]

Just like the former responsible *Amt für Fernmeldewesen* (Bureau of Telecommunications, AfmBw), the ANBw had no command authority at all over all of the Bundeswehr's various Sigint resources. Rather, it remained dependent on the army, navy and air force's Sigint yield, as well as on their competing interests. In 1966 the armed forces command staff had recommended the creation of joint-service direction-finding sites, but it was not until 1988 that the first Bundeswehr's direction-finding network, with five fixed HFDF sites linked via a digital communications system, was put into operation. The Navy's Direction-Finding Site I was at Husum, the Air Force's Direction-Finding Site II in Eriskirch, the Army's Direction-Finding Site II (codename 'Zitrone') in Diepholz, the Army's Direction-Finding Site III (Codename 'Winzer') in Mainz-Schwabenheim, and at the village of Übersee on Chiemsee was the Army's Direction-Finding Site V (Codename 'Weide'). DF Sites I and V were disbanded in 1992.[41]

The ANBw itself carried out some additional collection under the codename 'Grau' but in addition processed and evaluated intelligence information obtained by Sigint. The Fm radar site near the Army's Sigint processing analysis center 'Schwarz' in Daun, which was subordinate to the ANBw, dealt with the so-called 'black holes' in the military's Sigint coverage through the use of dedicated mobile Sigint intercept units. The three Corps-level Sigint battalions, Telecommunications Battalions 120, 220, and 320, operated their own intercept sites and parallel HFDF stations, designated 'Rot' at Rotenburg/Wümme, 'Blau' at Donauwörth, and 'Grün' at Frankenberg. An additional German Army HFDF station was located at the town of Meitingen. In 1994 all these corps telecommunications battalions were restructured and became Eloka (electronic warfare, Elektronische Kampfführung)] telecommunications battalions.[42]

From 1990 onwards, many German Sigint installations were downsized or closed following the end of the Cold War. The German Army abandoned its Sigint intercept towers along the East German and Czech borders, and the three Sigint companies, which had manned these sites (Telecommunications Companies 945, 946 and 947) were disbanded. The Sigint site at Dannenberg on the East German border, which had been jointly operated with the British Army's 13th Signal Regiment, was vacated. The same thing happened at the US military's Elint intercept station at Wobeck on the East German border.

One exception was the Sigint installation at Kötzting. It was kept in operation even after the restructuring of the Bundeswehr's Sigint infrastructure was completed in 1994. A measure of just how unexpected

142 SECRETS OF SIGNALS INTELLIGENCE DURING THE COLD WAR

MAP 5.1
LISTENING, DIRECTION-FINDING AND RADAR INSTALLATIONS OF
THE BUNDESWEHR 1989

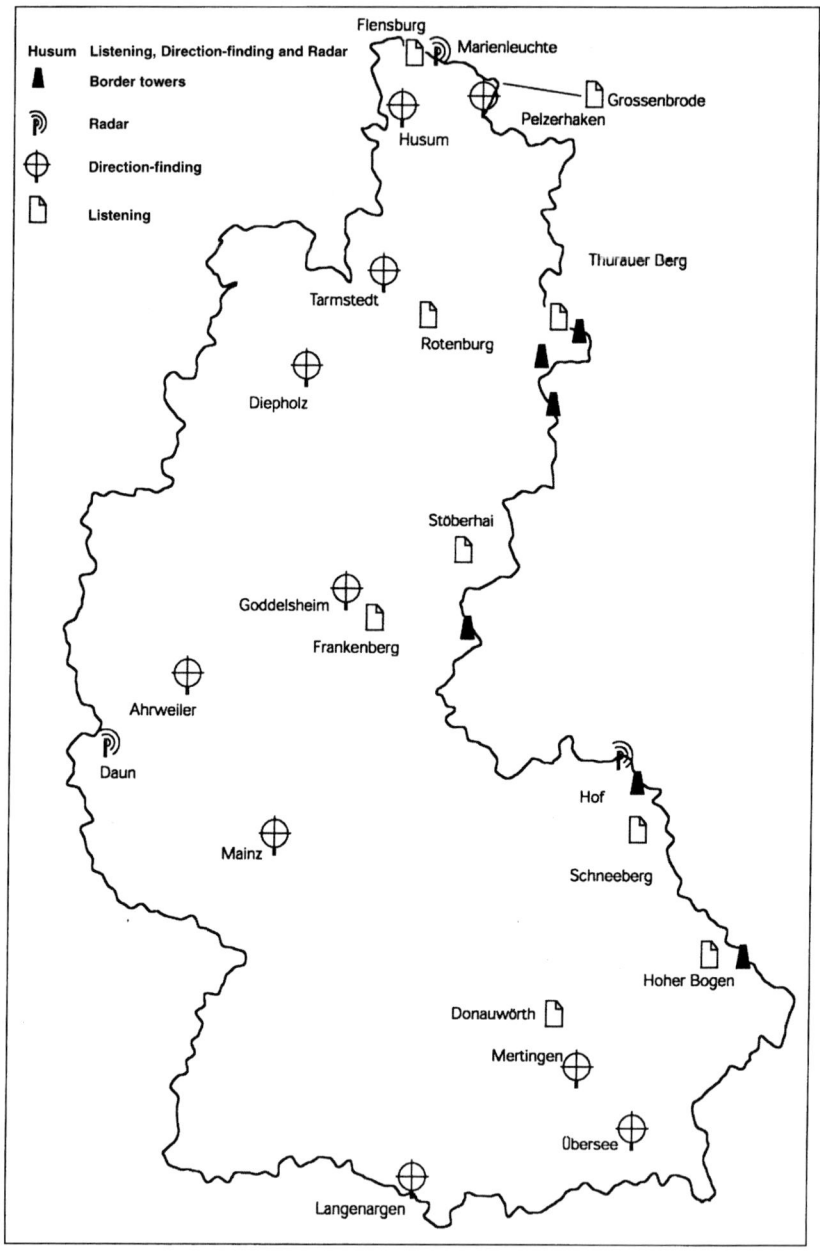

the collapse of the East German regime in November 1989 is demonstrated by the amount of money spent on the West German's border Sigint sites. Between 1987 and 1991, about 14 million DM was spent expanding the Sigint tower at Stöberhai in the Harz Mountains, and the expansion of the Sigint tower at Dannenberg, which was finished in 1991 (two years after the collapse of the East German government) cost about 7 million DM. Both sites had to be subsequently closed for lack of any eavesdropping operations for them to do.[43]

Since 1991 the German Air Force has been solely engaged in strategic Sigint collection aimed at the countries that formerly comprised the Soviet Union. This has required a substantial modernization of the Air Force's Sigint facilities. Some Sigint units, such as Telecommunications Sectors C and Q, were closed. Others were restructured and in the five provinces that formerly comprised East Germany, the former East German Sigint stations at Guetzkow and Zodel, managed by the newly created Telecommunications Sector D in Berlin, were created. At Telecommunications Area 70 in Trier, roughly 90 million DM was invested between 1992 and 1998 in the construction of a new Sigint processing and analysis complex buried inside bunkers at the General von Seidel Barracks. One sizeable German Air Force Sigint unit, Telecommunications Area 72 in Feuchtwangen, was restructured although not deactivated.

In December 1994, the German weekly magazine *Der Spiegel* wrote that the change of command at the ANBw from General Dieter Farwick to General Günther Wenger signaled a revival in the competition between the ANBw and the BND. The magazine wrote that: 'Farwick has incurred the wrath of the BND. Pullach was supposedly to relinquish some areas of expertise to Farwick's domain.' By 1993 there were already some indications that because of new overseas tasks the Bundeswehr's resistance against the BND's domination of the Sigint field was increasing. An agreement between the BND and the Bundeswehr signed in August 1992 gave the BND not only the Sigint coordination role, but also the responsibility determining for strategic Sigint requirements. This limited the Bundeswehr to operational and tactical Sigint missions. According to the 23 September 1993 agreement between the German Chancellery in Bonn, Bundeswehr and BND, the latter intelligence service was also given the exclusive right to exchange information with cooperating foreign intelligence partners.[44]

It was in this area that BND President Konrad Porzer acknowledged that tension existed during a June 1994 meeting of the Parliamentary Defense Committee. General Olshausen, Staff Section Director VI of the armed forces command staff (*Führungsstab der Streitkräfte*) pleaded at the meeting for a 'complete examination of political, commercial, sociological, technological-scientific, military-political, and military facts. Only in correlation, with inclusion of non-military, but militarily relevant facts is it possible to recognize crisis-prone developments early on.' It is of course manifest to all that the ANBw claim that it was responsible for handling a complete and worldwide potential crisis identification would sooner or later lead to a huge conflict with the BND.

Since the Bundeswehr, as happened during the 1999 war in Kosovo, poaches on the BND's traditional hunting grounds using reconnaissance aircraft like the Breguet Atlantique and the ECR Tornado, unmanned drones, and Sigint – the traditional domination of the BND over the military will eventually diminish. Despite the growing competition between the two parties, during the war in Kosovo the BND and the Bundeswehr were the epitome of harmony and cooperation in their relations with their foreign partners, forming a joint organization called the German 'National Intelligence Group'.[45]

SIGINT JOINT VENTURES IN THE FEDERAL REPUBLIC OF GERMANY

In early 1970, within the framework of its ambitious Sigint infrastructure development plan, the BND checked on the cost of building its own 'Wullenweber' antenna system, which is a huge 40-acre circular antenna farm that had been in use by NSA since the early 1960s. The cost of the new system was high. In 1972, BND officials estimated the cost of building the antenna system at '133 but probably closer to 160 million DM',[46] (about 36 to 44 million US-$) which was a substantial sum of money at the time. At the same time, BND officials asked the question as to whether political intelligence gathering would be really improved by building the new antenna, or if the system would be used mainly for intercepting military traffic, since the Bundeswehr would operate the planned antenna system, referred to within the BND as 'Object Komtesse', near the northern German city of Flensburg.

In April 1971, Minister of Defense Helmut Schmidt recommended to the German Chancellor's intelligence services coordinator, Horst Ehmke, that the planned antenna system be dedicated to political coverage rather than to military targets. But 'lasting discrepancies' between the BND and the Bundeswehr over whether the antenna would be used for military or political surveillance were not resolved. The result was that the costs involved and differences between the BND and the Bundeswehr forced the Germans to kill the plan to build their own giant antenna system, and instead turn to the US for assistance.

During the early summer of 1973 Vice President Blötz asked the CIA's Chief of Station in Munich, Arthur Stimson, in a letter if the BND could install 75 radio intercept receivers and 6 direction finding positions inside the American listening post at Augsburg-Gablingen in southern Germany. This station was equipped with the AN/FLR-9 'Wullenweber' antenna system. He also asked if it was possible to participate in NSA's HFDF network, which stretched from London to Turkey. However the competitor of the BND, the Bundeswehr, also wanted to move into the Augsburg listening post.

The battle between the two organizations came to a head during a meeting on 28 August 1973 between Blötz and the Deputy Inspector-General of the Bundeswehr Karl Schnell. Blötz was shocked to find that the Bundeswehr had invited several representatives of the National Security Agency (NSA) to the meeting to discuss the Augsburg station. The BND vice president was surprised about two issues.

First of all, why had the Bundeswehr not yet submitted its plan for the joint use of Augsburg to the BND? This had been agreed upon via an agreement signed in late July 1973.

Second, why did the Bundeswehr not stick to the 'from us urgently requested restraint to talk with the Americans about this subject'? Blötz regarded this as a manifest violation. Brigadier General Karl Heinz Page countered that a proposal had been submitted to the BND already quite some time ago. In addition he claimed that the Bundeswehr had not held talks with US representatives nor were they planning to do so in the near future. However, during a break in the meeting, one of the Bundeswehr generals present at the meeting showed Blötz the agenda for the next day's meeting of the Combined Group Germany (CGG), which was the joint NSA-GCHQ Sigint liaison office in Munich. The meeting was to be attended by British and American intelligence officers, and one of the items on the agenda was the 'Wullenweber' installation. The BND vice president learned then and there that General Page had lied to him about not talking with the Americans.[47]

Further negotiations were planned to be held at BND headquarters on 19 September 1973 concerning the joint use of the US Army Field Station at Augsburg (the covername for the Augsburg site was 'Drehpunkt'). However, the planned participation at the meeting by members of the Bundeswehr was cancelled at the request of the CIA and NSA. The BND then snubbed the Bundeswehr headquarters at Hardthöhe with the suggestion that senior BND officials from Pullach would separately brief selected military officers during the autumn of 1973 about missions, future plans, and capabilities in the context of the BND as the sole German foreign intelligence service.

On 21 January 1974, Blötz submitted the findings of a panel of experts to the Bundeswehr's Deputy Inspector-General Karl Schnell:

After integrating the German participants (BND and Bundeswehr) into the [Augsburg] station, the independence of the [German Sigint] processing control and mission implementation shall remain intact. Our own recording systems would be attached to the US antenna. Thus the joint use of the US mammoth direction-finding facility is possible.

Blötz also remarked that the Federal Republic of Germany would profit from this accord because it would give them access for the first time to NSA's worldwide Sigint collection and reporting system. However, he did not conceal the snag: 'To accomplish this, the first step was for the FRG's telecommunications eavesdropping by the BND and Bundeswehr be linked to the US switching center at Augsburg.'[48] On 5 February 1974, during a visit to NSA's headquarters at Fort Meade, Blötz signed the 'Drehpunkt' agreement in a 'festive ceremony'. This agreement gave the green light to begin two Bundeswehr construction projects at Augsburg: a new West German listening post inside the Augsburg station called Telecommunications Site South, as well as a similar program for a BND station at Augsburg.[49]

In October 1978, the BND noted that the results of its direction-finding work at Augsburg ('Drehpunkt') had worsened because the capacity of the American HFDF network to handle the increased workload was insufficient.[50] It seemed that the German requests for direction-finding plots and position reports were always placed at the tail end of the long queue behind the US HFDF missions. Even after the BND liaison staff with NSA had moved during the 1980s from a villa in the Munich suburb of Krailling to Gablingen, the Germans there remained second-class citizens, with most of the station's operational facilities remaining out of bounds to them.[51]

Sigint collaboration between the West German intelligence services and NATO Sigint units in Germany started slowly under Chancellor Brandt, but accelerated beginning in the summer of 1974 under Chancellor Schmidt. Towards the end of 1973, BND Vice President Blötz visited the American listening post at Teufelsberg in West Berlin 'to find out whether our joint endeavors with the Americans are paying dividends'. In July 1974, the BND and the Americans formed a joint working group concerning Sigint collection and processing by their intercept sites arrayed along the East German and Czechoslovak borders, followed by a second joint working group concerning Sigint collection against satellites and other space vehicles. The BND staff member 'Springmann' (cover name) headed the former working group.

But discussions about improving joint US-German Sigint efforts only took on a sense of urgency after US Secretary of Defense James R. Schlesinger (July 1973–November 1975) advocated improving and optimizing the efficiency of certain partnership relationships within NATO, including Sigint relationships with US allies in Europe. This led in 1974 to the initiation of secret talks at Pullach designed to create what BND officials described as a 'Sigint Trinity', comprising the BND, NSA and the British Government Communications Headquarters (GCHQ).

Blötz noted in a memorandum that: 'The director of GCHQ, John Burrough, already carefully hinted at such a trilateral discussion level.' According to Blötz this idea was very obviously talked over by Burrough and NSA Deputy Director Buffham who had known each other already for many years and were personal friends. According to Blötz a deal was made 'so that in each of their separate discussion with me, their willingness for trilateral top-level deliberations could be manifested'.[51] These trilateral discussions began in London in January 1975, focusing on cooperation with the BND on two of its Sigint programs, Operations 'Laus' and 'Roman', as well as collaboration in the field of space Sigint.

Concurrently, the BND staff at Pullach began coordinating for the first time ever with the NSA liaison officer to the BND in Munich, Bayard Keller, which American or German intercept stations would copy and process East German communications traffic.[53] In June 1975, Blötz expressed his thanks to Buffham for the good trilateral cooperation taking place in Augsburg and Munich, assessed these cooperative efforts: 'As the first step on the road towards steadily improving mutual Sigint cooperation'.[54]

However, behind the scenes the BND and the Bundeswehr were concerned about giving so much Sigint information to the US and Britain, with a BND official writing in November 1975 that 'caution [was in order because] of the eventual intentional hemorrhaging of Sigint information'. This resulted in the Germans delaying on improving Sigint relations with the US and Britain. But the Americans pushed harder, and requested in October 1976 better coordination of effort with German military Sigint authorities in the areas of intercept, processing and communications. At the same time, the Americans asked the Germans to agree in writing to provide NSA with copies of raw German 'Roman' Sigint reports, and not just the more general 'Laus' intelligence summaries based on Sigint. Even before a written agreement was drafted, BND Vice President Blötz agreed that the German Navy Sigint unit, Naval Telecommunications Staff 70, would supply NSA with the 'Skateboard' material collected by this staff on military activities in the Baltic Sea.[55]

The Americans had other requests. First, the Americans wanted any materials intercepted by the German Air Force Sigint towers along the East German border that referred to NSA's Teufelsberg listening post in West Berlin. Second, they desired an agreement with the Bundeswehr concerning cooperation in the field of VHF direction finding field. Both of these requests the Bundeswehr Deputy Inspector-General was willing to grant.[56]

In December 1976, the French Sigint agency, the *Direction Général de Sécurité Extérieur* (DGSE), entered the picture. Vice President Blötz noted a suggestion from 'Narzisse' (codename for the French intelligence) to broaden French participation of the BND Operation 'Laus'.[57] Although France was not militarily integrated into NATO, it maintained listening posts on German soil at Bahrsdorf in the Harz mountains, at Appen north of Hamburg, and a major intercept station at Landau. Next to its radio intercept site in Berlin-Spandau, the BND also operated a listening post jointly with its French partner at the Berlin Tegel airport. The BND concealed the site, called the *Arbeitsgruppe für Vergleichsuntersuchungen* (Working Group for Comparative Research) from the British and Americans,[58] and reports coming from the station were assigned the special codenames 'Eisberg' and 'Sandwüste'. Both of these listening posts were closed after the end of the Cold War.[59] Today, the BND's presence in the new German capital is limited to a few leased antenna aerials mounted on top of the Alexanderplatz television tower in downtown Berlin.[60]

Throughout the Cold War, the Federal Republic of Germany steadily took over new Sigint tasks from the Americans. In 1969 the US Air Force border intercept station in the Bavarian town of Hof was taken over by the *Amt für Nachrichtenwesen der Bundeswehr* (ANBw) as a Comint and Elint intercept station. This site operates an AN/FLR-12 antenna system and was originally established by the then CIA Office of Elint (OEL). The Hof collection site was sold for $1 in return for data exchange rights.[61] And in the mid-1990s the huge American Sigint collection station at Gablingen was taken over by the German military, and is now known as Telecommunications Station South of the Bundeswehr.[62]

In 1988, a new BND Sigint collection station was opened at Bad Aibling Station in southern Germany to intercept tropospheric-scatter radio links in Eastern Europe and the former Soviet Union. The station was officially known under the name of *Fernmeldeweitverkehrsstelle* (Telecommunications Site). Its secret codename was 'Object Seeland', which was later renamed as 'Object Orion'. It proved its worth in 1991 when the BND could provide the German government with important intelligence information about the coup d'état against Mikhail Gorbachev on the basis of Sigint collected by the station proving that the coup leaders did not have the support of the Soviet military.[63]

This joint US-BND station is now the second largest US eavesdropping installation in Europe after the NSA base at Menwith Hill in Britain. The German intelligence staff at Bad Aibling is not allowed access to the data collected by NSA at this station. This has nurtured suspicions on the part of

THE BUNDESNACHRICHTENDIENST 149

MAP 5.2
RECONAISSANCE INSTALLATIONS OF THE BND 1989

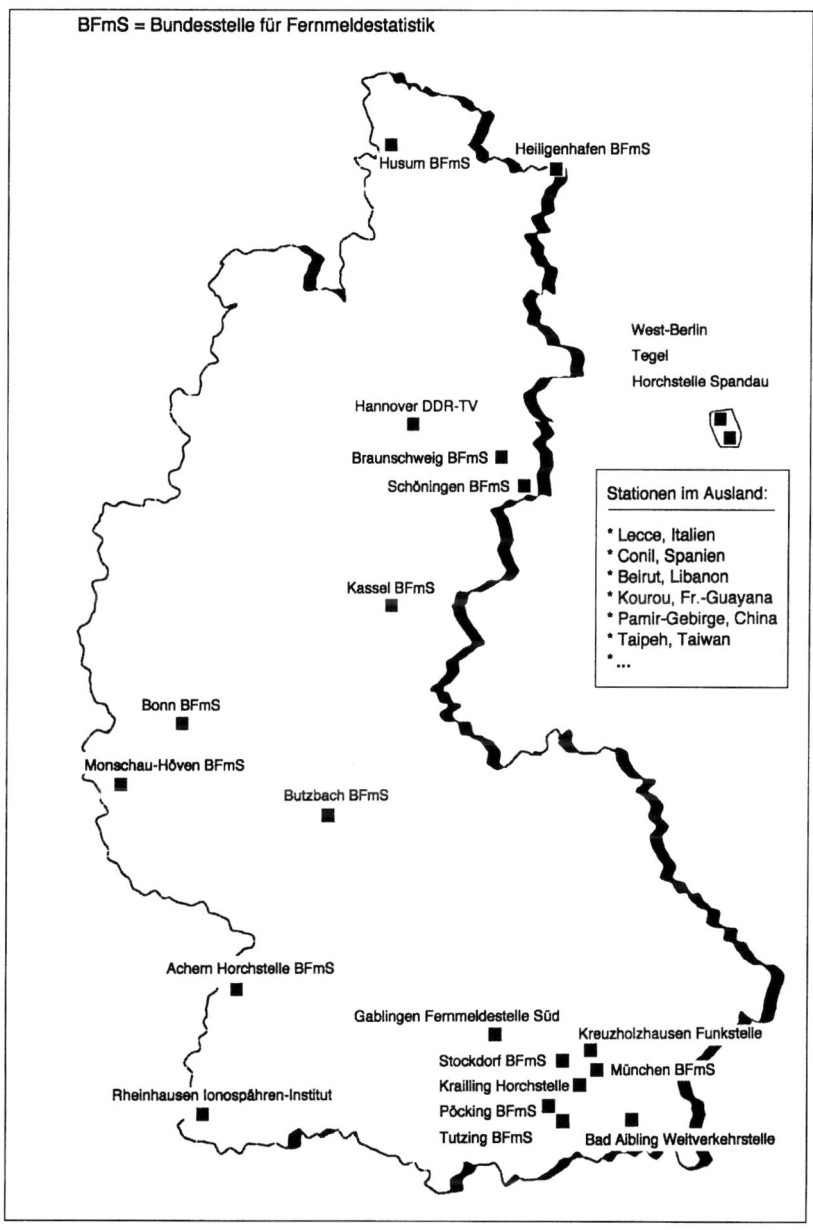

many German intelligence experts that the US is conducting industrial espionage against the Federal Republic of Germany from the base.[64] In November 1999, German Undersecretary for Intelligence Affairs Ernst Uhrlau and BND President August Hanning were allowed to visit Bad Aibling Station. After the visit, Uhrlau stated that NSA Director General Michael V. Hayden had given him a guarantee that no German interests were being violated by the station's intelligence work. Uhrlau told the German weekly magazine *Der Spiegel* that Bad Aibling Station conducted no economic espionage against German firms.[65]

However, an insider early in 2000 told the author of this article in *Der Spiegel* that while Uhrlau and Hayden's statements in November 1999 were correct, in fact NSA's economic espionage against Germany had been transferred to Menwith Hill Station in Britain. The main economic focus of the Bad Aibling listening post was now Switzerland and Liechtenstein and in particular banking transactions and money laundering.[66]

SATELLITE RECONNAISSANCE

On 30 June 1975 the contract of the BND with the Institute for Space Exploration in Bochum lapsed. For some years the university's institute had to study Soviet satellite communications and deliver the results to the BND. The intelligence service demanded the return of its devices worth 400,000 DM, which the Institute had been loaned by the *Zentralstelle für das Chiffrierwesen* of the BND.[67] This was because the German intelligence community had in the past not shown very much interest in the Institute's endeavors in the field of monitoring Soviet satellite communications. However, this did not mean a final farewell from eavesdropping through satellite emissions at all. However, as a place in time it was simply out of step.

In January 1975, the BND had presented its first plan for space reconnaissance to the joint German/American 'Asmara' working group.[68] This group that had been created under the direction of the BND staff member 'Kreipe' (covername) in July 1974.[69] In late July 1976, BND General Hubertus Grossler, chief of BND Department II, presented the first comprehensive study of West German space and satellite eavesdropping activities.[70] At the time, the BND was already operating a few ground stations that were intercepting foreign satellite transmissions. In Brunswick, a station of the Federal Bureau for Telecommunications of the BND concentrated on monitoring Inmarsat communications satellite communications, while a Trial Station of the Federal Bureau for Telecommunications, located in a quarry near the city of Schöningen, was also engaged in satellite communications intercept.

The most notable BND asset used for space Sigint collection was an installation disguised as a government scientific establishment called the 'Ionosphere Institute' in Rheinhausen, Germany. The total budget to build this 'Object Packhouse' was about 90 million DM (about $25 million), with the money coming out of the German Federal Postmaster's budget.[71] In 1971 this station, which within the BND was designated 'Service Department 525',[72] received a new B6 parabolic satellite antenna.

However, in the summer of 1971 the BND staff member responsible for the project, 'Dr Göing' (covername), wrote a report questioning the effectiveness of the antenna, stating that he foresaw *'unjustifiable costs'* in view of the rather modest use it had received up to that point.[73] The site's refurbishment dragged on.

Finally, in October 1976 NSA offered to supplement the station's equipment and provide other technical assistance until the BND was ready for the final conversion of the station.[74] 'Despite certain limitations in independence incumbent in the offer, it was decided to accept NSA's proposal of cooperation because it would give us access to the reconnaissance results of the world-wide [Sigint] collection and radar systems of this partner', stressed Blötz in a secret memorandum. Nonetheless, at the same time he had his staff check out all other possibilities to determine whether he could loosen the NSA ties 'by building an operational system on the national level' and 'through participation in organizations involved with reconnaissance satellite, i.e. civilian establishments such as Spacelab or Erts'.[75]

Then there was the question of West Germany building a reconnaissance satellite in conjunction with a European partner. The Inspector-General of the Bundeswehr Harald Wust asked the Vice President of the BND in August 1976 an important question. Could the Federal Republic of Germany do permanently without satellites as a reconnaissance vehicle? He had been inspired to ask this because of a discussion he had with the chief of station of the Japanese intelligence service in Bonn, General Saigo. The Inspector-General favored a joint effort with Tokyo because the Japanese approach was so different [and more beneficial] when compared with the French offer. Paris only seemed to be interested in a financial contribution by the Germany while having all the industrial manufacturing done for a proposed system at home.[76]

The agenda of the BND as regards the visit of the parliamentary control commission on 12 April 1989 drew up the balance sheet as regards the further developments up to the end of the East–West conflict. In 1985, an inter-ministerial working group chaired by Horst Teltschik, responsible for the foreign- and security policy in the Chancellery from November 1989 until the end of 1990. It had laid down the domestic requirements of the Bundeswehr and BND. The proposed German-French cooperation to build a spy satellite was rejected in late 1985. At the same time, the BND tried unsuccessfully to obtain the cooperation of the US intelligence services, seeking the use of commercial satellite systems such as the French Spot and the American 'Landsat'. In 1987, the German Minister for Foreign Affairs launched a new initiative to revive German-French cooperation to build a reconnaissance satellite, but Chancellor Kohl let the proposal founder (probably due to American pressure). At a 1988 Western European Union (WEU) conference, France formally proposed the construction of a European reconnaissance satellite. For the Germans, however, this proposal appeared to be too time-consuming and dependent on too many partners.[77]

After the WEU presented its report in November 1987, which included a cost estimate of 27 billion DM, the German government decided to participate in the initial or experimental phase of the project, but did not commit itself to future WEU project participation. A training center and ground-based processing station for the new satellite was completed at Torrejon Air Base in Spain, and since November 1995 imagery from the French-Italian-Spanish Helios I reconnaissance satellites has been processed and analyzed there.

In 1989, the BND had its own satellite reconnaissance goals clearly defined. These were verification and monitoring of the non-proliferation treaties on ABC weapons of mass destruction; crisis-proof reconnaissance of nuclear threats (strategic reserves, nuclear mission mechanisms, operational centers); and indications of crises and wars even in Third World nations. The service was also interested in early warning of natural and environmental catastrophes, as well as the accumulation of information about harvests, natural resources and infrastructures. The BND estimated that one ground station, one receiving and compiling station, two photo and two radar satellites, and one data relay satellite were sufficient. The service favored a binational cooperative solution in which France with its Helios II mounted photo sensors and the Federal Republic with radio sensors on Horus would work together.[78]

Four billion DM spread over ten years was the cost estimate of the German share at that time, but presently that has become at least seven billion. This high cost also motivated Defense Minister Volker Rühe to vote against the project. However, as a result of his negotiations with the French President, Chancellor Kohl did not leave any doubt that he preferred the German-French project above all those American offers for the sale of US satellites to Germany. With a mini-Department 40-X directed by its staff member 'Beigel' (covername) plus an analytical albeit embryonic unit, which in the future would have staff of up to 250 led by his colleague a 'Trendel' (covername), the BND had adapted itself by 1996 to the beginning of the active space reconnaissance.[79]

The cooperative efforts of Paris and Berlin with radar and photo satellites were only a first step towards the liberation from American dominance in space-supported reconnaissance. Whether it is the seed of a budding European service that will germinate into tight German-French teamwork, will be proven by the collective follow-up All Osiris radio reconnaissance project, mainly because its operators must allow all other partners mutual access to their individual reconnaissance priorities and interests. The Kosovo War has in this respect been a catalyst for greater French-German cooperation. After US ambassador, John C. Kornblum, said during the air campaign in a discussion on German TV (Talkshow Sabine Christiansen) that American generals had decided that the photo satellite reconnaissance results over the terrain in Kosovo should not be given to the governments of the other NATO member states, Walther Stützle, undersecretary at the Defense Ministry in Berlin, answered that this must be a reason for the German government to support the activities towards independent satellite intelligence capabilities.

INTELLIGENCE AGENTS' RADIO COMMUNICATIONS

In the 1974 amended edition of its mission statement, the 'Intelligence Service Definitions for the Bundesnachrichtendienst', referred to the Radio Transmission Service, which was the BND radio unit that either transmitted coded instructions at specific frequencies for 'secret service liaison' within Europe, or contacted agents using clandestine radio transmissions. For example, the BND station in Rome, the Command Center IV from Marquese de Mistura, in 1968 controlled its own radio station, called Radio Station Helios, which serviced the BND's 'Exchange and Security Region Mediterranean'.[80]

The *Schnellinformationsdienst* of the BND (Rapid Information Service) repeated coded messages on all frequencies for recipients outside of continental Europe, and the *Sonderblindfunkdienst* (Special Binary Telegraph System Service) transmitted worldwide coded instructions via Morse code to specific personnel. With a numbers rapid transmission device, an electronic memory and telecast mechanism, messages were swiftly transmitted via radio signals.[81]

The BND also perfected rapid telephone transmissions by developing the 'Schnellbahn' system, which could transmit compressed coded messages to agents in Eastern Europe. The BND also developed radios in which a 'Harpune' receiver was hidden, able to receive the compressed messages. The BND received DM 20 million by selling this system to some Western allies. For example, counter intelligence in Poland in 1986 found a 'Harpune' receiver arresting British agents.[82]

The BND operated a major radio transmission installation at Dachau/Kreuzholzhausen. Its secret designation being 'Object Mühle', this station was publicly referred to as a test site of the 'Federal Bureau for Telecommunications Statistics'.

During the 1980s, the BND also used a communications station called 'Object Alpina' near the town of Tutzing to transmit messages to clandestine agents working abroad, as well as send wireless traffic to BND foreign residences. Despite the fact that the service prevented East German agents from penetrating its operations, it could do little to protect its excellent reconnaissance and communications system from the eavesdroppers of the East German *Ministeriums für Staatssicherheit* (Ministry for State Security, MfS). From a listening post with the codename of 'Topas' in Czechoslovakia, the MfS was able to intercept the BND's uncoded 'Alpina' radio relay traffic to Munich.

Since the satellite communications links to the legal BND stations in Cairo and Amman, as well as to the illegal BND base in Damascus, also ran through this station 'Object Alpina' near the town of Tutzing, no secrets remained hidden from the MfS intercept operators about what was transpiring between BND headquarters at Pullach and its three Middle East stations.

For example, in 1984 the BND station in the Egyptian capital believed that they had discovered which flat the new East German military attaché would move into. Following the orders of their headquarters in Pullach in the Isar valley, the BND agents rented a nearby apartment and equipped it with plenty of eavesdropping equipment. But the MfS had warned East German military intelligence well in advance, and the Cairo flat was subsequently not occupied by the military attaché, but rather by a harmless foreign trade bureaucrat. The expensive eavesdropping equipment brought in from West Germany specifically for the operation was used for over six months before the BND eavesdroppers finally realized that their bugging operation was not monitoring an East German military officer. The MfS eavesdropping specialists in East Germany noted with great glee that Dr Becker, the BND chief of station in Cairo, reacted in maniacal rage when he learned of the disaster.[83]

In autumn 1995, the BND once again was enormously shocked when it turned out that a Southeast European intelligence service had succeeded in deciphering codes and passwords of the BND's data transmissions.

THE BND'S RELATIONS WITH FOREIGN INTELLIGENCE PARTNERS

The BND could always count on the secret export of German communications technology to foster good relations with partner intelligence services. In 1958, it sent the King of Saudi Arabia several mobile radio sets for use by his palace guards. In 1965 it helped the Indonesian military intelligence service with the overthrow of President Sukarno by sending radio sets. During the 1970s, the BND sent eavesdropping equipment to Uganda's dictator Idi Amin, and during the mid-1980s[84] it sent to Libya radar and surveillance technology, a telephone eavesdropping center, and provided training support.[85] During the late 1980s, telecommunications equipment labeled 'Made in Germany' also went to the guerrilla organization Renamo, which was trying, with the assistance of South Africa, to destabilize the legal Mozambique government.[86] In April 1986, the BND sent a Sigint station taken from the Bundeswehr's stores to Pakistan's intelligence service (known within the BND by the codename 'Eichkatze').[87]

Following the end of the Cold War, German Army intelligence agents reportedly supplied the Algerian Islamic Salvation Front (AIS) in 1994 with radio equipment and weapons. The BND supported the AIS of which the controlling echelon mainly found asylum in Germany in order to break the French economic monopoly in its former colonies in case of a takeover by the AIS in Algeria.[88] Only after a massive protest by Paris in 1994, was the secret support of the Algerian opposition abandoned. In 1991, the BND delivered intelligence related material as computers and cameras to the value of DM 100,000 to the intelligence service of the Islamic Republic of Iran, and in 1992, it trained members of the Iranian intelligence service in its Munich based training center on the use of the computer equipment.[89]

In 1991, the BND exported together with the MAD (*Militärischer Abschirmdienst*) to Albania telephone surveillance equipment taken from the stores of the defunct East German Ministry for State Security,[90] and they supported their new intelligence partner in Hungary with Sigint systems taken from the BND's abandoned Sigint towers along the former East German border.

The BND also made itself at home on foreign shores. For example, in the summer of 1988 the BND shipped to a Christian militia group in Beirut via Cyprus a high-tech Sigint intercept facility to expand the Agency's sphere of influence in the Near East. At the same time it ensured that it would have access to the raw data generated by the system. The BND also dispatched its Sigint specialists to the European Space Center at Kourou in French Guiana, from where Ariane rockets launched commercial satellites into orbit. According to French press reports, the BND together with the French have listened to American communications traffic sent by Inmarsat and Intelsat communications satellites from Kourou and from another Sigint station at New Caledonia in the South Pacific.[91]

Shortly after the creation of the West German government in 1956, and the subsequent integration of the Gehlen Organization into the German government under the new name of the Bundesnachrichtendienst, the intelligence service began establishing telecommunications and Sigint sites outside Germany. It still operates these kinds of sites today. Below are four examples covering the period from the 1960s to the present.

First, at the beginning of the 1960s the BND enhanced its already good liaison relationship with the Turkish secret service by helping it open radio-electronic eavesdropping stations, like the Turkish station at Samsun on the Black Sea. The BND also sent some of its staff, among them some personnel from the BND 'Homing Station Overseas' at Chiemsee, to Turkey to conduct Sigint operations against Russia.[92] During the 1970s a German supply aircraft flew from Munich-Riem airport to Turkey every week carrying supplies and other materials to support the Turkish technical and human intelligence programs.[93]

Second, in 1967 the BND and the Spanish secret service installed Sigint equipment at a ranch in La Mancha with the help of a Munich-based BND commercial proprietary company called Thum & Co. On 26 March 1971, the Spanish intelligence service, which was referred to within the BND by the codename 'Flieder', sent a telegram to Pullach in which it proposed taking over the BND Sigint installation in Spain in return for use of the site in case of war.[94] The BND, however, did not pull out of the ranch until after a modern, state-of-the-art listening post had been built near Cadiz in 1975.

The purpose of the new and complex Sigint station at Cadiz was to monitor the communications relay links for the submarine cable that had their land terminals on the Atlantic and Mediterranean coasts of Spain and Portugal. This would allow the BND to break into the communications hub connecting North and South America, West Africa, Britain and the Arabian Peninsula. In the spring of 1975, the BND put a five million DM price tag on this secret station,[95] which was called 'Object Eismeer'. In March 1975, the Spanish government gave its blessing to the project after its concerns about a violation of the Portuguese communications relay system were clarified.[96] An agreement between the BND and its Spanish counterpart, CESID, was signed in Spain in September 1975.[96a] FN This marked the beginning of a BND operation called 'Delikatesse', which did not end until the station was closed in December 1992.

Third, in the early 1960s the BND established a station on the island of Taiwan, and opened on the island what was then one of the most modern listening posts yet built. Pullach also helped the Taiwanese with staffing the station. For example, during a September 1973 working lunch in Bonn, the Vice President of the BND introduced the Taiwanese intelligence chief of station in Bonn, Ango Tai, to a BND officer named 'Prigge' (covername), who was to serve as his intelligence advisor. At the time, Prigge was working at the BND's Brunswick listening post. During the autumn of 1977, two BND cipher experts visited the headquarters of the Taiwan intelligence service, which was referred to within the BND by the codename 'Frettchen'. In December 1978, the chief of the German Chancellery agreed to expand the intelligence cooperation between the BND and Taiwan's military intelligence service in the field of satellite intelligence.[97]

During the late summer of 1992, the BND gave a computerized Sigint processing and analysis system worth several hundred thousands of DM to the Taiwan military intelligence organization, which was installed in a military barracks in the southern part of the Taiwan capital Taipei. This system systematically decoded and analyzed communications traffic obtained by Taiwan from their Sigint collection stations. In 1992, Taiwanese Sigint personnel were trained in new technology and BND-developed analytic software at the BND Telecommunications School at Pöcking (which was publicly known as the Federal Bureau for Telecommunications, Test Station). This school was affiliated with the Sigint collection and analysis site called 'Object Kleefeld.' In addition, the BND posted three of its Chinese-speaking staff members to Taiwan for four-year tours at the Taiwan Sigint processing and analysis center in Taipei.

Fourth and final example, for monitoring the hinterland of the Soviet Union, the BND built during the early 1960s an eavesdropping installation that monitored a huge area using interconnected Sigint intercept stations in Husum, Lecce in Italy, and Tehran, Iran.[98] However, this capability vanished when the Tehran intercept station was lost during the 1979 Iranian revolution. The BND, together with the CIA, looked for a replacement site in the People's Republic of China. In 1985, this station was opened in the Pamir mountains near the Afghan border in the presence of a high-ranking delegation from the German Parliamentary Control Commission for the Intelligence Services. A German Sigint liaison officer has been posted at the German embassy in Bejing since 1985.

The Afghan Ministry of Intelligence Affairs (WAD), until the Afghani regime fell in May 1992, carried out an operation against the station, which in the late 1980s provided intelligence support to the mujahedin fighters in Afganistan. WAD agents bought the listening post's garbage from Chinese civilians and found out that the American as well as German intelligence services at the station also collected intelligence against their Chinese hosts.[99] While the United States pulled out of these stations in protest of the Chinese human rights policy during the early 1990s, the BND Sigint station, designated Operation 'Lanze', still continues to this day.

August Hanning, the President of BND since November 1998, visited the town of Gudermes in the Russian breakaway province of Chechnya in early March 2000 to share intelligence with the Russian foreign intelligence service, the SVR, whose BND covername is 'Sequoia'. The intelligence information dealt with, among other things, with the support that the Chechen rebels were receiving from Islamic terrorists, mainly from Afghanistan, such as Osama bin Laden. It has been the information from the Pamir Sigint station, in addition to the BND's old connections to the Afghan mujahedin, which enabled Mr Hanning to deliver important and exclusive information to Russian President Vladimir Putin's foreign intelligence service. In a first reaction the BND said that its chief only had a look at the situation there. Two days later, the German government declared that there had been some cooperation with the Russians in the area of combating international terrorism within the framework of the G-8 nations, but that there had been no direct assistance given to the Russian army in Chechnya.[100] On 17 April 2000, Mr Hanning briefed the press on the role of terrorists in the region, but denied that the BND has a listening post in the Pamir mountains.[101]

Operation 'Lanze' was not the only joint venture between the German and Chinese intelligence services. Since the 1989 massacre at Tianamen Square in Bejing, 50 Chinese intelligence experts have visited Germany to negotiate agreements in areas of common interest. The Chinese wanted a quid pro quo, and claimed support in controlling the Chinese democratic opposition in China and abroad. As a consequence of this Chinese-German cooperation, in 1991 the Munich-based electronics firm Rhode & Schwarz delivered some PA 055 eavesdropping antennas and audio surveillance 'bugs' to China. These were used to monitor the movements of foreign diplomats and journalists in Beijing by car in order to determine where they would meet opposition members, most of whom were students.[102]

Intelligence personnel from smaller countries have also been trained by BND experts. For example, between 1982 and the end of the 1980s, German specialists traveled twice a year to meet with officials from the Brazilian foreign intelligence service, which the BND assigned the covername 'Biber'. Their job was to train their Brazilian colleagues in the correct use of audio surveillance bugs.[103] Nevertheless, the BND and its most important partners have carried out joint operations in the Sigint field. Bob Woodward in his 1987 book *VEIL* reported on the work of joint CIA-NSA Sigint units operating inside US embassies abroad, which were known as Special Collection Sites:

> The CIA and the NSA had developed techniques barely imagined by the host countries ... phone lines and rooms could be tapped or bugged without physical intrusion or connection. Room conversations could sometimes be picked up from the windows by electronically measuring the vibrations of the window glass with a small, invisible beam.[104]

One such example of close Sigint cooperation involving the BND happened in Libya. Since there was no American embassy in Libya, it was impossible for the CIA and NSA to place a SCS Sigint site in that country. The CIA asked the BND for help, and the BND responded by opening a station inside the German embassy in Tripoli in January 1987. From inside the German embassy, German intelligence specialists tapped and bugged the embassies of East Bloc nations as well as the communications facilities of the Libyan government, and delivered the results of these operations to their American partners.[105]

Four secret BND documents pertaining to a joint operation conducted with the Israeli secret service, the Mossad, in Germany raise questions about whether these special collection operations were really an American invention. On 8 September 1975, the Vice President of the BND, Dieter Blötz, met the Israeli Mossad chief of station in Paris, David Kimche, who told him that the Mossad had developed a new laser measuring system for conducting bugging operations. He also said that Mossad would like to use this technology for the first time in Germany for an attack on the Egyptian military attaché stationed in the German capital Bonn. A flat near his house had already has been rented for this purpose. Kimche also offered the proliferation and discussion on the new technology with specialists from BND. Blötz agreed to the proposed operation three weeks later.[106]

On 16 October 1975, Mossad and BND experts met. They agreed that a prototype of the system had to be built by January 1976, and then tested during an eavesdropping operation. The BND officer involved in the operation was a Mr Grad, who about a month later became BND chief of station in Morocco. Grad also found a target for a second test. Both tests were done together. However, the operation in Bonn was carried out exclusively by Mossad, who nonetheless shared the results of the operation with Pullach.[107]

CRYPTOGRAPHY

During a conversation with the Permanent Secretary of the German Chancellery, Franz Schlichter, the Vice President of the BND on 19 January 1978 complained that the Code Coordination Bureau (ZfCh), was completely omitted from the new description of tasks to be performed by the BND. But the Permanent Secretary thought that the omission was a better idea than if 'suddenly it should be made public what kind of contact exists between the BND and the ZfCH'.[108] However, the relationship between the BND and the ZfCH involved more than simple 'contacts', because until 1989 the Center for Ciphering, located at the Mehlemer Kreuz in Bonn-Bad Godesberg, was officially a unit (Sub-Department 62) of the

BND. The ZfCH consisted of five sections, namely: General Cryptography and Central Tasks, Mathematics, Crypto-Technology, Decoding A, and Decoding B. Its primary emphasis was not cryptanalysis, but rather the encoding of information, for which it acted as the Executive Agent for the entire German government. It was for this particular reason that the BND instructed the Bundeswehr in March 1975 'to eliminate all crypto-technological devices which could have been compromised by an officer in the (Bundeswehr) Bureau of Telecommunications suspected of having committed espionage', presumably for East Germany, and requested the help of the Director of the ZfCh in implementing this order.[109]

The genesis of the ZfCH can be traced back to 1947, when a former Wehrmacht cryptanalyst named Erich Hüttenhain brought together a small group of former German Army cryptologists at the US Army base at Camp King outside the town of Oberursel, which became an integral part of the Gehlen Organization's nascent Sigint unit, then called the Study Group for Scientific Investigation. In 1950, with the permission of the Allied High Commission, the Ministry of Foreign Affairs installed Section 114 as a cryptographic service under the direction of Adolf Paschke. The members of its scientific advisory committee were some of the best cryptologists in Germany, including Erich Hüttenhain, Kurt Selchow, Rudolf Schauffler, and Heinz Kuntze. With the decision to rearm the Federal Republic of Germany, this section was closed in 1955, and all of its coding and decoding responsibilities were transferred to the BND in 1956. The Bundeswehr limited itself to simple cryptographic (coding) responsibilities.[110]

The science of cryptology is characterized by a constant and never-ending race between the makers and breakers of codes, with usually the codemakers coming out on top. That is why BND Vice President Blötz warned in an 28 August 1973 speech at the Richthofen hall at Hardthöhe that it was becoming increasingly difficult to decipher encrypted radio traffic. He emphasized that communications security meant that more staff and fiscal resources were needed to keep up with the improvements in codemaking. However, the BND's dominant position in cryptography proved to be very convenient with regard to the export of cryptographic technology to Third World countries, since it enabled the BND to implant 'Trojan Horses' inside the exported technology. The BND was able to read encrypted radio transmissions from Argentina, Brazil, Mexico, Japan, Malaysia, India, Nigeria, and Israel because the cryptographic technology exported by Germany to these countries had an inherent technical weakness. At Pullach, the decrypts based on this technology were called 'Yellow Stroke Reports'. Insofar as which countries were monitored most intensely, priority was given to the Middle East and North Africa, with Egypt, Iraq,

Iran, Jordan, Lebanon, Libya, Morocco, Pakistan, Sudan, Syria, and Tunisia getting most of the attention.[111]

Because of information contained in one of these 'Yellow Stroke Reports', the BND knew, for instance, that in late November or early December 1975 there had been a meeting of Iranian, Saudi Arabian and Egyptian intelligence officers in Cairo.[112] In June 1976, the Pullach eavesdroppers also intercepted a radio message from the Iranian intelligence chief of station in Cairo to his headquarters in Tehran about an impending visit of a BND delegation to the Egyptian capital.[113]

NATO partners, such as Italy, Spain and Turkey, were not exempted from these 'Yellow Stroke Reports'. What General Gehlen had started during the Cold War regarding surveillance of Western countries was continued by his successors. In July 1973, the BND noted a substantial decrease in the number of 'Begonie-Narzisse Yellow Stroke Reports', which meant that Denmark ('Begonie') and France ('Narzisse') had improved the security of their cryptographic systems. The Vice President of the BND quickly ordered an inquiry to determine the reasons for this decline.[114] The results of this inquiry are unknown at present.

It should be noted that the German listening post at Monschau-Höven outside Bonn was used primarily by the BND to eavesdrop on the telegraph and radio traffic of all important embassies in Bonn.[115] After the German government's relocation to Berlin in the summer of 1999, another listening post in Germany assumed this function.[116]

Despite the above-mentioned setback, the professionalism of German cryptanalysts and cipher experts was an important reason why the BND had a good reputation with its foreign intelligence partners. For instance, it led in April 1973 to the 'exchange of coded raw material with "Hockey"',[117] which was the codename for Britain's GCHQ. But the ZfCh experts did the British Sigint service its greatest favor in 1982 when they were one of the several NATO intelligence services that could read the encrypted radio traffic between Buenos Aires and the Argentinian military on the Falklands Islands (the Argentinians referred to them as the Malvinas) during the Falklands War.[118] They did this thanks to a code key that the Abwehr had captured from the French during World War II. The cipher, still in use in Argentinia in the 1980s, had this way come to Gehlen.

The BND also expanded its communications intelligence (Comint) interests in Africa. A strategically important Sigint system called 'Advookat' was shipped to the Republic of South Africa by way of the German electronics firm AEG Telefunken, and began operations at the South African Simonstown listening post in 1975. In September 1977, a ZfCh staff member named Mr Rave suggested to BND management that specialists from his bureau should be sent to South Africa. The RSA partner service, which was known within the BND by the covername 'Panther', had already received eight decoding programs from the BND, but it appeared that Pretoria's intelligence personnel were not able to effectively use these software programs. Pullach recognized how politically explosive such support for the South African apartheid regime was since this nation was covered by a UN embargo.[119] Nonetheless, the South Africans received highly sensitive technological materials and logistics support from Germany. According to Gert Hugo, who was until 1991 the director of Ciskei's security service, in 1984 the BND gave South African agents three-month training courses in electronic surveillance skills.[120]

On 17 December 1990, the *Bundesamt für Sicherheit in der Informationstechnik* (Federal Bureau for Security in Information Technology, BSI) was established, which grew out of the BND's ZfCh. Initially, its headquarters was attached to the BND in 1989, and renamed it the *Zentralstelle für Sicherheit in der Informationstechnik* (Central Office for Security in Information Technology). Later, it became a federal bureau with a staff of about 300 personnel, and was placed under control of the German Ministry of Internal Affairs. Critics believed right from the start that the formation of the BSI was an attempt by the German government to create a mini NSA.[121]

The first director of the Central Office and of the BSI was Otto Leiberich, who had worked for the BND since 1957. In 1962, he was promoted to the position of top mathematician at the BND, and served as the director of the ZfCh until its closure in 1989. The transcripts of a secret German Parliament (Bundestag) budget committee meeting detailed the alliance between the BND and the BSI. In the documents, the Director of the BND's Department 2 (Telecommunications Reconnaissance) explained to the president of the Federal Auditor General's office why in 1994 the BSI wanted a mainframe computer costing several million DM for use in cryptanalysis. However, nothing changed which meant that the BSI only officially had the responsibility for decoding. In reality the BND had the decisive influence on the BSI and apparently the BND has retained the responsibility for cryptanalysis. In the 1998 organization chart for the downsized BND, the Agency's Department 2 (Telecommunications Reconnaissance) is once again responsible for 'deciphering'. The BND has been able to maintain its cryptanalytic proficiency with the help of American-made supercomputers, and it currently can solve the cipher systems of many nations, except Japanese and Italian ciphers.[122]

THE BUNDESNACHRICHTENDIENST 163

LEGAL FOUNDATIONS

The worldwide bugging of embassies and eavesdropping on diplomatic personnel is the type of special 'hospitality' that intelligence services give to the representatives of other nations. For example, the BND reported in 1973 that ten microphones had been discovered in the German embassy in Warsaw.[123] This was twice as many as had been discovered the previous year. By the end of the 1950s, the BND itself had already carried out eavesdropping operations directed against the Soviet trade mission in Bonn and apartments occupied by Russian diplomatic personnel in the greater Cologne-Bonn area. The BND also had set its sights on bugging the offices of the Chinese news agency Hsin-Hua in Bad Godesberg.

Faced with moral qualms about his duties, in 1963 an employee of the Federal Office for the Protection of Constitutional Rights (BfV), Werner Pätsch, publicly revealed some illegal joint ventures that the BfV and engaged in with several allied intelligence services. In the course of this affair, it became quite clear that the Allied victors of World War II, with no legal barriers to stop them, had engaged in the wholesale invasion of the privacy of German citizens by tapping telephones and reading personal mail. These powers were replaced on 13 August 1968 by the State of Emergency Law, or the G-10 Act. Under this act, the BND now assumed the responsibility for postal and telecommunications control in order to combat threats to the constitution, the security of the Federal Republic of Germany, and allied troops stationed on German soil.

In the summer of 1971, the jurisdiction to exercise these powers was transferred from the BND to the German chancellery in Bonn.[124] With the permission of the German Parliament's G-10 commission, the BND has the powers to monitor cable communications nets also in order to assist the other German intelligence and security services. The BND must always possess the most advanced technology to perform this task as, for instance, happened in February 1974, Blötz asked the chief of Department II what operations had to be executed in order to get access to new, high-frequency lines.[125]

In early 1975, the BND conducted Operation 'Flattermann', a G-10-sanctioned eavesdropping operation against the East German chief of protocol Günther Marsch, who was stationed in Bonn. The operation was conducted because the BND was interested in Marsch's activities against foreign diplomats stationed in Bonn. While Marsch talked on the telephone with 'models' and prostitutes, he asked if he could bring photographic paper with him on his 'dates'. Surveillance confirmed that despite his many dates, he never had any sex with the ladies of the world's oldest profession. Since this surveillance operation produced no information of intelligence value, in early 1975 the chief of chancellery discretely told East German Secretary of State Günther Vogel about Marsch's subversive activities.[126]

The case of nuclear scientist Klaus Traube is among the greatest secret service scandals in Germany. This occurred in 1975 during an illegal wiretapping of the telephone of a supposed terrorist sympathizer by the BfV. It eventually led to the dismissal of the Minister for Internal Affairs, Werner Maihofer. The BND only provided technical assistance to the operation, which it routinely did during eavesdropping operations for the military security service (MAD) and the BfV. The BND also provided technical and manpower support to the federal and provincial criminal bureaus for their eavesdropping operations.[127]

For example, in February 1974 BND President Gerhard Wessel gave permission for the BND to provide assistance to the Provincial Criminal Bureau (LKA) of Bavaria for an eavesdropping operation against Baader-Meinhof Group sympathizers. As a precaution, he ordered that the BND officers assigned to the operation were to be posted or seconded to the LKA office in Munich for the duration of the operation.[128] However, not only terrorists were targets of the BND: critical journalists were too. In November 1976, some Cologne *Bild* newspaper employees of the BND cadre have, according to the affected author, even provided 'agency assistance' while eavesdropping on Günter Wallraff.[129]

RECENT DEVELOPMENTS

The new Anti-Criminal Activities Act passed the joint parliamentary committee of both the Bundestag and the Bundesrat in the summer of 1994, and subsequently became law on 12 September 1994. The Green Party (*Die Grünen*) and former Communist Party (PDS) voted against the Act in the Bundestag. The Bundestag's Federal Commissioner for Data Protection explicitly warned in a public hearing on 11 April 1994 about the citizens' loss of personal freedoms and liberties through the inclusion of the intelligence services in anti-criminal activities. He stated that 4,000 conversations a day, or about 1.5 million conversations a year would be monitored pursuant to this Act. He recommended an 'authoritative instruction of the parliament and the public about the BND's endeavors and also how well the individual rights of the affected people were safeguarded, and whether the new means justified the final results'.

The aim of this act was to equip the BND for its new tasks of monitoring international terrorism, weapons and technology transfer, illegal drug trafficking, and cross-border counterfeiting and money laundering, as well as to facilitate the control of non-cable carried telecommunications. It also allowed the BND to execute where appropriate direct telephone monitoring and eavesdropping operations at the request of the Ministry for Internal Affairs and the G-10 Commission of the Bundestag.[130]

In July 1995, the Federal Constitutional Court issued a provisional directive which overturned a large part of the new enforcement rights. In July 1999, the highest German court declared parts of the anti-criminal activities legislation unconstitutional. Not surprisingly, the BND was not happy with this verdict. On 1 March 1996, the BND officially began its strategic monitoring campaign, and as of 1998 was using a computerized watch list containing 6,979 G-10 commission-authorized individual names, key terms, and telephone numbers. However, the BND stated that it was the limitations imposed by the Federal Constitutional Court which was responsible for the campaign's colossal failure in combating organized crime.[131]

In the two and a half years since March 1996, 4,600 hits on proliferation and about 6,000 on weapons trade were evaluated. In the same period international terrorism and war on drugs scored 100 each, resulting in the deactivation of the related keywords.[132] In the keyword database of the BND in 1998, 500 search subjects in relation with terrorism, 2,000 related to proliferation and 400 to the war on drugs were activated. On 23 December 1996, the President of the BND, Hans Jörg Geiger, presented in Karlsruhe the 'Report on the Implementation of Strategic Control Measures according to Article 1, Paragraph 3 of the G-10 Act'. The report found it inappropriate 'to question the purpose of the strategic control measures due to the small number of the provided reports'. The BND declared in its comments to the Constitutional Court that it had a daily limit of 15,000 communications being monitored because of limited capacity. Of this amount, after sorting and selecting by machines and evaluation by human operators, only five daily eavesdropping reports were analyzed.[133] But outside experts believe that that the BND is hiding a great deal about its monitoring operations.

With fax and telex messages, it is possible using existing technology to determine the identity of both the sender and receiver of a message. However, when it comes to telephone conversations, only one side of the conversation, the so-called 'down-link' portion, can be intercepted by the BND's eavesdroppers. The BND has stated that: 'of an oral satellite conversation, say between Europe and Asia, the BND can only monitor that part that is emitted by the Asian subscriber and then transmitted to Europe by satellite. Now, using this example, the corresponding remarks of the European subscriber cannot be recorded by the BND due to technological-physical reasons.' However, this is only partly true because Pullach also operates Satcom intercept joint ventures with Taiwan, as well as with the French at Kourou in French Guyana and New Caledonia, which theoretically can pick up the other end of the telephone conversation.

THE FUTURE OF SIGINT IN GERMANY

After World War II Reinhard Gehlen immediately put much effort in gaining a monopoly position for him and his organization in the intelligence service community. In 1947 he pushed aside his rival at the US intelligence services community, Herman Baun,[134] and was successful with his intrigue resulting in the downfall of the secret service of the 'Amtes Blank' (forerunner of the Ministry of Defence), the 'Friedrich-Wilhelm-Heinz-Dienst' by Adenauer's Minister of State Hans Globke.[135] He controlled, by means of his trustee Josef Selmayr, the foundation of the military counter espionage service in 1957 and took over from Wilhelm Flicke the American financed Sigint centre in Lauf. Until the mid-1950s, numerous attempts were made for an all-embracing West German security apparatus. In 1955, Franz Josef Strauss, at that time Federal Minister for Special Assignments, urged Globke to establish a state security service and two years later the CDU faction in the Bundestag deliberated upon a constitutional bases for the *Organisation Gehlen*.[136]

Beside this Gehlen, supported by Strauss, always hoped to become director of such a department enabling him to achieve the status of state minister. Even when this goal became unattainable, the BND was able to obtain a unique position in the field of Humint, including the tasking of the military attachés, and to create a dominating Sigint position over the Bundeswehr. A position that was strengthened by the internal competition within the armed forces between the Army, Navy and Air Force. In April 1971, the Vice President of the BND Blötz even tried to 'hand over to the BND the joint signal intelligence activities in the BMVg region (with conscripts) for which an arrangement should be made how to detach a communication intelligence section under the BMVg in case of employment'.[137] In that time the functional hierarchy of the BND in wartime, during which the staff should be withdrawn to Spain, was not yet clear.[138]

On 5 May 1971 the direction of Department II of the Bundesnachrichtendienst again deliberated about the question: 'Is it possible that the BND takes over the joint signal intelligence gathering of the Bundeswehr?'[139] These points of view should result in a cabinet decision, but were eventually rejected on juridical grounds.[140] The foundation of the intelligence service of the Bundeswehr (ANBw) by Defense Minister Hans Apel in June 1978, was only a half-hearted attempt to ensure the Bundeswehr, which for territorial defense focused on the East, had its own and independent Comint capability. At the time the intention to recruit military communication intelligence experts from the BND remained an empty threat.

Due to the agreement between the BND and the Bundeswehr, the first in 1958, followed by the Zugvogel treaty of October 1969 and finally under BND President Konrad Porzner in 1992, Pullach could repeatedly strengthen its dominating position in the field of Sigint. This offered the BND, apart from a huge amount of exclusive information, also numerous other advantages. Possessing the monopoly of exchanging information with NATO members and all other third parties, the service was able to benefit from the large amounts of Sigint material in the exchange with small intelligence services.

Besides this the BND also received support from the German industry because the service was able to promote the export of technical intelligence material in more than 100 countries.[141] And last but not least the weak BND spots in the field of Humint could be compensated because during the process of analysis, agent reports, open source information and useable Sigint material could be mixed unnoticed. Because there was no competing organization in West Germany, an objective review of the information and its source was denied to the politicians. At the most reports sent by the diplomats of the foreign ministry contained expanded facts and assessments.

However, in the spring of 1999 Chancellor Gerhard Schröder installed a committee chaired by the former Bundespresident Richard von Weizäcker that presented proposals for a reform of the Bundeswehr on 23 May 2000. Already its terms of reference: 'to discern timely the development of crisis and in order to be able to take decisions, the Bundeswehr has a requirement for an independent and unfiltered access to current and comprehensive information',[142] indicates a declaration of war against the US monopoly and against the Bundesnachrichtendienst.

Parallel to this Defense Minister Rudolf Scharping also ordered the development of proposals for the future structure of the armed forces. In many sectors competitive institutes produced various proposals, however there was unanimity about the new orientation on strategic intelligence. The Minister of Defense said that the Bundeswehr should 'contribute to the early detection of crises by using their own resources as well as by cooperating with allies and ensure full information is furnished to the political and military leaders at all times'. And in the Priorities he wrote: 'The Bundeswehr will acquire a spaceborne reconnaissance capability of its own to improve Germany's capacity to assess political and military situations and to supplement Alliance capabilities.'[143]

A serious threat for the old Sigint-monopoly of the BND is formed by the concrete transposition of this preamble in the authorized 'instruction of the Inspector-General of the Bundeswehr for the planning of the armed forces in the future' dated July 2000. 'Capabilities lacked so far – such as the capability of strategic deployment and strategic intelligence as well as an improvement of the inter-operability and performance of the control systems and -equipment becomes prominent.'[144] The preamble to take into account 'the common aspect of the armed forces ... particularly in the field of command, intelligence collection and reconnaissance' eliminates the present competition between the elements of the armed forces and therefore creates a good starting point for the battle about how to divide the responsibilities with the BND. Besides this the Inspector-General demands 'priority for facilities to improve the control capability emphasizing strategic reconnaissance, operational/tactical reconnaissance and superiority in the field of information'.

In the years to come this priority will have its effects in a technical modernization of intelligence tools, or in the words of the Weizäcker commission 'In the field of collecting intelligence and reconnaissance, new capabilities should be acquired for satellite supported reconnaissance and broad scale airborne regional reconnaissance, both imagery and Sigint.' It is not to be expected that the Bundeswehr will finance the expensive satellite capability and airborne intelligence reconnaissance assets and then hand over the responsibility to the BND in case of an operational deployment.[145] According to the plans of the Inspector-General, the communication elements for electronic warfare, the stationery and mobile communication, and Elint units of the Army and the Navy, the Army telecommunication in Germany as well as the mobile long-distance communication units not subordinated to the operational troops and facilities and elements of Communication Sector 70 of the Air Force in Trier will be united. The emancipation of the Bundeswehr from the BND reaches the top in the establishment of a Strategic Reconnaissance Command under control of the Joint Support Command.

At the future disposal of this reconnaissance holding of the Bundeswehr are the following elements: stationary HF communication interception, satellite interception (downlinks), interception of troop-, aircraft- and ray transmitter communication mainly by mobile elements, electronic interception from mobile platforms and weapon systems, airborne Sigint collection by Breguet Atlantic aircraft, seaborne interception with the present three survey vessels as well as with 'optronical' and radar data from AWACS aircraft (NATO Airborne Early Warning Forces).

The ANBw with its 620 employees was 'a customer of intelligence rather than a producer'. However, with the new intelligence center an effective supply service, based on many pillars is growing. The gathering of these information sources in one hand provides a much more differentiated picture than the BND and the often-uncoordinated efforts of the 'old' Bundeswehr could ever produce.

With large steps, the Inspector-General marches towards the fast transition, for the set-up and the reconstruction of 'all elements and facilities of the military intelligence service into the structure of the armed forces', which will, in chronological order, take place on 1 October 2001. Already in the summer of 2000, certain working groups within the ANBw, were already active. In the 1990s, the Bundeswehr already did its independent reconnaissance 'run on trial' in a potential deployment area, during the crises and war in the Balkan region. Already during the separation wars in Yugoslavia and lately in the Kosovo war, the Bundeswehr had carried out electronic interceptions independent from the BND using its direction finders and interception stations (reaching up to the Urals) in Germany. The ANBw was very satisfied with the results.[146]

Parallel to the concentration of interception facilities, the Ministry of Defense at the same time procures an analytical braintrust by uniting elements so far linked to various authorities. The scientific capacities of the Federal Academy of Security Policy in Strauszberg and of the universities of the Bundeswehr in Hamburg and Munich, the reports from the German military representative with the military committee of NATO and with the WEU will fall under the Inspector of the Armed Forces. He also takes direct operational command over the new intelligence center of the Bundeswehr in Bad Neuenahr-Ahrweiler and over the military counter-espionage service (MAD) in Köln as well as over the *Amt für Militärkunde*, that is, the military part of the BND. The proposals of the Bundeswehr are aimed at the withdrawal of the BND from the entire collection of military intelligence. They are therefore dynamite for Pullach, which in 1998 applied for 70 million DM[147] for the purchase and maintenance of technical equipment for its 10 interception stations in Germany and has to battle more budgetary cutbacks.

In May 1992, the French SR (*Service des Renseignement Militaire*) was restructured by General Jean Heinrich and became the DRM (*Direction du Renseignement Militaire*, 650 military personnel), obtaining monopoly for military communication intelligence including the satellite branch. The expertise was withdrawn from the French counterpart of the BND, the DGSE.[148]

Still such a step in Germany is not yet likely to take place but the trend goes in that direction. For example, the Weizäcker commission already ordered for intelligence 'the joint purchase and a cooperation of partners...originally collected information should be available for all partners'.[149] Herewith a judgement is passed over the old monopoly of the BND for the exchange of intelligence material not only with NATO partners. The conflict of authority between the BND and the Bundeswehr is pre-programmed when in case of an international deployment the

Bundeswehr cooperates with non-NATO members and consequently shares intelligence without using a long route via Pullach. In case of close allies outside NATO such as Israel and Japan, one could envisage that the Bundeswehr, looking for future beneficial talents, cooperates with the military intelligence services involved.

Another source of conflict lies in the limitation of the definition *military intelligence*. For the Bundeswehr, the defense policy, the defense economy and other political areas defining the endurance of an enemy state in times of war, are areas of interest. The fact that in the BND the military and defence policy sector becomes secondary means a severe detraction of its position within the political structure of Germany.

In the so-called new area of interest of the BND, in the field of the war against organized crime, the service has not achieved the intended success so far. Under pressure of developing priorities in the face of financial cutbacks, the BND already partially withdrew from these sectors. As soon as the Bundeswehr has postured itself in the new intelligence- and exploitation priorities, Pullach has to focus on political intelligence, international terrorism, arms trade and proliferation as well as on economic intelligence, including industrial espionage against foreign enterprises. It is doubtful that the civil foreign intelligence service will achieve in these difficult sectors such a wide acceptance, as the Bundeswehr will on its most peculiar terrain.

However, the main part of the new independence of the Bundeswehr will be the responsibility for the evaluation process in which also requirements for communication intelligence will be taken over. This will automatically exceed the restriction to Sigint. Apart from the knowledge derived from all own intelligence sources, the Bundeswehr has set eyes on the use of raw material (Sigint, Radint, Masint, etc.) of counterparts, on the presence on the Internet and information networks of target organizations, on intelligence collection by Humint with military attachés as well as with deployed elements of the Bundeswehr and of industrial enterprises, as well as on the evaluation of open sources such as specialized literature, scientific studies, etc. 'However, an improvement of the professional analysis and evaluation of the available information is necessary.' The Weizäcker commission concluded that 'the required evaluation expertise for this must be bundled and extended'.[150]

The commission recommends strengthening the national capacity for analyzing and evaluating received information by parceling and quality improvement of the available agencies (BND, the military intelligence service and intelligence services of the armed forces elements) and wants to facilitate the newly created organization in the vicinity of Berlin.[151] In an explanation it was stated that the present capabilities are divided up, that the

Bundeswehr suffers from poor coordination and that the official lanes are long and winding. It is true that the BND in 2001 will move its intelligence processing unit of about 1,000 employees to Berlin in order to be closer to its clients, but certainly not to let them merge into a new analysis organization independent from Pulach. In future, the Bundeswehr will be able to play the key role in an area overlapping the evaluation process. It is not yet clear which department will take the responsibility here. Leaving the responsibility to the BMVg would politically be too ticklish.

Already in 1993, this author suggested transferring all Sigint activities to the armed forces, to assign the war against the organized crime to the Bundeskriminalambt exclusively, to bring the integrated intelligence analysis under the Ministry of Foreign Affairs and to limit the BND to Humint.[152] However, because in a coalition government, the Ministry of Foreign Affairs traditionally is assigned to the smaller coalition partner, the placing of such a braintrust under the department of the Bundeskanzler, that traditionally patronizes foreign affairs, is more probable.

At the end of this development, automatically the need for a coordination commission on a political level, for instance in the committee of the Secretary of State in the Federal security counsel will grow. Besides direct cooperation with the BND and the Bundeswehr, this commission will take care of the production of intelligence requirements from the top. These requirements will bring together and review information from various sources. Eventually this will lead to a quality control of the various intelligence services. However, because the ANBw is not controlled by the PKK (Parliamentary Control Commission), it does report to the Bundestag Defense Committee (*Verteidigungsausschuss*).[153]

The armed forces show little tendency to submit their apparatus, emerging into a regular intelligence service, to the parliamentary control of the PKK. Looking at the new capabilities for intelligence collection by secret services, including the use of satellites, the involvement of Humint in the intelligence process and because of their appearance on the international stage, the same parliamentary control for the BND and the *Verfassungsschutz* should be required here for the sake of the democratic principle. But neither in the center of power in the red-green government, nor within the opposition is there a majority for such a solution.

NOTES

1. Erich Schmidt-Eenboom, 'Empfänglich für Geheimes', in Klaus Beyrer (ed.) *Streng Geheim. Die Welt der verschlüsselten Kommunikation* (Heidelberg: Umschau 1999).
2. Hermann Zolling and Heinz Höhne, *Pullach intern* (Hamburg: Hoffmann und Campe 1970) Chapters 5–10. For the CIA connection: Mary Ellen Reese, *General Reinhard Gehlen: The CIA-Connection* (Fairfax, VA: George Mason UP 1990).
3. Zolling and Höhne (note 2) pp.124ff. and 197.
4. Georg Meyer, 'General Adolf Heusinger und die Organisation Gehlen', in Wolfgang Krieger and Jürgen Weber (ed.) *Spionage für den Frieden? Nachrichtendienste in Deutschland während des Kalten Krieges* (München: Olzog 1997) pp.225–46.
5. For the decisive influence the OG had on Germany's rearmament see Erich Schmidt-Eenboom, 'Der innenpolitische Einfluß des BND', in Hans-Jürgen Lange (ed.) *Staat, Demokratie und Innere Sicherheit in Deutschland* (Opladen: Leske & Budrich 2000) p.187ff.
6. A former officer of the Wehrmacht, who had worked since 1950 at the OG Sigint center at Butzbach, claimed that this study was not done by General Heusinger, but rather by another former Wehrmacht intelligence officer named Wilhelm Flicke.
7. Source: Amt Blank, Military Section, File Heusinger, Federal archives-military archives BW 9/3574, pp.26–39.
8. D. Urwin, *Western Europe since 1945* (London: Longman 1990) pp.110–19 and 126–35.
9. Zolling and Höhne *Pullach* (note 2) p.110, using the diary of H. Baun.
10. According to one former officer of the Wehrmacht's espionage service who used to operate out of Finland. He declined the offer and choose a career as an architect. Confidential interview, Feb. 1986.
11. Rudolf Grabau, *Ideen und Planungen für eine militärische Funkaufklärung in Westdeutschland nach Ende des 2. Weltkriegs* (Much: private study 1999) p.3.
12. Biographical information on the members of German Sigint are in the National Archives, NARA 3452.
13. Erich Schmidt-Eenboom, *Schnüffler ohne Nase. Der BND* (Düsseldorf: Econ 1993) p.223.
14. Grabau, *Ideen und Planungem* (note 11) p.6.
15. Ref. surveillance areas of the military listening post Lauf, Confidential, document held at the Fernmeldemuseum of the Bundeswehr in Feldafing.
16. Within the framework of this contribution, it is not possible to clearly define the telecommunications branch of the Bundesgrenzschutz and the customs criminal branch (ZKA). In 1991, the ZKA in Cologne obtained a surveillance mandate under the Eighth Act for the Amendment of the Foreign Trade Law and became the fourth intelligence service at federal level. The mandate includes mail and telecommunications control.
17. Ref. draft of the federal budget for the financial year 1957, separate plan 04, chapter 04 in the Federal Archives at Koblenz, reference Bundeskanzleramt B 136/4885.
18. For 1976 letter Blötz to General Wust, 12/03/76, Secret. For 1997: Udo Ulfkotte, *Verschlußsache BND* (München: Koehler & Amelang 1997) p.116. Although Ulfkotte pays attention to Sigint in his book, he hardly deals with the Sigint activities of the BND. Also in other German publications Sigint is playing a marginal role and is hardly mentioned. A book solely dealing with the BND and Sigint is missing.
19. Grabau, *Ideen und Planungen* (note 11) p.33.
20. The history of the army's telecommunications units is presented in detail by Rudolf Grabau, *The Army's Telecommunications Corps*, Volumes 1–4 (Bonn: Fernmeldering e.V. 1995–99) *passim*.
21. For the Sigint of German Air Force in detail Rudolf Grabau, *Der Neubeginn der FmEloAufkl und der Eloka der Luftwaffe ab 1956 und das dem Neuaufbau zugrundeliegende Konzept* (Much: private study 2000) *passim*.
22. Franz Josef Strauss, *Die Erinnerungen* (Berlin: Siedler 1989) p.358.

23. Grabau (note 11) p.34f.
24. Erich Schmidt-Eenboom, *Undercover* (Cologne: Kiepenheuer & Witsch 1999) p.156f.
25. BND, Grundlegende der 'Bundesstelle für Fernmeldestatistik', Stand vom 1 Aug. 1970, 4 pages.
26. Blötz memo, Secret, 2 Oct. 1990, p.1.
27. Blötz note of a discussion with BND president Wessel, Secret, 27 Jan. 1971, p.1.
28. Letter Blötz to the Chief of Department II, Secret, 25 April 1972, p.2.
29. Blötz note of a discussion with Wessel, Secret, 1 Sept. 1971, p.1.
30. Letter Blötz to the Chief of Department II, Secret, 25 April 1972, p.1.
31. Blötz note, Secret, 14 Feb. 1975.
32. Letter Blötz to Arthur Stimson (CIA), Secret, 5 April 1974, p.2.
33. Letter Blötz to General Wust, 12 March 1976, p.2.
34. Blötz note, Secret, 18 Dec. 1974, p.3.
35. Letter Blötz to General Dr Schnell, Secret, 30 Jan 1975, p.1.
36. Blötz note, Secret, 8 Oct. 1976, and letter Blötz to Buffham, Secret, 22 Oct. 1976, Item 4.
37. Letter Blötz to General Dr Schnell, Secret, 8 Sept. 1975.
38. Agreement between Blötz and General Wust, Secret, BND document 125/1976.
39. Blötz note, Secret, 8 Oct. 1975, p.5.
40. Schmidt-Eenboom, *Schnüffler ohne Nase* (note 13) p.236.
41. Schmidt-Eenboom, *Empfänglich für Geheimes* (note 1) p.164.
42. Ibid. p.165.
43. Private information from the secret Bundeswehr building programs 1991.
44. Schmidt-Eenboom, *Empfänglich für Geheimes* (note 1) p.165.
45. Private information by an involved Bundeswehr officer.
46. Blötz note of a discussion with deputy minister Wilhelm Berkhan, ministry of defense, Secret, 4 July 1972, p.3.
47. Blötz note, 29 Aug. 1973 of a discussion with Schnell on 28 Aug. 1973, p.7f.
48. Letter Blötz to Schnell, Secret, 21 Jan. 1974.
49. Memo Blötz, Secret, 12 Nov. 1974.
50. Blötz note, Secret, 8 Oct. 1976.
51. Schmidt-Eenboom, 'Empfänglich für Geheimes' (note 1) p.167; After Gablingen was closed early in the 1990s the liaison staff went to Bad Aibling.
52. Blötz note, Secret, 12 Nov. 1974, p.1.
53. Blötz note, Top Secret, 16 Jan. 1975, p.2.
54. Letter Blötz to Burrough, 4 June 1975.
55. Letter Blötz to Buffham, Secret, 22 Oct. 1976, Item 6.
56. Blötz note, Secret, 8 Oct. 1976.
57. Blötz note, Secret, 14 Dec. 1976, p.4.
58. Blötz note, Secret, 24 Jan. 1978, p.7.
59. Schmidt-Eenboom, *Schnüffler ohne Nase* (note 13) p.224.
60. Private information from an employee of German Telekom, 1999.
61. Jeffrey T. Richelson and Desmond Ball, *The Ties that Bind: Intelligence Cooperation Between the UKUSA Countries* (London: Allen & Unwin 1990) pp.170–1.
62. This happened after in 1991 the Naval Security Group and the 6913th Electronic Security Squadron and in 1993 the 701st Inscom's Military Intelligence Brigade were deactivated.
63. Schmidt-Eenboom, *Schnüffler ohne Nase* (note 13) p.226.
64. IWR Daily Update, Vol. 7, No. 137, 21-07-2000 quoting *Washington Post*, 24 July 2000.
65. *Der Spiegel*, No. 47/1999, p.32f.
66. Private information by a BND employee, Dec. 1999.
67. Letter Blötz to Karl Liedtke, member of parliament from Bochum, 11 Feb. 1974.
68. Blötz note, Top Secret, 16 Jan. 1975, p.1.
69. Letter Blötz to Dr Louis Tordella, 22 July 1974.

70. Blötz note, Secret, 22 Oct. 1976.
71. Schmidt-Eenboom, *Schnüffler ohne Nase* (note 13) p.270.
72. Blötz note, Secret, 5 May 1971, p.3.
73. Blötz note of a discussion with Wessel, Secret, 13 Oct. 1970, p.9f.
74. Blötz note, Secret, 8 Oct. 1976.
75. Letter Blötz to General Wust, Secret, 6 Feb. 1976, p.2.
76. Blötz note, Confidential, 20 Aug. 1976, p.4.
77. Agenda of the BND as regards the visit of the parliamentary control commission on 12 April 1989.
78. In detail Bernd W. Kubbig and Tillmann Elliesen, *Wozu sollen die europäischen Satelliten Helios II und Horus dienen?*, Hessische Stiftung für Friedens- und Konfliktforschung, HSFK-Report 3/1997 (Frankfurt 1997) *passim*.
79. Private information from a former BND officer, 1996.
80. Juegen Saupe and Frank P. Heigl, *Operation EVA* (Hamburg: Konkret 1982) p.114.
81. BND, *ND-Begriffsbestimmungen* (Pullach 1974).
82. Schmidt-Eenboom, *Schnüffler ohne Nase* (note 13) p.222f.
83. Confidential interview with a former officer of the Ministry for State Security of the GDR, 1994.
84. Schmidt-Eenboom, *Schnüffler ohne Na*se (note 13) pp.137f., 179 and 270.
85. Erich Schmidt-Eenboom, *Der Schattenkrieger. Klaus Kinkel und der BND* (Düsseldorf: Econ 1995) p.108ff.
86. Juergen Roth, *Die Mitternachtsregierung* (Hamburg: Rasch und Röhring 1990) Chapter 5 and private information from an interview with a former GDR military attaché in Maputo, 1992.
87. Schmidt-Eenboom, *Der Schattenkrieger* (note 85) p.53.
88. Strauss, *Die Erinnerungen* (note 22) p.358.
89. Erich Schmidt-Eenboom and Jo Angerer, *Die schmutzigen Geschäfte der Wirtschaftsspione* (Düsseldorf: Econ 1994) p.97.
90. Private information from an interview with a former MAD officer.
91. Schmidt-Eenboom, *Empfänglich für Geheimes* (note 41) p.180 and also Jean Guisnel in the weekly news magazine *Le Point*, 6 June 1998.
92. Private information from a wife of an involved BND officer.
93. Schmidt-Eenboom, *Der Schattenkrieger* (note 85) p.28.
94. Note Blötz of discussions with Wessel, Secret, 26 March 1971, p.3 and Secret, 16 April 1971, p.7f.
95. Blötz note, Secret, 5 March 1975, p.5 and p.8.
96. Blötz note, Secret, 16 July 1975, p.3.
96a. Blötz note, Secret, 8 Oct. 1975, p.6.
97. Blötz note, Secret, 11 Dec. 1978, p.2.
98. Blötz note of a discussion with BND president Wessel, Secret, 15 Feb. 1974, p.7.
99. Telephone call with a Afghan WAD captain in summer 1996 then living in Tadjikistan.
100. 'BND soll Russland unterstützt haben', *Süddeutsche Zeitung*, 8 April 2000; 'Aus Feinden werden Freunde', ibid., 10 April 2000 and 'Deutschland und der Islamismus im Kaukasus', *Neue Zürcher Zeitung*, 10 May 2000.
101. 'Geheimdienst befürchtet Ausweitung des Krieges in Tschetschenien', *Berliner Zeitung*, 17 April 2000.
102. *Der Spiegel*, No. 24/1992, p.16.
103. Schmidt-Eenboom, *Der Schattenkrieger* (note 85) p.36.
104. Bob Woodward, *VEIL. The Secret Wars of the CIA 1981–1987* (London: Headline Book Publishing 1988) p.314.
105. Schmidt-Eenboom, *Schnüffler ohne Nase* (note 13) p.274.
106. Blötz note, Secret, 29 Oct. 1975, p.1.
107. Blötz note, 10 Sept. 1975; note, 29 Oct. 1975, note, 16 Oct. 1975, note, 17 Nov. 1975, all Secret.

THE BUNDESNACHRICHTENDIENST

108. Blötz note, Secret, 24 Jan. 1978, p.2.
109. Letter Blötz to General Dr Schnell, 1 April 1975.
110. For the first years of cryptography in the Federal Republic of Germany: Michael von der Meulen, 'Cryptology in the Early Bundesrepublik', *Cryptologia* (July 1996).
111. Schmidt-Eenboom, *Der Schattenkrieger* (note 85) p.68.
112. Blötz note, Secret, 15 Oct. 1975, p.3.
113. Blötz note, Secret, 10 June 1976, p.5.
114. Blötz note of a discussion with Wessel, Secret, 19 July 1973, p.6.
115. *Der Spiegel* 52/1998, 'Auf einem Ohr blind', p.33.
116. Private information from a former BND employee, 1999.
117. Note Blötz, Secret, 23 May 1973.
118. Schmidt-Eenboom, *Schnüffler ohne Nase* (note 13) p.246; Dutch Editor's note: also the Dutch Sigint service could read this traffic and forwarded it to GCHQ. See the contribution on Holland by Wiebes.
119. Blötz note, Secret, 7 Sept. 1977, p.1.
120. *Freitag* (German weekly newspaper), 18 Sept. 1992.
121. Schmidt-Eenboom and Angerer (note 89) p.224f.
122. *Der Spiegel* 52/1998, 'Auf einem Ohr blind', p.33.
123. Blötz note, Secret, 9 July 1973, p.2.
124. Blötz note, Secret, 22 April 1971, p.2.
125. Blötz note, Secret, 11 Feb. 1974.
126. Blötz note, Secret, 5 March 1975, p.3.
127. For the Traube case: Richard Meier, *Geheimdienst ohne Maske* (Bergisch-Gladbach: Lübbe 1992) pp.16–30.
128. Blötz note of a discussion with BND President Wessel, Secret, 15 Feb. 1974, p.5.
129. Schmidt-Eenboom, *Undercover* (note 24) p.82ff.
130. Karl-Ludwig Haedge, *Das neue Nachrichtendienstrecht für die Bundesrepublik Deutschland* (Heidelberg: Kriminalistik 1998) p.209ff.
131. *Der Spiegel* 52/1998, 'Auf einem Ohr blind', p.32f.
132. *Frankfurter Rundschau*, 7 Dec. 1998.
133. Bundesnachrichtendienst – The president, *Report on the Implementation of Strategic Control Measures according to Article 1, Paragraph 3 of the G-10 Act*, Pullach 23 Dec. 1996.
134. Zolling and Höhne (note 2) p.112.
135. Susanne Meinl, 'Im Mahlstrom des Kalten Krieges', in Krieger and Weber (note 4) pp.258–61.
136. Bodo Wegmann, 'Vergleichende Untersuchung der Entstehung und Entwicklung geheimer Nachrichten- und Sicherheitsdienste in Deutschland-Ost und – West 1945 bis Ende der fünfziger Jahre', Diplomarbeit Universität Berlin 1998, p.41.
137. Blötz note, Secret, 21 April 1971, p.3.
138. Blötz note, Secret, 28 April 1971.
139. BND II 1, Tagebuch Nr. 666/1971, Secret.
140. Confidential information from a former high ranking officer of the Bundeswehr.
141. Schmidt-Eenboom, *Schnüffler ohne Nase* (note 13) p.270.
142. *Gemeinsame Sicherheiz und Zukunft der Bundeswehr*. Bericht der Kommission an die Bundesregierung, Berlin 22 May 2000, p.49, im folgenden Weizäcker-Kommission.
143. Ministry of Defense (ed.) *The Bundeswehr – Advancing Steadily into the 21st Century, Cornerstones of a Fundamental Renewal* (Berlin: MoD 2000) Nr.48.
144. Bundesministerium der Verteidigung – Generalinspekteur der Bundeswehr Fü S VI2, For official use only, July 2000, *Weisung zur Ausplanung der Streitkräfte der Zukunft*, p.10.
145. Weizäcker-Kommission (note 142) p.94.
146. Confidential interview with a Bundeswehr officer, July 2000.
147. *Der Spiegel* 52/1998, p.33.

148. Erich Schmidt-Eenboom, 'Frankreich', in idem (ed.), *Nachrichtendienste in Nordamerika, Europa und Japan*, CD-ROM (Weilheim: Stöppel 1995).
149. Weizäcker-Kommission (note 142) p.49.
150. Ibid.
151. Ibid. S.84.
152. Schmidt-Eenboom, *Schnüffler ohne Nase* (note 13) pp.446–53.
153. Shlomo Shpiro (ed.) *Guarding the Guards – Parliamentary Control of the Intelligence Services in Germany and Britain* (Sankt Augustin: Konrad-Adenauer-Stiftung 1997) p.33.

6

France, Sigint and the Cold War

ROGER FALIGOT

Unlike the Anglo-Saxon signals intelligence community, the French Sigint effort has never attracted much attention from historians. Is it because these French 'electronic warriors' worked alone – for linguistic and political reasons – and did not share as much information as typically took place between the Sigint organisations of Britain, her former colonies (Canada, Australia, New Zealand) and the US? Or is it because the official archives on the subject are sealed in France? The answer is almost certainly – both.

Despite these obstacles, it is possible to piece together much of the French Sigint jigsaw puzzle thanks to open and private sources, as well as from the recollections of the actors who participated in this secret war. Taking part in this joint study with authors from many countries that specialise in Sigint is a new and stimulating experience. It provides an opportunity to present the French side of the story with a limited goal: to open the door for others – scholars and historians – who will, no doubt, perform a more exhaustive study of this subject in future.

THE EARLY YEARS OF FRENCH SIGNALS INTELLIGENCE

During World War I, the French performed their signals intelligence work within the framework of the Allied cause against the Central Powers, led by Imperial Germany. They capitalised on the inventions of French engineers such as Edouard Branly, and the fact that the French had a long tradition of cipher work dating back at least to the fourteenth century.

At the beginning of the twentieth century, France possessed an excellent team of cryptologists as well as some of the best radio intercept equipment available, such as the radio intercept station located at the top of the 300 meter-high Eiffel Tower in Paris, which could copy radio transmissions coming from thousands of kilometres away. In 1905, during the war in which the Japanese Empire defeated Tsarist Russia, the French intercepted the communications of the Japanese ambassador in Paris, Monoto Ichiro, and the legendary French cryptologist Commandant Etienne Bazeries deciphered them and passed their contents on to France's Russian allies. Despite this effort, the Japanese still won the war, thanks to their modern wireless radio transmission system, which allowed them to concentrate their naval forces and sink the Russian fleet at the Battle of Tsushima.[1]

During World War I, French communications intelligence (Comint) achieved impressive results during the fighting on the Western Front, which are described more fully elsewhere.[2] During 1914–18, the French operated several radio intelligence centres, such as in Chartres and Bordeaux. But the most important was the radio intercept station atop the Eiffel Tower in Paris, which was connected to the Comint intercept and processing station at Le Mont Valérien outside Paris, which was then the most important Sigint site in France, and remained in service through the end of the twentieth century.

The French also placed intercept stations throughout their overseas empire. For example, there was an intercept station inside the French Concession in Shanghai, which worked for the French foreign intelligence service, the Service de Renseignement (SR). It played an important role monitoring the activities of the Moscow-based Comintern (Communist International) intelligence service in Asia, led by the former French Communist leader Jean Cremet and the famous German spy Richard Sorge. Both friends worked for the Red Army's intelligence service, the GRU, and operated a network of clandestine radio sets in Hong Kong, Macao and Shanghai, which sent short wave messages to the Comintern's Far Eastern Bureau (FEB) at Vladivostok. This net also worked with the radio network of the Chinese Communist Party, led by the special service chief and radio technician Pan Hannian, which was under the technical supervision of the Comintern International Liaison Office. The French and British security services in Hanoi, Shanghai and Singapore were eager to intercept messages and tried to dismantle this system.[3]

Although the Cold War officially did not start until more than a decade later, it was in 1937 that most of the French intelligence services set up anti-Soviet intelligence units, as well as anti-Nazi sections. Soviet intelligence operatives from NKVD, the Comintern and the GRU were extremely active in France, as exemplified by the kidnapping in Paris of White Russian General Yevgeniy Miller in 1937, who the Soviets wanted because he had plotted against USSR.

Under the left-wing Popular Front government headed by Léon Blum that came to power in 1936, several important French intelligence organisations were set up. In March 1937, a joint organisation called the 'Contrôle de la Surveillance du Territoire' (CST) was established, which was the predecessor to the current French counterintelligence unit, the DST. The CST supervised a nationwide 'radio-police' network, which was in charge of identifying and locating clandestine radio transmitters in France that were being operated by foreign spies. The DST performs the same function today.[4]

The second new unit was the Service de Renseignement Intercolonial (Intercolonial Intelligence Service, or SRI), which performed intelligence collection from, and monitored potential threats to France's Asian and African colonies, which included operating listening posts from these French-controlled overseas territories.

In 1937, two new technical sections were created within the SR intelligence service: Captain Paul Arnaud set up a radio transmission section known as Radio-Chimie-Photo (RCP); while a separate unit, called simply Section D, intercepted and decoded foreign communications traffic. An expert in cryptography then working at the French Army GHQ's Cipher Section, Commandant Gustave Bertrand, joined the SR and was given command of Section D.

Finally, a new unit called Section 'I' (which stood for Intercept) was established along with a Naval Intelligence Cipher Section. Radio intercept operators belonging to an organisation called the Réseau d'Ecoute et de Radiogoniométrie des émissions radio-électriques étrangères (Listening and Direction-finding Network for Monitoring Foreign Radio-Electric Communications, or REG for short) were stationed at various intercept stations throughout France, the most of important of which was Le Mont Valérien outside Paris.

THE BIRTH OF THE RADIO-ELECTRONIC COMMUNICATIONS GROUP

When Hitler's army occupied France in 1940, French military intelligence analysts and Sigint technicians from the disbanded REG regrouped in Vichy to create a new Sigint organisation called the Groupement de communications radioélectriques (Radioelectronic Communications Group, or GCR), headed by Colonel Paul Labat and Lieutenant Colonel Gabriel Romon. Although they were located in the capital of the newly-created Vichy collaborationist regime, they kept the communications of the German Army in the northern occupied zone of France under surveillance.[5] Colonel Romon also provided much valuable information to the 3,000-person 'Alliance' intelligence network in France, which was led by one of the most important and successful of the leaders of the French Resistance, Marie-Madeleine Fourcade. The Alliance Network provided Britain's MI6 with invaluable intelligence information throughout the war.[6]

Working officially for the Postes & Télécommunications (P&T), the GCR ran six intercept stations and operated an analysis center in the town of Hauterive near Vichy. They decoded some 3,000 messages from the German Luftwaffe and Wehrmacht. Commandant Bertrand (whose alias was 'Gaudefroy' in the Resistance), worked with the GCR, bringing the results of this cryptanalytic work to a clandestine base in Cadix, from where they secretly sent to London to be used by Allied intelligence analysts. But Bertrand was arrested by the Nazis, and was sent to a concentration camp.[7]

When the Nazis occupied France's southern zone ('La Zone Libre') in late 1942, Colonels Romon and Labat were captured and sent to a German concentration camp, where they both died. Both men are remembered today within the French intelligence community as heroes of the Sigint war against Germany. Before being captured, Colonel Labat had managed to send to the US data on the GCR direction-finding equipment that he had developed in order to locate clandestine Nazi transmitters in the non-occupied zone of France. These technical documents helped the Americans develop the best anti-submarine detection system used during World War II.[8]

Outside of occupied France, other French cryptologic organisations were created. In Algiers in 1942, the SR had a Technical Control Service led by Albéric de Maistre, which intercepted Axis communications traffic. In London, General Charles de Gaulle's Free French secret service, the Bureau Central de Renseignement et d'Action (Central Bureau for Research and Action, or BCRA) was founded by André Dewavrin, whose codename during the war was 'Passy'. The BCRA also had a Technical Section (TCF) under a Captain Lecot (codename 'Drouot'), which was in charge of coding and decoding messages. Lecot operated jointly with the British experts from the Government Code and Cipher School (GC&CS) and the Special Operations Executive (SOE).

Before the war, the chief of the BCRA Intelligence Section, Jean Fleury ('Panier'), had been a technical director with Radio-France, Radio-Paris and Radio-Saigon from 1927 to 1939. It should therefore come as no surprise that he was instrumental in organising clandestine transmitters for the Resistance inside France, such as the 'Electre' network near Lyons in 1942; and made sure that BCRA agents parachuted into France could avoid being located when sending shortwave radio messages by the German FunkAbwehr's mobile direction-finding vehicles (which the French referred to as 'voitures-gonio', or direction finding cars).

French and British intelligence officers at this time also invented a new form of clandestine communications called 'personal messages', by which secret messages could be sent to operatives in occupied France buried inside innocuous messages broadcast over the radio. This form of secret communications was used throughout the Cold War, and is still in use today around the world. According to French counterintelligence officials, foreign intelligence agencies are today burying secret messages inside otherwise innocuous e-mail messages.[9]

In December 1942, after the Allies captured Northern Africa, General de Gaulle asked Colonel Jean Joubert des Ouches to set up a central cipher and cryptographic department in Algiers. A better choice could not have been made since Joubert des Ouches had already served in a similar position in 1919 under Prime Minister Georges Clemenceau and Marshal Foch after the end of World War I.

By August 1943, a decree from the French Committee for the National Liberation (CFLN) signed by Generals Henri Giraud and Charles de Gaulle, both rivals at the time for the position of head of the French government, gave birth to an inter-ministerial organisation called the Direction Technique des Chiffres (Cipher Technical Directorate), which was formed to separate the special Comint operations conducted by the secret service BCRA from the future French government's cipher and communications section, which was controlled by the DTC.

In 1943 and 1944, the French conducted a series of special operations in Algeria and Tunisia to locate and arrest Muslim nationalists who had been trained by the Abwehr and parachuted into North Africa. These operations were called 'Atlas' and 'Anti-Atlas'. Once aircraft from the German Gartenfeld squadron had parachuted these Algerians back into their homeland, they were supposed to collect intelligence and radio back to Berlin (and later to Siegmaringen) the information they had picked up. Almost all of the agents, however, were located by French radio finding operations, and after they were captured they were given a simple choice: either take part in a radio deception operation against their former employers or else be shot.

One of the agents who won some fame, although he was never very talkative about the incident, was a young Kabyle from Algeria named Mohammedi Saïd, alias Si Nacer. Because he agreed to play the French game under the supervision of the French Radiogoniometric Control Section, his life was spared and he was jailed for ten years. He was freed from prison in 1954, just in time to join the Algerian insurrection against the French, and as a leading figure of the Algerian maquis, always wore a Nazi helmet as a gesture of defiance. Once Algeria was freed, he became the Minister of Independence War Veterans. In the 1990s, before he died, Mohammedi Saïd became a senior *éminence grise* of the Algerian Muslim fundamentalist party, the Islamic Salvation Front (FIS). In 1991, the Algerian Army staged a coup d'état when it appeared certain that the FIS was about to win nationwide elections, and then banned the FIS. By 1992, a full-fledged civil war was raging in Algeria between the Algerian military and Muslim fundamentalists. This bloody war continues to this day.[10]

As France was being liberated during the summer of 1944, the French government began creating a new intelligence organisation by fusing together the pre-war services, the former Free French intelligence services from Algeria and Britain, and the internal Resistance. In November 1944, this organisation became the Direction Générale des Etudes et des Recherches (General Directorate for Studies and Research, or DGER). Its first director was Jacques Soustelle, who was subsequently replaced by André Dewavrin ('Passy'). Colonel Jean Fleury naturally became head of the DGER Technical Directorate (cipher and radios). The newly Service Technique de Recherche (Technical Research Service, or STR) was organized by Commandant Gustave Bertrand, who as detailed above, had played a key role as head of Section D in the prewar SR and in the Resistance.

The DGER, with 10,000 officers and NCOs, many of whom were latecomers to the Resistance, became notorious for its corruption and inefficiency. Its Directorate of Technical Control (DCT) interfered with telephone communications within France for political motives. The US forces in France also took the opportunity to set up some discreet intercept bases in places like Marseilles in the south of France. Lieutenant Stasia O'Neill, an Irishwoman who had served in the Resistance in Southern France, was attached to the US Counterintelligence Corps unit in Marseilles in 1944, and later recalled in an interview with the author:[11]

> Our service had as its cover name Signal Communication Installation of Delta Base Section. I worked under Colonel Ira P. Doctor of the Signal Corps, and Captain Fitzpatrick of the CIC (Counterintelligence Corps). Our entire set-up in the Rue de la Bourse at Marseilles was linked to the telephone exchange. Sixty-two persons were then working day and night at the exchange. We listened in to people suspected of collaboration during the war, and also to Resistance groups or French political groupings. The section for civilian enquiries (CID) was concerned above all with the 'purge' of collaborators, and the Information Service with the provision of intelligence to the secret services in Washington.

As time went by, a good working relationship developed between the US and French intelligence services. At the time that the Central Intelligence Agency (CIA) was formed in 1947, its Paris station, headed by Philip Clark Horton, began distributing selected American Sigint materials to the French intelligence services.[12]

FRANCE, SIGINT AND THE COLD WAR 183

THE BIRTH OF FRENCH POST-WAR SIGINT

After the end of the war, the wartime BCRA leader, Colonel Dewarin ('Passy') was asked by General de Gaulle to reorganise the DGER and thwart Communist infiltration within the French intelligence community. By order of a ministerial decree dated 28 December 1945, Dewarin founded a new French foreign intelligence service, the Service de documentation extérieure et de contre-espionnage (SDECE).

The SDECE's headquarters was established inside the Caserne Mortier on the Rue Tourelles in downtown Paris. All who worked inside SDECE headquarters referred to the place as 'La Piscine', which in English translates to 'The Swimming-Pool'. Within the SDECE there were two units that were responsible for the service's communications and signals intelligence activities.

The first was the 'Service des matériels techniques', or the SDECE Technical Services (known within the SDECE as 'Service 26') under the command of Colonel Paul Arnaud, who ran this department until 1964. Arnaud supervised sections such as 26/2, under Navy Captain Ducorps, which was in charge of communications between SDECE headquarters and its overseas stations. Section 26/4 under two civilians named Massport and Christin, was responsible for protecting the communications of the French deputy military attachés and 'chargés de mission' in French embassies overseas, who were in fact SDECE undercover operatives. Service 26 had 12 powerful radio transmitters at the SDECE station in Alluets-le-Roi near Paris, which were responsible for handling the service's overseas communications traffic, as well as occasionally engaging in Sigint intercept.[13]

Service 26 worked closely with 'Service 28', or Technical Research Service (Service Technique de Recherches, or STR), which was the SDECE unit responsible for intercepting and deciphering foreign diplomatic communications traffic coming in and out of the embassies in Paris and elsewhere. The section's headquarters was located at 9 Avenue Maunoury in downtown Paris, but its main technical station remained at Le Mont Valérien outside Paris, where the linguists and technicians worked. Service 28 was led by the legendary Gustave Bertrand, who had since been promoted to General. It was Bertrand who, in October 1948, through the head of Swiss intelligence, Colonel Max Waibel, established the first contacts with General Reinhard Gehlen, the former German military intelligence officer who was to found the Bundesnachrichtendienst (BND) in 1956. Shortly thereafter, the two former enemies began exchanging Sigint intercepts from the Soviet bloc. The Cold War may have begun in 1947, but both men had fought the Soviets long before.[14]

Bertrand's deputy was Lieutenant Colonel Black, who had organised the 'I' Naval Station in Morocco in 1937, which monitored Axis and Spanish naval and shipping traffic. A cipher specialist, he had stayed in North Africa and helped in the radio wireless war as head of the Centre Interarmées d'interception radiogonio et de décryptement (Inter-Army Center for Radio Direction Finding, Intercept and Deciphering, or CIIRD) in Rabat before rallying with his Center to the Free French cause in 1943. When Bertrand, then SDECE deputy director in overall charge of the agency's Technical Department, quit in 1949, Black replaced him. Unfortunately, Black had to resign when the French internal security service felt that French codes may have been compromised by Black's secretary, who was suspected of being a Soviet agent, recruited in Algiers by the NKVD in 1942.[15]

But the SDECE's Sigint responsibilities were limited to diplomatic communications intelligence. All other Sigint collection and processing functions were assigned to a new organisation. On 15 March 1946, a provisional government decree revived the former Vichy French Comint organisation, the Radioelectronic Communications Group (GCR). It was placed under the Président du Conseil (the French Prime Minister), and had a budget of FFr 450 million. The GCR reactivated the old Sigint collection and processing center at Le Mont Valérien near the town of Suresnes, where hundreds of French Resistance fighters had been executed by the Nazis during World War II. The former special radio school at Hauterive near Vichy was also reopened to train French radio intercept technicians for work with the GCR in France and abroad.[16]

The first Director General of the GCR was a French Army officer, Colonel Bodin. All senior management positions within the GCR were held by French military personnel, although a significant portion of the GCR's professional staff were civilians. The GCR was a rather large agency by the standards of the day, even when compared with GCHQ in Britain at the time. On 3 October 1948, according to the well-informed satirical weekly magazine *Aux Ecoutes*, a law was passed earlier that year that provided for a supplementary appropriation of FFr 542 million to pay the salaries of the 2,900 military and civilian staff who worked for the GCR.[17]

The GCR operated a world-wide network of radio intercept and direction-finding stations. According to a declassified report, as of October 1947, the GCR was operating listening posts in Paris and elsewhere in Metropolitan France, as well as overseas at Algiers (Algeria), Ajaccio (Corsica), Saigon (Vietnam), Noumea (French Polynesia), Tunis (Tunisia), Rabat (Morocco) and Dakar (Senegal). Also at about this time, the SDECE opened a small listening post within the French sector of West Berlin.[18]

The self-reliant GCR worked almost exclusively for the SDECE. The GCR was described in a May 1950 French National Assembly financial report as a 'service that centralises in principle all radio listening and direction finding (DF) resources in France and the French Union. Placed under the Prime Minister's office, this service works roughly 80 to 90 per cent of the time to fulfill the SDECE's needs.' The remaining 10 to 20 per cent of the GCR's intercept operations were for the French Army General Staff.[19]

Another important aspect of SDECE radio operations consisted of sending coded messages to French agents operating behind the Iron Curtain. At this time, the SDECE began parachuting agents into Yugoslavia, Czechoslovakia and Albania. Messages destined for these agents were sent from a SDECE radio station hidden within a mansion at Montmorency, northwest of Paris. But the agent insertion operation, called Operation 'Minos', failed and the police and internal security operatives of the Eastern European nations caught the agents as they landed. A subsequent internal inquiry by the SDECE concluded that the operation had been betrayed by someone within the small radio team at Montmorency, or elsewhere within the SDECE leadership.[20]

From the time that French troops began occupying a portion of Germany in 1945, the SDECE maintained an intelligence collection unit in its occupation sector called the Direction de la Recherche en Allemagne (Research Directorate in Germany, or DRA), which included a large and important Sigint intercept station at 'Camp Napoléon' in West Berlin, whose antennae were directed at communications traffic coming from East Germany. By the mid-1950s, the SDECE Berlin listening post was successfully intercepting a wide range of East German military and security communications traffic.

This station worked hand-in-hand with specialized French Army intelligence units in the French occupation zone of West Germany, such as the French Army listening post outside Landau that was operated by the 44th Régiment de Transmissions (44th RT), which monitored Soviet military radio transmissions coming from East Germany and Czechoslovakia; as well as an HFDF outstation at Appen, northwest of the city of Hamburg. By 1954, the French Air Force's Sigint organisation in West Germany, Group Electronique 30/450, activated an HF listening post on top of a mountain near the town of Achern opposite the French city of Strasbourg. HFDF outstations were established at Freiburg and Furstenfeldbruck near Munich. Also in the early 1950s, the French Army's 42nd Régiment de Transmissions and a detachment of the French Air Force's Groupe Electronique 30/450 established a listening post in a specially constructed tower on the grounds of Berlin-Tegel Airport in West Berlin.[21]

The SDECE, the GCR and the military were not the only French government agencies engaged in Sigint collection. On the other side of the intelligence spectrum, the civilian counterintelligence agency, the Direction de la Surveillance du Territoire (DST), which fell under the control of the Interior Ministry, had its own counterintelligence radio-electronic monitoring system. On the fourth floor of the DST headquarters on the Rue des Saussaies in Paris, a former lawyer named Paul Berliat set up a unit called the Police de contrôle radioélectrique (PCR), which operated intercept stations in Boullay-les-Trous (Yvelines Prefecture) and Limours (Essonne Prefecture) that monitored foreign radio programs as well as clandestine radio transmissions.

With a few so-called Centre d'Ecoutes et de Radiogoniométrie (Listening and Direction Finding Centers, or CER) spread throughout France (specifically at Nancy, Marseilles, Rennes, Toulouse and Lille), the DST intercepted coded messages between foreign radio stations and clandestine agents operating in France. Among their many successes, the DST intercepted the radio transmissions between the Czech capital of Prague and the underground Spanish Communist Party in France. The DST under Roger Wybot, whose actions and psychological make-up bore a striking similarity to his American counterpart, FBI director J. Edgar Hoover, also managed to bug the French Communist Party headquarters in Paris. In 1949, the DST was concentrating on monitoring the activities of the French Communist Party's Secret Radio Communications Section, which had been run since the Nazi occupation by Auguste Lecœur. Lecœur was later expelled from the French Communist Party in the wake of a Stalinist purge in the 1950s because of allegations that he was a 'police informer'.[22]

THE 'DALAT MONKS' OF THE SDECE IN INDOCHINA

In the 1950s, the decolonisation process led to the increased use of Sigint as the French Army sought to combat armed nationalist movements in French colonies around the world. At the end of World War II, French military and special forces returned to Indochina, initially to fight the Japanese forces occupying the former French colony in Southeast Asia. After Emperor Hirohito's surrender in September 1945, the French forces became involved in a bloody and protracted war against Communist guerrilla forces in Vietnam, who called themselves the Viet Minh. The political leader of the Viet Minh was Ho Chi Minh, a formidable strategist who received at different times military support from the Chinese, the Soviets and even the Americans in his fight against the Japanese. His military commander was a talented soldier and extremely capable strategist named General Vo Nguyen Giap.

Because of the difficulties inherent in managing and coordinating a large-scale military campaign against the colonial French forces, the Viet Minh depended to a great degree, but not entirely, on HF radio communications to direct their conventionally-structured military forces deployed throughout North and South Vietnam. Much of Vietnam's terrain was hilly or mountainous and heavily vegetated, and the Communist forces were mobile guerrilla units who usually operated far from their parent unit. The Viet Minh regiments and lower echelons of command therefore used lightweight, low-powered radio sets for their communications which they had either captured from the French or been given by the Chinese. Because of their low power, these radio sets could not be heard in Hanoi, much less the neighbouring province, which made the job of intercepting their transmissions exceedingly difficult. The Viet Minh also employed at an early stage sophisticated cipher systems to protect their most important radio traffic from prying ears. Despite these precautions, the French were able to solve many of the cryptographic systems used by the Viet Minh.

Vietnam was an extremely difficult environment for conducting effective Sigint collection, not just for the French, but also for the American forces who followed them. Despite the Viet Minh's communications security procedures, the French Comint intercept service in Indochina, the GCR was able to collect large amounts of Viet Minh radio traffic, and the French cryptanalytic agency in Indochina, the Services Technique des Recherches (STR), managed to solve the Viet Minh logistical code, as well as part or all of several Viet Minh operational codes used between 1947 and 1957. The French were also able to track the movements and activities of Viet Minh combat units because the Vietnamese radio operators violated every tenet of good communications security procedures by indiscriminately using their radios. A 1952 French report stated that: 'The study of enemy radio is by far our best source of intelligence.' But the success and the amount of material produced by the French Comint organisations in Indochina was so great that it outstripped the ability of the available French intelligence analysts to adequately process the Viet Minh intercepts.[23]

Codebreaking was not the only French success story in the intelligence war against the Viet Minh. As early as 1950, the French Air Force had mounted World War II-vintage radio intercept receivers and direction finding equipment in an American-made PBY Catalina amphibious patrol aircraft, which the French used to locate Viet Minh targets for parachute assaults. In 1953, the French abandoned the slow and ungainly Catalina flying boat, and put radio direction-finding gear on Nord NC.701 Martinets and a Beech C-45 twin-engined transport aircraft in order to monitor Viet Minh radio transmitters. Although a small effort, the French ARDF program was sufficiently successful that captured Viet Minh documents ordered radio operators to maintain radio discipline in order to counter the French ARDF threat.[24]

The successful French Comint collection effort against the Viet Minh in Indochina by the Services Techniques de Recherches (STR) and the GCR was directed by a specially created organisation in Indochina called the Direction générale de la documentation (DGD), headed by Colonel Maurice Labadie from his headquarters in Saigon. The SDECE in Indochina, led by Colonel Maurice Belleux, formed its own special Sigint organisation at Dalat in the Vietnamese Highlands with personnel seconded from Service 28 in Paris. By the mid-1950s, the SDECE Sigint organisation in Indochina had become a vast organisation, supported by radio direction-finding materials from listening posts run by France's foreign partners in Hong Kong and Taiwan. As Belleux explained to the author: 'At Dalat, dozens of men, grouped into an interception unit, devoted themselves to the task. For every code, I set up two-man teams. They lived like monks; they were separated off, working feverishly, shut away until they had succeeded in breaking "their" code. Then, and only then, did I send them off on leave to relax and rest. And through this system we obtained a 60–80 per cent success rate: the messages of the Viet Minh General staff were decoded. Then we passed on the reports, so that they could be acted on. Unfortunately, at GHQ there were many recipients who did not believe the reports. It was absolutely incredible. General Raoul Salan [commander of French forces in Indochina] was one of those who gave us due credit for our work.'[25]

Thus, at the time of the Cao Bang disaster in 1950, the SDECE was able to report on the basis of Sigint that eight Viet Minh battalions were being trained in the newly-formed People's Republic of China. However, in high places within the French military in Indochina, the enemy was underrated. In the end, in spite of the fact that the SDECE was able to successfully monitor and exploit Viet Minh communications, the superior forces and the brilliant skill in 'People's Warfare' by General Vo Nguyen Giap ensured the encirclement and ultimate destruction of a French expeditionary force at Dien Bien Phu in 1954. Just before Dien Bien Phu, General Giap had learned of the successful French Sigint operations and had changed his operational codes.[26]

Lieutenant General Wang Zheng, head of the Chinese People's Liberation Army 3rd Department and the Special Nanyang Radio School, helped the Vietnamese develop their own Sigint effort. But at Dien Bien Phu, despite Soviet and Chinese support and assistance, the Viet Minh could not understand the radio transmissions by the French radio operators, many of whom came from Brittany and used their Celtic language when talking on the radio.[27]

FRANCE, SIGINT AND THE COLD WAR						189

In spite of the mistrust between the French General Staff and the US, partly due to OSS support for Ho Chi Minh ten years earlier during World War II, there was a limited intelligence sharing agreement between the French forces in Indochina and the US intelligence community. Early in 1951, the CIA provided the French with 100 radio intercept receivers for use in Indochina, and an informal agreement was reached whereby the French, British and Americans split up the responsibility for Comint coverage of China and agreed to share Comint product from these operations.

There also appears to have been an unwritten Comint-sharing agreement between the French and American armies, whereby in return for HFDF bearings from the US Army Security Agency (ASA) listening post at Clark Air Base in the Philippines, the French STR passed to the American embassy in Saigon copies of finished Comint reports gathered by French military listening posts in Vietnam and Laos.

After the French defeat at Dien Bien Phu in 1954 and the subsequent withdrawal of French forces from Indochina in 1956, a 40-man French Army Comint unit remained in Indochina to run five intercept and HFDF posts. But in mid-1957 the French Comint personnel left the country after disagreements arose between the French and the new president of South Vietnam, Ngo Dinh Diem, who had formerly been an agent in the pay of the dreaded Japanese secret police, the Kempeitai, during World War II. Three of the French HFDF sites, at Rula Ray near Quang Ngai, Con Son Island, and Ha Tien were taken over by the South Vietnamese in July 1954. These sites used old American AN/SCR-255 HFDF sets which the Vietnamese inherited from the French, but these were World War II-vintage pieces of equipment which had long since been phased out of service in the US because they were inaccurate, unreliable and difficult to maintain. The South Vietnamese were never able to get these sites operational after the French departed and they soon fell into a state of disrepair.[28]

THE COMMUNICATIONS WAR FOR ALGERIA

At the same time that French forces were being defeated in Indochina, in late 1954 another insurgency movement flared up in a French colony, this time in French North Africa. Some of the Algerian nationalist leaders had been recruited by the Nazis during World War II, as we have seen in the case of Mohammedi Saïd. Others, like the future President of Algeria, Ahmed Ben Bella, had fought in the French Army in Italy and helped liberate France. After the war, they did not understand why their own country was not given its independence by France. This was the root-cause of the November 1954 insurrection in Algeria by the 'Special Organisation', which became the Algerian National Liberation Front (FLN), which fought an eight-year war against the French, without mercy shown on either side.

Four years later, on 21 September 1958, two young women belonging to the FLN tried to blow up the Eiffel Tower. They knew that on the third floor of the tower's observation deck, the DST had installed radio transmitters to relay messages to DST mobile surveillance units, as well as jam shortwave FLN communications, including radio transmissions such as the 'Voix de l'Algérie combattante' (Fighting Algeria's Voice) that was beamed at the Paris region for consumption by Algerian immigrants living in the French capital. Besides, the blast would have been very demoralising for the French if it had succeeded. But the DST acted swiftly thanks to insider information and crushed the plot.

This was all part of what was known at the time as the 'war of the radiowaves'. The SDECE organised 'black radio' stations, such as the Arabic-language 'Voice of Reform', which was meant to support a pseudo-group called the Movement for Algerian Renaissance (MRA) that had split from the main Algerian resistance organisation, the 'Mouvement national algérien' (MNA), which was a left-wing Algerian party founded by the 'father of Algerian nationalism' Messali Hadj, which opposed terrorism as a means to gain independence.

In August 1958, another 'black' radio transmitter known as Shark al-Adna began operations from the French Riviera. Another station, codenamed 'Kléber', was set up in the town of Jouy-sur-Eure near Chartres, and broadcast programs in the Arabic and Kabyle languages under the supervision of François Bistos, the director of the SDECE archives. Bistos was later expelled from La Piscine in 1970 when it was discovered that he was a 'Romanian spy'.

Until 1962, the SDECE supplemented this with radio programs broadcast on medium waves towards Egypt using a 300-kilowatt radio transmitter in Marseilles (which is today the local State Television FR3 station), with the help of one of Egyptian leader Colonel Gamel Abdul Nasser's former friends who had been 'turned' by French intelligence.

The war in Algeria led to an expansion of the SDECE's Sigint efforts. In the early 1960s, the SDECE Sigint unit, Service 28, was renamed Section III, under the command of a Mr Blanchard. Section III's Sigint efforts were managed by the SDECE Technical Division (referred to within the SDECE as 'Service 5'), headed by Colonel Paul Arnaud. Despite the change in names, the unit's mission remained the same – the interception and decryption of foreign diplomatic communications, including those of the FLN. During the early stages of the war in Algeria, from 1954 to 1959, the SDECE Sigint organisation proved to be very efficient in intercepting and reading FLN radio traffic.

Then in 1959, as Constantin Melnik, then the French Prime Minister's intelligence adviser acknowledged, the FLN changed their codes and made them impossible to decipher. Despite their inability to read the encrypted FLN traffic, French human intelligence (Humint) together with traffic analysis of FLN radio messages revealed a political split within the FLN leadership, which was reflected in a fracturing of FLN radio networks inside Algeria, as well as between FLN bases in the neighboring states of Morocco, Egypt and Tunisia.[29]

During the Franco-Algerian war, a military intelligence unit called Services Techniques (Technical Services) was set up on the Rue Michelet in downtown Algiers under the command of a Colonel Tesseyre, who had formerly been a 2ème Bureau intelligence officer in Indochina. He supervised a GCR department in Algeria consisting of two Sigint units: (1) an intercept and deciphering unit; and (2) fixed and mobile radio direction-finding units, such as T26 direction-finding aircraft. This organisation worked under the direction of the Centre de Coordination Interarmées (Inter-Army Coordination, or CCI) in Algiers, led by Colonel Léon Simoneau, which worked closely with the SDECE's Northern African 'delegate', Jean Allemand (alias 'Colonel Germain'). Allemand, thanks to communications intercepts, was able to organize the first airplane hijacking in human history. He diverted to Algeria an Air Atlas DC-3 plane flying from Rabat to Tunis. That plane carried the most important leaders of the Algerian revolution: Ahmed Ben Bella, Muhammed Boudiaf, Hocine Aït Ahmed and others, all of whom remained in prison until Algeria obtained its independence in 1962.[30]

Meanwhile, Charles de Gaulle returned to power in 1958, and he asked the French intelligence services to 'act on all fronts', which included a mandate to spy on the American and British agents who supported Algerian independence. Long before France withdrew from NATO in 1966, the DST's American Section, codename 'Mejean', eavesdropped on the US embassy in Paris and other American diplomatic interests in France.[31] This was made easier by the modernisation that took place within the French intelligence community.

In the 1960s, the SDECE's Technical Service, led by Colonel Pelletier and then by Colonel Eugène Caillaud, was still responsible for coding and decoding messages between SDECE headquarters and its agents abroad. It worked closely with the GCR, – which still reported to the Prime Minister's office – whose 1,000 military personal continued to perform 'surveillance of the radio waves and the unmasking of clandestine transmissions to foreign countries'.[32] Many of the radio intercept operators came from the French Navy, which historically has produced the best-trained radio and intercept operators. This was exemplified by the career of Navy

Commander Michel Steichen, who became the operational head of the GCR at Le Mont Valérien during the 1960s, and remained in this position for a decade until the 1970s. During the early 1960s, 54 GCR Sigint operators and analysts were integrated into the SDECE headquarters as a separate 'Analysis Staff'. They worked so hard that that one of the chiefs of the SDECE Technical Department, a mathematician and cipher expert, died of a heart attack in his office at SDECE headquarters on the Boulevard Mortier trying to crack FLN codes.[33]

In 1959, at the height of the Algerian War, the French Intelligence Coordinator, Constantin Melnik, was given 'carte blanche' to organise a centralised phone-tapping system in France. To perform this mission, a new organisation called the Groupement interministériel des communications (Inter-ministerial Communication Group, or GIC) was formed, commanded initially by Colonel Labrosse, and later by Colonel Caillaud who had been seconded to the GIC from the SDECE. Working on requests from the SDECE, DST and other French police units, the GIC engaged in a massive eavesdropping effort within France, but engineers from GIC and the GCR worked occasionally on joint foreign intelligence projects, and signals technicians frequently transferred from one organisation to the other.

In October 1965, the Moroccan opposition leader Mehdi Ben Barka was kidnapped in Paris and probably murdered by King Hassan's secret service with the help of French policemen, 'spooks' and criminals. He was never seen again, and even today his ultimate fate remains an unsolved mystery in the history of the French Fifth Republic. The SDECE was indirectly involved, insofar as one of the men involved was a part-time agent (the term used was 'honorable correspondent') of the SDECE's Service 7. This was responsible for the acquisition of intelligence by surreptitious means, such as stealing or opening diplomatic bags, bugging hotel rooms, etc. As a consequence of a trial and scandal that affected the Gaullist regime, on 19 January 1966 responsibility for the management of SDECE was withdrawn from the Prime Minister's office – then Georges Pompidou, and instead placed under the auspices of the French Ministry of Defense. In fact, since then the chain of command has usually been short-circuited by the presidential adviser on intelligence and security matters. The fallout from the Ben Barka scandal momentarily disorganised the SDECE-GCR arrangement.

Another Sigint problem further undermined the SDECE's operational intelligence activities. It came from the Anglo-American UKUSA Pact and their decrypts of Soviet KGB communications traffic during World War II, the so-called Venona decrypts. Extracts from the Venona documents relating to France were handed over by the British internal security service, MI5, to their French counterpart, the DST. In this organisation an obsessive 'red mole hunter' comparable to James J. Angleton in the CIA, investigated the SDECE leadership as well as some of de Gaulle's close collaborators. This person was Marcel Chalet. That DST team felt, thanks to the Venona information and the confession of the KGB defector Anatoly Golitsin, it would be possible to unearth the Russian 'moles'. The end result was quite different, but pleased equally the KGB, the CIA and the DST: it disrupted the SDECE for at least five years, and created ill-will and distrust towards the French on the part of the NATO intelligence community for a long time.[34]

The irony was that because they trusted their US sources, the French security services genuinely tried to uproot the Soviet 'moles' in their midst. But by 1964, because of security considerations, the US had cut the French off from further access to American Sigint information.

ALEXANDRE DE MARENCHES SWALLOWS THE GCR

The man who purged the SDECE of suspect elements, and at the same time reforged the alliance with the US intelligence services was Count Alexandre de Marenches, who headed the SDECE for almost 11 years, from June 1970 to December 1981. One of De Marenches' first acts was to take control of the GCR. In July 1970, the some 1,000 GCR staff members were absorbed into the SDECE, which was officially placed under the command of the Ministry of Defense on 13 January 1971. Following the takeover, for a few years the GCR remained an autonomous service under the supervision of the SDECE Technical Division (known as 'Service 5'). The commander of the GCR at this time was Colonel André Benoît, who was also the director of the large Le Mont Valérien Sigint processing center and additional outstations, such as the station at the fort in Noisy-le-Sec outside Paris, which was also the headquarters of the SDECE Action Service.

Later in the 1970s, the GCR was fully merged into the SDECE Technical Division, which then consisted of all of the former SDECE and GCR Sigint assets. But the new division remained responsible for conducting strategic signals intelligence collection and maintaining various Sigint collection stations around the world.[35] The GCR fell under the 'Moyens techniques', or Technical Department of the SDECE, which in 1971 was commanded by the Colonel Delamalle and his deputy Colonel Dardot. At the time of the GCR's integration into the SDECE in 1970, the GCR consisted of about 1,200 technicians, 800 of whom worked directly for the SDECE, with the others employed by various other French intelligence units. The GCR was described in a 1973 newspaper report as follows: '... the Le Mont Valérien staff are monitoring all radio-electric signals [in order] to decipher them'.[36]

Another part of Alexandre de Marenches' reorganisation plan was, because of Marenches's close relationship with the Anglo-American intelligence community, he felt France could be part of the UK-USA pact without officially joining. Under Marenches, NSA, GCHQ and the reorganised SDECE/GCR restarted the process of exchanging Sigint information. The integration of the GCR into the SDECE created problems, principally because of the clash of cultures between the civilian-dominated GCR and the military-dominated SDECE management. Many in the GCR staff, which was reduced to 500, disapproved of the new administrative regime of the SDECE, which was much more restrictive than under the previous system when they worked under the Prime Minister's office and were treated as civil servants, just like teachers or postmen. In a battle reminiscent of the British GCHQ situation in the mid-1980s, the right to belong to trade unions was restricted. To ice the cake, a special SDECE investigative team felt that some GCR personnel had been recruited by the KGB in Berlin, which led to what the trade unions representing GCR employees described as a 'witchhunt'. The head of counterintelligence within SDECE – Colonel Chopin de Janvry – and DST security personnel interviewed the suspected GCR staff in such a brutal manner that some later committed suicide.[37]

The CGT trade union made public the suicide of three GCR engineers in the 1970s, one in Berlin (a senior officer named Georges Blanc), and the other two (Claudine Sokolowski and Edmond Dessille) who worked at the Sigint station at Domme in the Dordogne region. Georges Blanc had spent 35 years in the GCR, having joined the GCR in 1941. He was a longtime trade union member – just like Bernard Collin, the Berlin GCR controller – and he refused to formally join the SDECE and abandon his union benefits. Under constant pressure, he committed suicide in Berlin, but not before writing a damning letter addressed to SDECE's leaders. The SDECE security section had obviously tried to sidetrack people who could have been labeled as sympathetic to the Communist cause who were at the forefront of the anti-Soviet Sigint battle. In July 1978, the 'deep malaise' became a matter of public record when the daily newspaper *Le Monde* reported that it was affecting France's espionage services, whose activities were increasingly directed towards operations within France itself.[38]

Following the GCR scandals, in 1979 the staff finally was given the same status and benefits as the agents working for the Interior Ministry's Transmission Service (STI). Until the fall of the Iron Curtain in the late 1980s, the special SDECE Sigint unit in West Berlin, the 'Antenne du service technique de recherche avancée de Berlin' (Technical Service Station for Advanced Research in Berlin, or ASTRAB) was successful in intercepting high-level East German communications. However, since the

STASI was able to infiltrate the GCR and the local DST, the East Germans were able to protect their communications and even manipulate what ASTRAB was listening to. Later, in the early 1980s ASTRAB also monitored the communications of the Libyan Mission in East Berlin, and was able to determine that the Libyans were responsible for the 5 April 1984 bomb attack against US military personnel in the La Belle discotheque in West Berlin.

In spite of these problems, the Sigint efforts of the SDECE Technical Services met with success. For instance, the SDECE was able to forecast the 1973 Yom Kippur War thanks to the intercept of Egyptian military radio transmissions. During the Vietnam War, the GCR listening post at Larçay, near the city of Tours, was responsible for monitoring signals traffic coming from Southeast Asia, and it is reported to have monitored US communications traffic in the region. De Marenches also had a GCR intercept station built close to his holiday home on Saint-Barthelemy Island near Guadeloupe, which was conveniently located near the United States.

The Middle East was an important French Sigint target during the 1970s. A GCR intercept station at La Courtine in the Creuse region covered the Middle East. For example, the SDECE monitored communications between terrorists and the control tower of the Tehran airport where an Air France plane had been landing.[39] In 1979, De Marenches was given transcripts of Soviet army communications deployed along the Afghan border by his friend Brian Tovey, who was the director of Britain's GCHQ. The French Count immediately sent men to reinforce the SDECE Kabul station inside Afghanistan, and was able to confirm to French President Valéry Giscard d'Estaing and to his British friends that the Soviets were about to invade Afghanistan.[40]

Other French clandestine radio stations run by SDECE were sending 'personal coded messages' to underground agents on the other side of the Iron Curtain. One such SDECE unit broadcast Tyrolian music on 6425 Khz between 11.30 and 11.40 a.m. from the Chartres region, but it ceased its activities in 1975 after an article was published in a French magazine specializing in radio matters called *Interférences*.[41]

By the end of the 1970s, the intercept and direction-finding stations belonging to the GCR Sigint network had been expanded and reinforced, such as the construction of the new Poncharramet listening post in the Haute-Garonne region, and the expansion of the independent intercept site at Domme in the Dordogne region. The Alluets-Feucherolles intercept station in the Yvelines region west of Paris, which specialised in the intercept of international telephone calls and faxes, was also modernised in the 1970s. Another station in a lovely seaside resort in northern Brittany was located opposite the GCHQ Morwenstow station in Cornwall, and interfered with British communications. However, it was subsequently closed down and the station's buildings were subsequently used as a holiday home for retired SDECE agents.[42]

In spite of the improved relations between France and the UKUSA nations, in 1973, under President Georges Pompidou, an important step was taken towards attaining a greater degree of autonomy when Paris decided to initiate a national cryptographic industrial effort – thanks in large part to the efforts of the Thomson-Sagem company – to manufacture electronic equipment in order to preserve France's technological independence. Under President Valéry Giscard d'Estaing (1974–81), several successful covert operations were conducted by the French in Africa thanks to advanced reconnaissance work by the long-range reconnaissance personnel of the French Army's 13th Regiment of Parachute Dragoons (13e RDP) and Sigint collected by French Breguet Atlantique and Mirage IV reconnaissance aircraft. This was the case during the 1977 Shaba revolt in Zaire, and during military strikes against the Algerian-backed Polisario guerrillas, who wanted the Western Sahara to secede from Morocco.

From that period onwards, French military Sigint collection units called Détachements autonomes de transmission (DAT) became very active in Africa, especially under the leadership of a SDECE Sigint expert named Colonel Jubault. They also benefited from intelligence information collected by the French Sigint listening post at Bouar in the Central African Republic, where that nation's ruler, 'Emperor' Bokassa, was President Giscard's personal friend.

REORGANISATION UNDER FRANÇOIS MITTERRAND

In May 1981, François Mitterrand was elected the President of France, and a new reorganisation of the SDECE was started. By presidential decree, dated 4 April 1982, the SDECE was renamed the Direction generale de la securite exterieure (DGSE). Four years later, on 10 July 1985, DGSE Action Service operatives sank the Greenpeace ship *Rainbow Warrior* in the harbor of Auckland, New Zealand. The subsequent scandal did not solely disrupt the Action Service. The DGSE's chief Admiral Pierre Lacoste was forced to resign, and his successor, General René Imbot, launched yet another purge of the service, which 'La Piscine' seems to undergo every ten years or so.

One of the victims of the purge was Henri Serres, a telecommunication engineer and head of the SDECE Technical and Computer Department, who was largely responsible for expanding and modernising the GCR during the 1970s and 1980s.[43] Thanks to substantial increases in the DGSE budget (from FFr 62 million in 1981 to FFr 310 million by 1988), new hardware and software was acquired which improved French Sigint operations.

Among the improvements was the introduction of a new collection system designed to intercept transatlantic satellite communications; and the development of new computerised intelligence collection systems that permitted the intercept of foreign phone calls, telexes and faxes on the basis of pre-selected and programmed keywords. It should be pointed out that this technology was nothing new for the specialists from the American NSA or British GCHQ.

Under President Mitterrand, just as it had been under his predecessor Giscard d'Estaing, French special operations in Africa continued unabated. In 1987, the operation to combat Colonel Gaddafi's Libyan troops in Chad, Operation 'Epervier', was supervised by elite intelligence personnel from the French Army's 11è Choc. Also involved in these operations were French Navy Breguet Atlantique Sigint collection aircraft. As early as November 1984, a Breguet Atlantique aircraft was being used by the French intelligence services as a mobile headquarters in Chad, carrying portable equipment for communications and electronic intercept. The intercepted Libyan transmissions monitored by the aircraft were immediately decoded and translated by Arabic linguists, which permitted more accurate and efficient attacks on the Libyan forces operating in Chad.[44]

In January 1988, French Air Force General François Mermet became the new Director-General of the DGSE. This former flight commander was very keen to further modernise 'La Piscine's' communications system. Rumour had it that he wanted to set up a [global Sigint collection] system similar to that of NSA. This was utterly impossible from a financial point of view, although he did manage to get a substantial increase in the DGSE's budget for technical intelligence. Likewise, the idea of merging the GCR with the DST's counterintelligence Sigint intercept unit, the PCR, was dropped.

However, it is true that new Thomson-CSF equipment for deciphering codes was bought under his tenure as chief of the secret service. The main target still remained the Soviet Union and its Eastern European satellite countries. Yet these communist countries were to survive for only three more years. The Middle East seems to have been another priority target. Indeed in February 1988, as General Mermet was returning from a visit to the United States to strike up a new partnership with the heads of the CIA and NSA, two of his DGSE operatives were shot dead in Beirut, Lebanon.

FRENCH ELECTRONIC WARFARE ASSETS

During the 1970s and 1980s, French Sigint operations were performed mainly by the DGSE, and to a far lesser extent by a French military intelligence unit called the Military Intelligence Exploitation Center (Centre d'exploitation du renseignement militaire, or CERM), which had been founded in December 1971 by Colonel Paul Bourgogne. The French military, however, did possess several extremely valuable special electronic warfare assets.

Since 1977, the French Air Force has operated a single four-engine DC-8 Sigint collection aircraft known as 'Sarigue' (Système aéroporté de recueil d'informations de guerre électronique), which was based at Evreux Air Base under the command of Escadron electronique 51 'Aubrac' (EE 51). This electronic aircraft, which was the only aircraft assigned to EE 51, was flown by the French civilian airline UTA before it was equipped with signals intercept and electronic countermeasures equipment.[45]

Since 1989, two Transall C-160 'Gabriel' electronic warfare aircraft have operated from Metz-Frescaty Air Base in eastern France. They belong to Electronic Squadron 54 'Dunkerque' (EE 54), which was formed in 1964. Able to fly at altitudes of between 7,000 and 10,000 meters, the C-160 Gabriel can intercept communications and radar emissions within a radius of 800 kilometers around the aircraft. One of these aircraft was based in Saudi Arabia in the autumn of 1992 to help reinforce the United Nations operation that banned flights over southern Iraq and enforce the embargo against the Saddam Hussein regime. It also supplements the operations of French AWACs aircraft, four of which had been purchased by France.

Prior to acquiring the 'Gabriel' aircraft in the 1980s, EE 54 used to fly eight Nord 2501 Noratlas transport planes equipped with special electronic countermeasures equipment to both collect electronics intelligence and confuse enemy forces.[46] According to veterans of these units, some of planes and helicopters belonging the EE 51 and EE 54 have been at times flown by a special aircrew detached from the most secret squadron in French aviation, Groupe Aérien Mixte 56 'Vaucluse' (GAM 56), located at Evreux Air Base, which is the DGSE's special air missions unit.

Likewise, the CERM and DGSE jointly operated a spy ship named the *Berry*. Built in 1958 under the name *Médoc*, she was bought by the French Defense Ministry in 1964 and based at Toulon. After ten years of service as a supply ship in the South Pacific, in 1974 the 2,700-ton *Berry* was modernised and converted into a spy ship. In the years that followed, the *Berry* sailed in the Mediterranean Sea off the coastline of the Middle East and in the Red Sea, all the while operating as a mobile Sigint intercept station.[47] In the summer 1987, the *Berry* sailed to the Persian Gulf and intercepted Iranian communications. During the 1990s, a crisis erupted in Algeria, prompted by the rise of Islamic fundamentalism in the North African country, which led to a military coup d'état in 1991. Throughout the 1990s, the *Berry*, under the command of Captain Jean Le Pivain, was often

FRANCE, SIGINT AND THE COLD WAR 199

positioned off the shores of North Africa intercepting Algerian military communications. To perform these Sigint collection operations, French officers and NCOs with special linguistic skills in Farsi or Arabic were embarked in the ship. These linguists were trained at the Strasbourg Intelligence and Linguistics School (EIREL). The intelligence collected by this French ship allowed analysts in Paris to draw a clear picture of the military clashes between the Muslim fundamentalist guerrillas and General Mohamed Lamari's Algerian government security forces.

Besides the *Berry*, the French Air Force's Sarigue and Gabriel spy planes, together with French listening posts in Arabic countries, such as Djibouti and the United Arab Emirates, were also important intelligence collectors on events in North Africa and the Middle East.

REDEPLOYMENT AFTER THE FALL OF THE BERLIN WALL

The fall of the Berlin Wall in 1989 forced the redeployment of French Sigint resources formerly deployed in Germany back to France. The Berlin GCR station (ASTRAB) was dismantled. The 44th Transmission Régiment (44 RT) at Landau in Bavaria, which formerly engaged in Sigint collection along the borders of East Germany and Czechoslovakia with its 'roving transmission teams' and 'technical analysis teams', was also returned to France. Its new home station is at Mutzig in eastern France.[48]

It usually worked hand in hand with the 54th RT at Hagueneau on the French side of the Rhine River, which was an electronic warfare unit. The 44th RT had a strategic mission, in charge of intercepting high-frequency radio communications from the Soviet Bloc. By the 1990s, the French Army Signals Corps included 23,000 soldiers, among whom were 2,000 personnel working directly in the field of electronic warfare, such as the 738th Electronic Warfare Company.

The Escadrons électroniques au sol (EES) in Bad-Lauterberg (near the Iron curtain), Berlin-Tegel (EES 02), Goslar (EES 03) (Harz) with an Elint Puma helicopter (HET), Furth-im-Wald (near the Czech border) and Achern (Schwarzwald) regrouped a hundred technicians who performed ELINT and Comint operations.

FRENCH POST-COLD WAR SIGINT

In 1989, exposés in the French press revealed that the DGSE had been spying on the United States, particularly collecting economic intelligence against American multinational corporations such as IBM, Corning Glass, and Texas Instruments. Some of these operations were conducted by agents assigned to the DGSE stations in New York City and Washington DC. But the majority of the intelligence collected was derived from Sigint intercepts of American commercial communications.[49]

Some press reports suggested that because of these revelations, the Director of the DGSE, General Mermet, was forced to resign. In fact, the changes at the top of the DGSE were more political in nature, prompted by the 1989 electoral defeat of Jacques Chirac and his conservative government. In March 1989, the new Socialist Prime Minister, Michel Rocard, replaced General Mermet with a civilian named Claude Silberzahn, who was just as keen as his predecessor to expand intelligence operations against the Americans, and he explicitly said so in the first press interview ever given by a French secret service chief to the Paris daily newspaper *Le Monde*.[50]

From this point onwards, the DGSE was managed by a collective leadership headed by a newly created Director of Intelligence (Directeur du renseignement). The first incumbent was a former DST official named Michel Laccariere, who was also the deputy director of the DGSE. His deputy, the diplomat Jean-Claude Cousseran, held the newly created position of Director of Strategy (Directeur de la strategie). In 2000, Cousseran became the head of the DGSE.

Under Silberzahn, the DGSE was once again reorganised, with Jérôme Ventre becoming the head of the DGSE Technical Department, which continued to supervise the service's Sigint collection operations. Silberzhan wanted to further develop French Sigint capabilities, especially against Africa. As he later explained: 'For the DGSE, technical intelligence chiefly entails the radiogoniometric [direction-finding] network, the system of communications interception (Comint), radioelectrical signals (Elint), photo interpretation (Imint) and penetration of computer systems.'[51]

Expansion of the DGSE's Sigint infrastructure continued in the 1990s. In 1991, it was publicly revealed that the DGSE intended to build a major Sigint station in the Camargues region in Southern France, with planning for the station having begun in 1987, was made public. In August 1991, however, following pressure from both local politicians and environmentalists, who claimed that wild Camargues ecosystem needed protection, a court stopped the construction, and the project was finally terminated in 1993.[52] But four years later, the DGSE was allowed to set up a 50-man new intercept station on the Plateau d'Albion in Haute-Provence, which would monitor communications coming from Africa, together with a 60-man intercept station at Saint-Laurent-de-la-Salanque, in the East Pyrenées.[53]

Meanwhile, Silberzahn's DGSE had to confront two major crises that changed the face of the world, as well as the traditions of the global intelligence community: the 1990 the invasion of Kuwait and the subsequent 1991 war against Iraq; and the same year the abortive coup d'état in Moscow that signaled the end of the Soviet Union. The Gulf War underlined for French intelligence officials US supremacy in the field of technical intelligence. For instance, shortly after Operation 'Desert Storm' began in 1991, the senior US defense attaché in Paris, Admiral Philip Dur, had an appointment with President Mitterrand to show him satellite pictures of the area where the offensive was taking place. This was helpful but humiliating for the French, since their Spot satellite was of little use in this type of a large, high-intensity military environment because of its lack of high-resolution capability. Besides, the French General Staff was always wary of information that they could not confirm with their own technical means and human intelligence sources. Some officials who happened to be pro-Iraqi even feared that that the US would try to manipulate their allies.

Yet the French played an important part in the overall Allied technical reconnaissance effort. Some of the Spot imagery was used by the Pentagon complementing the work of the US Keyhole imaging satellites, and the magazine *Armed Forces Journal International* later acknowledged that SPOT satellite imagery and photographs from the French F-1-CR Mirage reconnaissance aircraft had been used to good effect.[54] One of the French Air Force's new C-160G 'Gabriel' tactical Elint aircraft and the sole DC-8 'Sarigue' Sigint aircraft were sent to Al Hasa Air Base in Saudi Arabia in 1990–91 to take part in Operation 'Desert Storm'. There they liased with the GCR's Djibouti intercept station, but operated under the command of the CERM Special Electromagnetic Intelligence Cell.[55] The spy ship *Berry* also was sent to the Persian Gulf to monitor Iraqi communications.[56]

The Gulf War had revealed major holes in France's intelligence system. The Socialist Defense Minister Pierre Joxe and his intelligence adviser, General Philippe Rondot, a former SDECE officer who later attained worldwide fame by kidnapping the terrorist Carlos in the Sudan and bringing him to France for trial, decided to triple the size of the French military intelligence system in order to cut France's reliance on American technical intelligence information.[57]

One of the results of the war was that the military's small intelligence unit, the Military Intelligence Exploitation Center (CERM), then commanded by Colonel Maurice Castagne, was deactivated and replaced with a new and much larger military intelligence department, the Direction du renseignement militaire (DRM). The DRM would have a large Sigint capability, which would be augmented by the launch of the first French spy satellite, Helios.[58]

The former Director of Operations of the DGSE, General Jean Heinrich, became the chief of the newly created DRM, which immediately started recruiting some 300 intelligence specialists from the armed services and universities. In April 1992, the DRM was activated with two headquarters, one in Paris within the Chief of the General Staff's office, which handled administration; and the other at Creil Air Base in the Oise region, where the DRM's technical and operations directorates were located.

The DRM's main operations center is at Creil Air Base, where the following units are based: the Inter-service Helios Unit (Unité interarmées Helios), which is in charge of exploiting imagery intelligence collected by the Helios reconnaissance satellite; the Inter-service Imagery Interpretation and Information Center (Centre de formation interarmées d'interpretation de l'imagerie, or CF3I); the Centre d'information sur les rayonnements électromagnétique (CIREM), which processes and analyses Sigint; as well as various military intelligence communications units. The DRM also drew personnel and technical support from the Inter-service School for Intelligence and Linguistic Skills (EIREL) in Strasbourg.[59]

In 1995, General Heinrich was replaced as head of the DRM by General Bruno Elie, who commanded the unit until 1998. As of 2000, the DRM, which consisted of 1,710 military and civilian personnel, was headed by Vice Admiral Yves de Kersauzon, who formerly was the French military attaché in London. The DRM's Technical Sub-Division (Sous-direction Technique, or SDT), which supervises the command's Sigint operations, was commanded from 1992 until February 1999 by Geoffrey d'Aumale. His replacement was Jacques Bongrand, a longtime aeronautical engineer and former Defense Ministry official.[60]

The DRM Technical Sub-Division worked closely with the Special Electronic Warfare Center (Centre Electronique de l'Armement, or CELAR) in the small town of Bruz, near the city of Rennes in Brittany. CELAR, which is described as the French 'nerve center for digital warfare', consists of about 900 military technicians.[61] The DRM's Sigint intercept resources have grown dramatically since the command's formation in 1992.

Most of the DRM's Sigint intercepts come from 11 radio intercept detachments deployed overseas, called Détachements avances des transmissions (DAT), which are manned by some 300 personnel.[62] In Africa, two DATs are located in Djibouti and on La Reunion Island in the Indian Ocean.[63] In the late 1990s, the DRM built two new DAT intercept sites in French Guyana in South America and on the island of New Caledonia in the South Pacific.[64]

Another modernisation program dealt with replacing the spy ship *Berry* with a more modern vessel. In 1991, a new development project called '*Berry*-NG' (*Berry* – New Generation) was initiated. It was decided within the French intelligence community that by 1998, the *Berry* would be replaced by a brand new surveillance ship, whose codename was 'Argos'.[65] In 1999, after being modified for intelligence work in Brest, the new French spy ship, the 4,870-ton *Bougainville* replaced the *Berry* as France's primary sea-based intelligence collector. The *Berry* was decommissioned in May 1999. *Bougainville* is now homeported at the Toulon naval base in the Mediterranean. In the near future, *Bougainville* will set sail from Toulon on its first operational Sigint collection mission off the North African coast.[66]

As far as the public was aware, the *Berry* was officially the only spy ship operated by the DGSE. But in fact, the DGSE secretly operated another spy ship, the 500-ton *Isard*, which had operated from 1978 onwards along the African coast and in the Indian Ocean conducting Sigint collection as well as clandestine operations in support of friendly African heads of state.[67]

Also located at the Brest naval base was another French ship called the *Monge*, under the command of Captain Gérard Etienne, which was not a spy ship as such. Yet the 21,000-ton French Navy missile and space tracking ship worked for the DGA and the DRM on scientific projects and performed space-monitoring missions. For example, in March 1996 she received a special mission: to calculate precisely the splashdown point in the South Atlantic of the lost Chinese spy satellite 'Jianping'.[68]

The French Navy's nuclear-powered attack submarines also possessed a limited Sigint collection capability. During the 1999 war in Kosovo, the French attack submarine *Amethyste* was deployed off the Yugoslav port of Kotor collecting intelligence on the activities of the Yugoslavian Navy.[69]

In 1994, prior to the genocide that was to sweep Rwanda and result in the murder of hundreds of thousands of people, the DGSE's Bouar station in the Central African Republic was able to intercept the communications of the Rwandan Tutsi leader Paul Kagamé as he spoke over telephone circuits carried by Inmarsat communications satellites.[70] In 1997, French Sigint pieced together information showing attacks by Tutsi supporters of the future leader of Zaire, Laurent-Désirée Kabila, against Hutu refugees who had come from Rwanda.[71] However, the Bouar station was closed in 1997, and the Djibouti base had to be beefed up in order to maintain adequate Sigint coverage of Africa.[72]

The DGSE's Sigint efforts have become increasingly high-tech in the last decade. Today, in the basement of the DGSE headquarters on the Boulevard Mortier in Paris is a new Cray 1 supercomputer that is used to process and analyse intercepts. Two computer systems, called 'Taiga' and 'Noemia' allow DGSE analysts to screen large numbers of intercepts for intelligence comparable to NSA's larger Echelon system. One intelligence source has quoted the figure of 8,000 keywords being recorded in the 'Noemia' computer program, which involves 'hacking' into the computer data systems of foreign companies. 'Taiga' (Traitement Automatisé de l'Information Géopolitique d'Actualité) was developed in 1987 by a researcher named Christian Krumeich at the firm Thomson-CSF for the

DGSE, which initially provided essential computerised data on the former USSR. The military intelligence organisation, the DRM, has also bought a dozen 'Taiga' stations in order to further expand its Sigint effort.[73]

An imagery reconnaissance satellite called Helios was developed by the French defence contractor Matra, in conjunction with the Italian and Spanish governments at a cost of FFr 9.6 billion. A ground station was constructed at Torrejón Air Base, Spain to analyse the pictures taken by the Helios satellite. In the summer of 1996, the DRM at Creil delivered the first pictures from the Hélios-1-A satellite to the photo interpretation unit of the Western European Union at Torrejón. After its launch in August 1995, Helios was often used to provide imagery of military developments in Bosnia.

Helios also carried into orbit a Sigint intercept cartridge that allowed it to intercept communications, as revealed by the French investigative journalist Jean Guisnel.[74] But the French kept the Sigint intercepts obtained from the satellite for themselves and did not share them with their European partners.[75] Germany decided to withdraw from a program to develop and build a second European spy satellite, called Helios-2, (the total cost of the system was FFr 11.6 billion) after considerable pressure from the United States, including a visit from CIA director John Deutch to dissuade German Chancellor Helmuth Kohl from joining the program following the signing of the French-German joint-development deal in the autumn of 1995.[76]

Therefore, the French-German divorce in 1995 was an obvious setback. It also financially weakened other French intelligence projects, such as the Zenon Sigint satellite, which was then in the early stages of development. And as a consequence, without telling the Italians and Spaniards, the French place a Sigint cartridge built by Dassault on the satellite at a cost of FFr 1 billion.[77] The irony was that during the Cold War, the US and French intelligence services collaborated in the fight against the Soviet Union. But as soon as the Cold War was over, France turned its Sigint collection system against its former partners.

In a separate program, a special agreement was struck between the German BND and the DGSE on a Sigint exchange program. Under the aegis of the DGSE, the two intelligence organisations agreed to intercept communications traffic being relayed through civilian communications satellites in orbit over American territory.[78] All of these new developments allowed the DGSE and the DRM to offer a response and a viable alternative to the Anglo-American UKUSA global spying programs, such as Echelon. A good example was given when it was revealed that the in 1993, the French had intercepted the communications of a US ambassador aboard a C-141 Globemaster military transport plane as it flew over the Russian province of Nagorno-Karabakh.[79]

It was not the first time that the Anglo-Saxon Sigint community was forced to realise that the Cold War was really over, and that 'friendly spies' were quite active in a field, which until then had been publicly described as their main speciality. At least they, unlike the public, knew that the French had managed in the last ten years to develop a widespread and effective Sigint system that could compete with their former allies and play a leading role in the building of a European intelligence community.[80]

NOTES

1. The author published extracts of these intercepts in Roger Faligot, *Naisho, enquête au cœur des services secrets japonais* (Paris: La Découverte 1998). He was given access to them thanks to Colonel Cattiew, former chief of the SCSSI, who among other contributions published a 'Rétrospective de la cryptologie de 1928 à nos jours', published in bulletin No. 26 (1998/1999) of the Association des Réservistes du Chiffre et de la Sécurité de l'Information (ARCSI – Association of Cipher and Information Security Veterans).
2. In her book *La France gagne la guerre des codes secrets, 1914–1918* (Paris: Tallandier 1998), Sophie de Lastours describes, thanks to the long-overdue opening of the archives, the successes of French cipher specialists during World War I, such as Generals François Cartier and Maurice Givierge, as well the brilliant cryptanalyst Georges Painvin, who broke the famous 'Victory radiogramme' code, which allowed the Allies to defeat the last German offensive in 1918. When Ms de Lastours published her well-researched book, many in the international Sigint community demonstrated great interest, and NSA and GCHQ, both of whom are always interested in solving historical enigmas, ordered dozens of copies of the book.
3. In 1932, Joseph Ducroux, a French Comintern agent, was sent in Asia to find Cremet, who had vanished the previous year while smuggling arms to Deng Xiaoping's guerrillas in Guangxi Province in southern China. An opponent of Stalin, he faked his own death and subsequently lived under an assumed name in Belgium. For his part, Ducroux was arrested by British police in Singapore. Some messages in his possession were deciphered, and as a result the whole Shanghai network was dismantled and Sorge narrowly escaped arrest. In Hong Kong however, the leader of the Vietnamese revolution, Ho Chi Minh, was arrested as a result. All of these men, Cremet, Sorge, Ducroux, and Pan Hannian had been trained at the OMS Mount Lenin School, near Moscow, where clandestine cipher and radio techniques were taught. Interview with the Ducroux and Cremet families, 1988. See also Roger Faligot and Rémi Kauffer, *The Chinese Secret Service* (London: Headline 1989) and *As-tu vu Cremet?* (Paris: Fayard 1991).
4. For a more detailed discussion of the history of the DST and its predecessors, see Roger Faligot and Pascal Krop, *DST, Police secrète* (Paris: Flammarion 1999).
5. The best description of French Sigint following the fall of France is Martin Thomas, 'Signals Intelligence and Vichy France, 1940–1944', *Intelligence and National Security* 14/1 (Spring 1999) pp.176–200.
6. Interview with Marie-Madeleine Fourcade, 13 Feb. 1988.
7. A colorful history of Bertrand's cryptologic work during World War II can be found in his memoirs, Gustave Bertrand, *Enigma ou la plus grande énigme de la guerre 1939–1945* (Paris: Editions Plon 1973).
8. Col. A. de Dainville, *L' ORA, la résistance de l' armée/guerre 39–45* (Paris: Ed.Lavauzelle 1974). See also on Romon, *Revue des Transmissions*, Oct. 1946.
9. Confidential interview.
10. Written communication with the author from Maurice Vial, who was the French counterintelligence case officer of Mohammedi Saïd in Algeria, 22 Feb. 2000. See also on the North African nationalists' alliance with the Nazis: Roger Faligot and Rémi Kauffer, *Le*

Croissant et la Croix Gammée: Les secrets de l'alliance entre l'islam et le nazisme d'Hitler à nos jours (Paris: Albin Michel 1990).
11. Interview, Stasia O'Neill, April 1984. Ms O'Neill, who lived in Lyon, remained a staunch supporter of the Irish Republican cause until she died in 1998.
12. Interview with John Bruce Lockhart, former MI6 station chief in Paris.
13. Interview with Col. Paul Arnaud, 13 Nov. 1986.
14. Reinhard Gehlen, *The Service: The Memoirs of General Reinhard Gehlen* (London: Collins 1972) p.258.
15. Interview with former SDECE officer Marcel Chaussée, 18 July 1984, and confidential interviews. This story is still controversial in France since Black was considered a good cryptologist. For instance, a French cipher specialist, engineer Charles Eyraud later thanked Lt. Col. Black for his help while writing his book *Précis de Cryptographie moderne* (Paris: Editions Raoul Tari 1953).
16. The sports journalist André Passevant, who later worked for the Communist daily newspaper *L'Humanité*, recalled in his memoirs how he was trained as an intercept operator at Hauterive in 1946. Hereafter he was then was sent to the 8th Transmission Regiment to help set up a GCR intercept station in New Caledonia in the Pacific, which was commanded by a French resistance hero, Captain Delhommelle. André Passevant, *Même si ça dérange* (Paris: Robert Laffont 1976).
17. *Aux écoutes*, No. 1.264, 15 Oct. 1948.
18. *European Security and Police Organisations*, 15 Oct. 1947, RG-38, Translations of Intercepted Enemy Radio Traffic, 1940–1946, Box 2740, US National Archives, College Park, Maryland.
19. The French Union then included France's colonies in Asia and Africa. Financial Report on the SDECE by M. Guy Petit, French National Assembly, 12 May 1950.
20. Conversation with Joël Le Tac, former BCRA operative, then SDECE officer, and later a French member of Parliament. See Franck Renaud, *Le Breton de Montmartre* (Rennes: Editions Ouest France 1994).
21. Nicolas Fournier and Edmond Legrand, *Dossier E... Comme Espionnage* (Paris: Editions Alain Moreau 1978), and communication from Dr Erich Schmidt-Eenboom (see his contribution in this work on German Sigint).
22. Faligot and Krop, *DST* (note 4); Interview with Auguste Lecœur in 1984.
23. Douglas Porch, *The French Secret Services* (NY: Farrar, Straus and Giroux 1995) pp.299, 313–14; John Prados, *The Hidden History of the Vietnam War* (Chicago: Ivan R. Dee 1995) p.193 and Alexander Zervoudakis, 'Nihil Mirare, Nihil Contemptare, Omni Intelligere: Franco-Vietnamese Intelligence in Indochina, 1950–1954', *Intelligence and National Security* 13/1 (Spring 1998) pp.203–5.
24. David Donald, *Spyplane* (Osceola, WI: Motorbooks International 1987) p.53; Col. Alfred F. Hurley, *Project CHECO Report: The EC-47 in Southeast Asia*, 20 Sept. 1968, p.2, Air Force Historical Research Agency, Maxwell AFB, Alabama; MS, US Army Security Agency, *ARDF in Vietnam, 1962–1972*, p.31, via Matthew M. Aid.
25. Interview with General Belleux, 12 July 1984.
26. See Roger Faligot and Pascal Krop, *La Piscine: The French Secret Service Since 1944* (London: Basil Blackwell 1989) p.89. Raoul Salan had been an intelligence specialist in Asia before World War II as head of the Service de Renseignement Intercolonial (SRI).
27. Unpublished MS, Roger Faligot, *Biographical Dictionary of Chinese Intelligence Operatives*.
28. John D. Bergen, *Military Communications: A Test for Technology* (Washington DC: US Army Center for Military History 1986) p.20; US Army Security Agency, *When The Tiger Stalks No More: The Vietnamization of Sigint: May 1961 – June 1970*, p.1, NSA FOIA, via Matthew M. Aid.
29. Constantin Melnik, *Services 'très' secrets* (Paris: Editions de Fallois 1989).
30. Interview with Jean Allemand, alias Col. Germain, in 1984. Allemand died in 1994. See also Roger Faligot, 'Comment le colonel Germain a enlevé Ben Bella', *Historia*, No. 573, Sept. 1994.
31. Faligot and Krop, *DST* (note 4).
32. Bernard Chantebout, *L' organisation générale de la défense nationale en France depuis la*

fin de la seconde guerre mondiale, Librairie général de Droit et Jurisprudence (Paris: R. Pichon & R. Durand-Auzias 1967).
33. Melnik, *Services 'très' secrets* (note 29).
34. Faligot and Krop, *La Piscine* (note 26) pp.210–29; Interviews with Colonels Hounau, Lannurien and Delseny, all of whom were SDECE department heads in the mid-1960s.
35. Interview with Alexandre de Marenches.
36. *Paris-Match*, 30 June 1973.
37. Faligot and Krop, *La Piscine* (note 26) pp.254–55.
38. *Le Monde*, 24 Feb. 1978.
39. Confidential source.
40. Interview with the author. See also, Roger Faligot and Rémi Kauffer, *Les maîtres espions, Histoire mondiale du renseignement* (World Intelligence History, Vol. 2, foreword by Alexandre de Marenches) (Paris: Robert Laffont 1994).
41. *Interférences* No. 3, Autumn 1975.
42. The author knows the location of this station, but because of the current wave of Breton nationalist bomb attacks on French government targets in Brittany, the author has chosen not to reveal its location here.
43. Henri Serres later worked for Matra, the French affiliate of Computer Sciences Corporation, and in March 2000 he became the head of information systems security within the Prime Minister's office, the Secrétariat général de la défense nationale (SGDN).
44. Confidential source who participated in the operations.
45. Donald, *Spyplane* (note 24) pp.87–8; 'Les avions-espions francais Gabriel et Sarigue', *Air & Cosmos/Aviation International*, No. 1460, 21–27 Feb. 1994, p.26.
46. Donald, *Spyplane* (note 24) p.88; Paul A. Jackson, *French Military Aviation* (Earl Shilton: Midland County Publications 1979) p.53.
47. Paul Michaud, 'New French Spy Ship', *Defence*, March 1991, p.10.
48. For details on these French special intelligence units see Jean-Jacques Cécile, *Le renseignement français à l'aube du XXIe siècle* (Paris: Editions Charles-Lavauzelle 1998).
49. Michael Wines, 'French Said to Spy on OU Computer Companies', *New York Times*, 18 Nov. 1990. Jean Lesieur, 'Le scandale des espions français', *L'Express*, 25 May 1990. See Peter Schweizer, *Friendly Spies* (NY: Atlantic Monthly Press 1993). The author of this book reveals, among other things, that in the economic war, GCR facilities at Mont Valérien 'began intercepting commercial traffic between German companies' in 1967 (p.103).
50. *Le Monde*, 13 March 1993.
51. Claude Silberzahn and Jean Guisnel, *Au cœur du secret, 1500 jours aux commandes de la DGSE, 1988–1993* (Paris: Fayard 1995).
52. Jean Guisnel, 'DGSE: chantier secret-beton en Camargue', *Liberation*, 12 July 1991; 'La DGSE agrandit ses antennas en pleine Camargue', *Le Monde*, 13 July 1991.
53. Jacques Isnard, 'La DGSE installera un center d'ecoute sue le plateau d'Albion', *Le Monde*, 9 May 1997.
54. Jean-Paul Dufour, 'Le satellite français Spot a guidé les raids américains dans le Golfe', *Le Monde*, 26 May 1991.
55. Jeffrey M. Lenorovitz, 'France Uses C-160G Aircraft to Perform Elint, ESM Missions' *Aviation Week & Space Technology*, 21 Jan. 1991, p.62.
56. Michaud, 'New French Spy Ship' (note 47).
57. Percy Kemp, 'The Fall and Rise of France's Spymasters', *Intelligence and National Security* 9/1 (Jan. 1994) p.16.
58. Jean Guisnel and Bernard Violet, *Services secrets: Le pouvoir et les services de renseignements sous François Mitterrand* (Paris: La Découverte 1988).
59. Kemp, 'The Fall and Rise of France's Spymasters' (note 57) p.15.
60. 'France: Jacques Bongrand', *Intelligence Newsletter*, 25 Feb. 1999.
61. Roger Faligot, 'Inside the nerve centre for digital warfare', *The European*, 28 Nov. 1996. See also Jean Guisnel, *Cyberwars: Espionage on the Internet* (NY: Plenum Trade 1997).
62. Senat, Project de loi de finances pour 2000, Tome IV – Defence, Section II. Le Renseignement: Une amélioration des moyens qui doit être renforcée.
63. Confidential source.

64. Senat, Avis No. 88 – Tome IV – Project de loi de finances pour 1998, Section B: Les moyens des differents services de renseignement; Senat, Project de loi de finances pour 1999, Section B: Les moyens des differents services de renseignement; Senat, Project de loi de finances pour 2000, Tome IV – Defence, Section II. Le Renseignement: Une amélioration des moyens qui doit être renforcée.
65. Confidential naval information.
66. Jacques Isnard, 'Bougainville and its Big Ears Sail, Destination Unknown', Le Monde, 10 May 2000.
67. Confidential source. See also Jacques Isnard, 'La France admet officiellement l'existence d'un second bateau-espion au service de DGSE', Le Monde, 5 Jan. 1994.
68. Télégramme de Brest, 18 Sept. 1997.
69. Jacques Isnard, 'NATO Denies Keeping France in the Dark', Le Monde, 11 April 1999; idem, 'Franco-US Naval Battle During Kosovo War', Le Monde, 5 Jan. 2000.
70. Guisnel (note 61).
71. Confidential source.
72. Guisnel (note 61).
73. Ibid.; Alain Joannes, 'Dossier. Les nouveaux agents secrets de l'economie', Le Télégramme, 19 Aug. 1999.
74. Jean Guisnel, Les pires amis du monde: Les relations franco-américaines à la fin du XXè siècle (Paris: Stock 1999).
75. The head of the DRM, Admiral de Kersauzon, acknowledged in a press interview that the tension between France and its partners reached the stage where the DRM refused to provide the Germans with imagery of Kosovo and the Yugoslav Federation during the 1999 NATO operations in Kosovo. See, 'We do not show our pictures to our German allies', Liberation, 3 Dec. 1999.
76. John Fitchett, 'Help With a Spy Satellite: Germany Poised to Join French Project', The International Herald Tribune, 19 Oct. 1995.
77. Le Monde du renseignement/The Intelligence Newsletter, Paris, 30 March 1995. Website: www.intelligenceOnline.com.
78. Jean Guisnel, 'Espionnage: les français aussi écoutent leurs alliés', Le Point, 19 June 1998.
79. Guisnel, Les pires amis du monde (note 74).
80. Ibid.

7

Scandinavia, Sigint and the Cold War

ALF R. JACOBSEN

One brief look at the map of Scandinavia will immediately make clear why the Nordic countries during the Cold War became of increasing importance to Signals Intelligence, and why in certain areas, it was at the really sharp end of the business. Finland, Sweden, Norway and to a lesser extent Denmark are all tucked away in the far north-eastern corner of Europe, leaning against the former Soviet Union from Leningrad (now St Petersburg) in the south to Murmansk in the north – Finland and Norway sharing a thousand-mile-long common border with the Russian superpower. The importance of Scandinavia to Signals Intelligence therefore was, and still is, a function of geography.

Norway and in particular Finland, but also Sweden sat ringside to many of the main events, and indeed to the main enemy during the Cold War. Proximity to the Soviet Union transformed the Nordic countries into excellent platforms for the collection of intelligence about the Soviet military machine. This of course became even more acute as the Cold War heated up during the 1970s and 1980s and the Kola Peninsula was turned into one of the largest military bases in the world. It was home of the huge Northern Fleet with its surface ships and nuclear submarines. The large island of Novaya Zemlya, the White Sea and the eastern parts of the Barents Sea became the main testing ground for new Soviet weaponry, especially Submarine Launched Ballistic Missiles (SLBMs) and cruise missiles.

The odds seemed overwhelming. Among the forces confronting the Nordic countries in north-western Russia, one had the forward elements of the Sixth Army with headquarters in Petrosavodsk outside Leningrad, including one infantry division and numerous Army and Navy assault brigades and amphibious forces spread across the territory.

On the Kola Peninsula itself more than 20 airfields with hundreds of bombers, helicopters and fighters lay within striking distance of the borders.

Along the coast, stretching from Gremikha in the east to Zapadnaya Litsa in the west, was the enormous Northern Fleet base complex growing. Hundreds of ships, including the *Kiev*-class carriers, the *Kirov*-class battlecruisers and the feared 'Typhoon' and 'Delta' -class rocket-firing nuclear submarines were at its centre, constituting the main Soviet submarine strategic deterrent.

Further to the east lay the White Sea and Archangel area with its Red Navy shipyards, the Nenoksa naval missile testing range and to the south of Archangel, the Plesetsk space centre, from where Intercontinental Ballistic Missiles (ICBMs) were test fired across the Soviet continent and into the Pacific.

Further to the south lay Leningrad with the Baltic Fleet and the whole Western theatre stretching through the Baltic States, Poland and East Germany right down to the Wall. One enormous military camp with its guards armies, its air fleets and its missiles ready to strike. It was a formidable and fearsome arsenal only minutes away from the Scandinavian countries, the nightmare scenario being a swift attack cutting off northern Norway in the north and sealing off the Danish straits in the south in the opening stages of a conflict.

A BALANCING ACT

As border states, squeezed between the two superpowers, the United States in the west and the USSR Evil Empire in the East, the Scandinavian countries had to conduct a very careful balancing act. On the one hand doing their outmost *not* to provoke the Soviet giant and on the other hand always sweeten and flatter Whitehall and Washington, which in the end provided the only guaranty of security. This policy was expressed in the Norwegian 'base declarations' and the Swedish concept of neutrality.[1] As one element of this balancing act, it was important to keep the big powers out of the border regions. In particular was this the case as regards their intelligence services, which were perceived as gung-go and trigger-happy without any deeper understanding of local affairs. To keep them *out*, and at the same time keep the Allies *in*, the Nordic intelligence services had to do the job themselves and to prove that they could deliver the stuff the Americans and others wanted. The Scandinavian countries had to build up big and technologically-advanced intelligence services but that could only be done with access to US hi-tech and US dollars.

It is therefore fair to say that the Nordic countries, with the exception of Finland and to some extent Sweden, more or less became American client states in the field of intelligence, large scale purveyors of intelligence raw materials. In exchange they would be placed under the US atomic umbrella

SCANDINAVIA, SIGINT AND THE COLD WAR 211

and receive huge and secret transfers of equipment and money. In this respect one may for example argue that the secret daily stream of intelligence data across the Atlantic from Northern Norway alone was one of Norway's main export articles during the Cold War, and also a considerable earner of much needed foreign exchange, millions of dollars being put into the secret accounts of the Defence Intelligence Staff in the Bank of Norway.

Today, Scandinavia's old and longsuffering neighbour Russia is down on her knees, her immense resources having been sold off at give-away prices to her new gangster elite, a tragedy which to some extent happened on the basis of advice by Wall Street. Accordingly the United States has lost much of her interest in the intelligence data she once got from Scandinavia. This raises another question of even greater importance to small nations. To what extent are the Nordic countries now able to conduct their own independent foreign policy, based on sound and high quality political and military information from their intelligence services, when these services have less to offer the US, and therefore, one may suspect, not any longer get back the intelligence from other and more important areas of the world which they used to do?

SOURCES

I am a writer, not a historian. Although an investigative journalist applies more or less the same methods as a historian, I shall in this contribution approach my theme with the liberty of a contemporary journalist. For one reason mainly: much of what we know, especially from later years, are learnt from oral sources; officers and politicians who are still bound by strict secrecy laws, and whose identities therefore cannot be revealed. This again means that what they say cannot be properly tested against written sources. I will also dwell mostly on developments in Norway, which I know best, and where a number of historians and parliamentary investigators have been given access to the officers and archives of the Defence Intelligence Staff.[2] It is fair to assume that Signals Intelligence in Scandinavia more or less developed along similar lines, with Finland forced to downplay her role due to her closeness to Russia, and Denmark playing a somewhat different role because of her closeness to Germany and Poland.

THE ROOTS

The 1920s and 1930s saw an awakening in Scandinavia to the possibilities and dangers of the telegraph and the radio. As news spread about the success of Sigint during World War I, in which the Nordic countries remained on the sidelines, a new consciousness developed among scientists, military officers and diplomats as to the need for secure communications. The pioneers were greatly inspired by Herbert Yardley's book *The American Black Chamber*. This book was followed in 1931 in Sweden by a similar study, *The Contributions of the Cryptographic Bureaus in the World War*, written by the hot-tempered and multi-lingual French/Swedish businessman Yves Gyldén whose other great interest was rugby.[3]

In neighbouring Norway, which was a young nation, having only gained independence from Sweden in 1905, all communications were enciphered by an instrument, invented by a Norwegian painter in 1883, called Strømdahl's cryptograph. Absolutely secure, said the inventor. However, anyone with an interest in cryptography could read the supposedly secret traffic on military and diplomatic channels, without any problem, because this cryptograph was based on the Vigenére cipher, which already had been known for 500 years.

The first man who understood the precarious nature of Norwegian ciphers was a young artillery captain and economist, Alfred Roscher Lund, who must be regarded as the father of Norwegian Signals Intelligence. On his initiative a secret cipher committee was set up in 1935 to establish secure communications and, as the other side of the coin, to explore how the communications of other countries could be intercepted and attacked by cryptanalysis. In fact, one can date the birth of Norwegian Signals Intelligence to Saturday 23 November 1935, when a rather pedantic story appeared in a popular evening paper, inviting readers to solve a number of puzzles. The main one was the title, where the words 'Secret writing' were hidden in a simple transposition cipher. Roscher Lund wrote the article, but not as entertainment for the readers. It was the first conscious effort to recruit puzzle solvers to his new 'Black Chamber', which he tried to establish with a government grant of only 200 dollars. By getting access to the addresses of the people who sent their solutions to the paper, Roscher Lund found 22 recruits who became the first members of his new Cryptology Club.[4]

Roscher Lund got his inspiration from Sweden where two eccentric brothers, Arvid and Ivar Damm, in 1915 set up the company Cryptograph Limited with financial support from Olof Gyldén, head of the Swedish Naval Academy and Yves Gyldén's father. In 1919 Arvid Damm invented the rotor wheel, essential to the new generation of electromechanical cipher machines, at about the same time as Koch in Holland, Hebern in America and Scherbius in Germany.[5]

The brothers were true enthusiasts. The last word Ivar, a teacher in mathematics, uttered on his deathbed was 'crypto'. Arvid reached his high point as a womaniser when he staged a fake wedding with a fake priest in a rented church only to be able to bed a Hungarian circus princess he had pursued since their first meeting in Helsinki. His money-making abilities did not match his passion for beautiful women and in 1921 his company had to be saved from bankruptcy by money from the famous Nobel family who had returned to Sweden after the Bolshevik revolution with a fortune from the Baku oilfields.

When the Nobels grew tired of waiting for a return on their investment, one of their employees, Boris Hagelin, born in Russia of Swedish parents, took over. Hagelin turned the company's fortunes around when he, in the 1930s, put a developed version of Damm's invention into mass production. Among Hagelin's first customers in this early period were the Nordic countries. He eventually sold 140,000 cipher machines, model Hagelin M-209, to the US Signals Corps during World War II and thus became a very rich man. While Damm and most of the other geniuses of Scandinavian Sigint died in relative poverty, Hagelin in 1952, to avoid the punitive taxes and regulations of the new Swedish welfare state, took his company from Sweden to tax-free Switzerland. There it became the Crypto AG, in later years a world renowned producer of modern cipher machines selling different models of the old Swedish invention to 120 countries world-wide.

FIRST ATTEMPTS

The scientific development spearheaded by Damm, Gyldén, Hagelin and others, plus the various attempts by the Swedish Armed Forces to collect signals intelligence from naval vessels in the Baltic Sea, led in July 1937 to the establishment of the Signals Intelligence Department of the Swedish Defence Staff, from 1942 until today known as the Defence Radio Institute (*Försvarets Radioanstalt*, FRA). In fact many of the mathematicians and officers who started with Sigint in 1937, could be found on the FRA payroll up until the 1970s, giving Sweden a unique tradition through World War II and most of the Cold War.

Roscher Lund became a personal friend of Hagelin and other Swedes and tried to copy their achievements when he finally, in September 1939, pushed by the outbreak of war, managed to persuade the Norwegian Labour government to put up 10,000 US dollars to finance the first Norwegian Black Chamber, camouflaged as the Information Office of the Department of Defence.

Twelve years earlier, on 18 June 1927, the Finnish Army had established its own radio intelligence service, headed by the gifted and ruthless Reino Hallamaa who had fought on the White side against the Bolsheviks in the civil war that followed Finland's declaration of independence from Russia in 1918. With help from Germany, Poland and Estonia, Hallama slowly built the small 'Statistical Office' of the General Staff into Scandinavia's most advanced Sigint service with listening posts and a highly qualified staff of cryptanalysts, among them the former head of Estonian Sigint, the brilliant Andres Kalmus, and the equally brilliant Russian linguist and actuary, Erkki Pale, who was recruited in late 1939 and led the attack on Soviet military systems.[6]

Despite a shortage of resources, Scandinavian Sigint had significant success in this early period. Both the Finns and the Swedes became experts in solving Soviet two to five digit additive codes. The Swedish mathematical genius, Arne Beurling, and his partner, the multi-talented Åke Lundquist, later Grand Master of chess, acknowledged botanist and translator of German classical poetry, and their associates among the cryptanalysts, solved the Soviet Baltic Fleet code, various Soviet Army and Air Force codes, before their wartime efforts culminated in the solution of the German Geheimschreiber, a cipher machine with the complexity of Enigma.[7]

From 1940 to 1943 the FRA intercepted and read tens of thousands of German military and diplomatic messages, passing through Stockholm on landlines from Oslo and Helsinki to the Nazi High Command in Berlin, thereby giving the Swedish government nearly complete insight into the German dispositions in occupied Norway and along the Murmansk front, where German and Finnish troops fought their common Soviet enemy.

To a certain extent information about codebreaking was exchanged on an informal basis between Stockholm, Hallama's battlehardened veterans in Finland and Roscher Lund's recruits in Norway.[8] The Finns, who shared a long and bloody history with both Sweden and Russia, were the undisputed frontrunners, mastering a number of Soviet ciphers, which greatly helped them during the heroic struggles of the first Winter War, from November 1939 to March 1940.

They continued their efforts when Finland joined forces with Nazi Germany in its attack against the Soviet Union in June 1941, and a significant number of German and Finnish cryptanalysts worked together on the Murmansk front. Through the pioneering work on the Abwehr by the Norwegian historian, Tore Pryser, we also know more about how the Germans and the Finns got their hands on the famous Petsamo codebooks.[9] It has been widely believed that one partially burnt codebook was taken by the Finns. The truth seems to be that a platoon of German Abwehr troops, supported by Norwegian Nazi collaborators, in the early morning hours of 22 June 1941 staged a surprise attack across the Norwegian border on the Soviet consulate in Petsamo where they successfully captured the Soviet codes. The resulting information was shared with the Finns, and also Japan, and later became passports to freedom for the German and Finnish experts who had been on the wrong side in the war.

Indeed, the wartime telegrams exchanged between Tokyo and its envoys in Berlin and Helsinki, discussing the combined Axis success against Soviet ciphers, were intercepted and read by the codebreakers in Washington and seem to have inspired the US Venona effort in 1943. There is therefore a clear connection between the early work of the Nordic codebreakers in the frozen wastelands of the Arctic and one of the deepest Allied secrets of the Cold War.[10]

In Norway Roscher Lund's Black Chamber existed for only six months, as he and his codebreakers were forced to evacuate Oslo and burn their archives when Hitler's Germany attacked out of the blue in the early morning hours of 9 April 1940, having occupied Denmark earlier that same night. During those six months Roscher Lund's small staff had concentrated on German and British ciphers with surprising success in one field. Through traffic analysis and codebreaking they had helped to unmask practically the whole Abwehr spy network along the Norwegian coast. Most of the radio agents were just about to be arrested by the Norwegian police when they were saved by the German occupation. Lund fought alongside the Norwegian and Allied forces as they retreated northwards during April and May until Norway, on 10 June 1940, surrendered after the Allied troops withdrew in an desperate attempt to stop the German advance through the Low Countries and France. That same day Lund and two of his men skied across the border to neutral Sweden, having been ordered to continue their work in Stockholm. In 1942 he became head of a new Military Intelligence Staff set up in London by the Norwegian government-in-exile, equipped and trained by MI6 and MI5.[11]

THE WAR ENDS

This brief sketch of the history of Scandinavian Sigint was necessary for these early years are important in understanding the post-war developments in Signals Intelligence. Finnish Sigint had been, from a technical viewpoint, a success, but at least doubtful politically. Much more needs to be done by historians to assess the effect the combined German/Finnish codebreaking on the Northern Front had on Allied shipping losses in the Barents Sea, for example. We know that Soviet coded messages, from aeroplanes and ships sent out to meet and protect Allied convoys, were intercepted and broken. They were then transmitted on landlines through Sweden to the Nazi commands in Berlin and occupied Norway from where submarines, surface ships and planes were directed against the convoys. The Swedes of course knew, because they read the teleprinter traffic.[12]

When the Murmansk front broke in October 1944, the Finns surrendered while the Germans applied scorched earth tactics as they withdrew through

Northern Norway. More than 50,000 people were driven from their homes; every single building in an area the size of Denmark burnt down. In the autumn of 1944, in order to escape the wrath of the Soviets, 350 Finnish Sigint staff members escaped into Sweden with their immediate families, and with 700 cases of crypto equipment in the well known Operation 'Stella Polaris'. Copies of the materials seem later to have been sold off to Sweden, France, Britain and the United States. About 15 of the codebreakers were employed by Sweden's FRA, while the anti-Soviet head of Finnish Army Intelligence, Colonel Aladar Paasonen, Hallama, Kalmus and several others sought their post-war fortunes in France and elsewhere. The rest of them eventually went back home to Finland, a few to face prison sentences, among them Erkki Pale, while others later rebuilt modern Finnish Sigint of which, so far, little has been revealed.[13]

It is less well known that the German partners of Finnish Sigint, organised in the so-called *'Meldekopf Nordland'*, retreated to southern Norway, where they surrendered in May 1945. The remaining group of 36 German codebreakers, well aware of the value of their knowledge of Soviet ciphers, crossed the border with their archives and offered their services to Sweden. However the Swedish government, already embarrassed by the influx of the Finnish 'stellists', and with the victorious Soviet Union closely watching their actions, turned them down and put them on a plane to Germany, where they were welcomed by the American Office at Strategic Services (OSS), in Wiesbaden in June 1945.[14]

WAR LESSONS

All the Nordic countries had learnt bitter lessons from the war. Once the immediate post-war jubilation had died down and the Cold War started, one thing seemed quite clear: if the small democracies facing the Soviet military colossus were to survive, they needed their own professional intelligence services. The elite, which had started cryptanalysis before the war, also realised that such new and better services needed to incorporate Sigint as the vital element.

In Denmark and Norway people said: 'Never again the ninth of April. Next time we'll be better prepared.' Both countries had been occupied by Nazi Germany for five years, but the situation was slightly different. In Denmark the King and the government had stayed on, for three years walking the difficult path of collaboration with the German occupiers, even signing the Anti-Comintern Pact in 1941. In occupied Copenhagen the pre-war Army and Navy Intelligence Staffs, led by Lieutenant Colonel Einar Nordentoft, continued to function as did the other state institutions until the crisis of August 1943, when the Nazi German plenipotentiary, Dr Werner

Best, tightened the screw and demanded tougher measures against the rising number of protest strikes and Resistance-led sabotage actions. Mass demonstrations followed and, on 29 August, both the government and the parliament stood down after having rejected Best's demands. The Nazi answer came swiftly. German troops attacked the Danish Army barracks. Fighting broke out, and the Danish Navy scuttled its ships. During the following state-of-emergency, the Danish King and the Crown Prince were interned as prisoners of war, while the running of the country was left to civil servants. The Resistance movement, being directed from neutral Stockholm, grew to a well-armed underground army of nearly 50,000 women and men, and put up, for the remaining 21 months of the war, a brave and vigorous struggle against the increasing Nazi terror.

On that fateful 29 August 1943, the Danish intelligence headquarters in Copenhagen closed down. Nordentoft escaped to Stockholm with his right-hand man, Navy Commander Poul Adam Mørk, head of the Naval Section since 1932. The third main character, Army Colonel Hans Lunding, was arrested by the Gestapo, transported to Berlin and sentenced to death by a Nazi court after several months of hunger and harassment. He spent the last year of the war in the Flossenbürg concentration camp near Munich, waiting for the sentence to be carried out. In his next-door death-cell, the legendary former chief of the Abwehr, Admiral Wilhelm Canaris, was held, having been arrested and tortured as one of the officers responsible for the failed attempt to kill Hitler on 20 July 1944. By exchanging signals through the wall, the Danish Army intelligence colonel was the last man to have contact with the German intelligence maestro. For Lunding, having to witness so many executions, the feared nightly knock-on-the door never came. When the Germans surrendered, he was still waiting. Two months later, in July 1945, he was able to return to Denmark as a free man.[15]

THE NAVAL RADIO SERVICE

In 1946, Lieutenant Colonel Nordentoft and Commander Mørk, having run Danish intelligence successfully from exile in Stockholm, entered into negotiations with the British and the Americans to secure funding and technical assistance for new and more efficient intelligence services. The visionary Mørk soon became the driving force behind the Danish Navy's first systematic attempts to collect Sigint. He had in the 1930s built up a highly successful shipwatching system in the Baltics and the Danish straits, based on a network of contacts among Danish shipowners, harbour masters, merchant marine captains, shipbrokers, shipping agents and fishermen. In his book, *The Colonel and the Commander*, the Danish historian, Wilhelm Christmas-Møller, writes this:

Over all Denmark's position, at the entrance to the Baltic Sea, was and still is the main reason behind the Danish intelligence services' ability to deliver genuine products of interest to others...the Straits have always had their strategic importance, and Denmark has been especially well suited to survey, if not control, the traffic passing in and out of the Baltic ...Mørk's maritime network was re-built after WW2, and became, from the end of the 1940s, a vital part of what Danish intelligence had to offer its partners abroad.[16]

Indeed, shipping intelligence, expanded to cover the seven seas as the Danish merchant marine fleet grew, has until today been a major product of Danish intelligence. When the United States, Great Britain and their UKUSA partners in the 1950s established their huge and world-wide program to monitor Soviet and Eastern Bloc merchant ship movements, the Joint Naval Intelligence Collection Organisation (JNICO), Denmark became a key member, alongside Norway, Holland, Sweden and other seafaring nations.[17] The shipping intelligence Mørk could produce and information from the Danish possession of Greenland seem to have been important when he secured from the Americans in 1947 the first deliveries of radio equipment for his brainchild, the new Naval Radio Service (*Søværnets Radiotjeneste*, SR). This service became the foundations upon which modern Danish Sigint would rest, and also worked with captured direction-finding equipment from the huge Comint intercept station the Germans had run in the north of Jutland during the war. When Mørk many years later briefed Christmas-Møller, he was 'visibly proud' when he told him how he had been able to develop Danish Sigint on financial contributions from outside ordinary budgets, and on 'donations from the Americans'.

THE VILLA

Since just before the war, the Danish Navy had been intercepting German naval radio traffic and made limited efforts at codebreaking. The small codebreaking unit was re-established in 1946 under Lieutenant Commander Finn Haugsted who again reported to Commander Mørk. Two years later, in the spring of 1948, time had come to move on. In a fashionable suburb of Copenhagen, Mørk found a huge marble villa, which he thought suitable as a new Sigint headquarters. However, when he presented his plans, including a $5,000 bill for one year's rent, to the Minister of Defence, Rasmus Hansen, the Labour Party stalwart protested: it belonged to a capitalist! Only when the cigar-smoking politician was presented with a box of fine Cuban cigars, bought for the occasion, did he budge. On 1 May 1948,

Commander Mørk and SR's 15 employees moved in. 'The Villa' was born: the name under which Danish Sigint is known even today.[18]

The service operated interception and direction-finding stations at Løgumkloster in the south of Jutland, later also at Hjørring in the north, and at Åkirkeby on the island of Bornholm, close to Poland in the Baltic Sea. From 1948, another station opened at Danmarkshavn on the north-eastern coast of Greenland, operated by one single radio telegraphist, which intercepted Soviet radio traffic. At the same time, the SR continued its interception of open and coded diplomatic communications to and from foreign embassies in Copenhagen, in cooperation with the state-owned Danish Telecom and Postal Service. In 1965, the interception of embassy telex traffic was moved to the basement of the East Asian Institute in the grounds of the University of Copenhagen. When discovered by students four years later, it erupted into a scandal known as the Kejsergade affair.

The Naval Radio Service achieved its first success covering a large Soviet naval exercise near the island of Bornholm in the Baltic in 1948. Soviet aircraft overflew the island, which caused anxiety both among the politicians and the population. However, analysis by SR of the intercepted communications proved that the Soviets only used Bornholm for navigational purposes, and did not have other intentions, to the relief of the government. Despite his achievements, and to his lasting bitterness, Mørk was bypassed when the Danish government, in 1950, merged and reorganised its intelligence services, appointing Colonel Hans Lunding, who was a supporter of the Labour Party, as Danish intelligence chief.

As second-in-command, Mørk concentrated himself on the running of Sigint. The SR changed its name to *Försvarets Central Radio* (FCR, the Defence Central Radio) and it moved in the autumn of 1950 with its 80 employees from 'The Villa' to a new secret headquarters complex at Aflandshage in a southern suburb of Copenhagen, from where it still conducts its main eavesdropping operations. As in the other Nordic countries, the onset of the Cold War saw a huge expansion of FCR, in close cooperation with NSA and the German Bundesnachrichtendienst, having been assigned the Danish straits, the Baltic and Poland as its main targets.

RETURN TO NORWAY

The Norwegian King and Government had chosen exile in London, from where they continued the fight against the Nazis. The various intelligence organisations had in 1945 several hundred employees at their London and Stockholm stations; and a secret army of agents and informers behind German lines in Norway, may of whom had been trained by the British Secret Services. When Colonel Roscher Lund left Norway in 1946 to become military adviser to Trygve Lie, the first UN Secretary-General, his successor Colonel Vilhelm Evang could build his operations round a substantial manpower base of war veterans.

The British had not allowed Roscher Lund to run Sigint operations from within the United Kingdom. However the Norwegians had kept in touch with the Swedes during the war. They also had upheld the scientific study of the Hagelin machine which were used on most of the exile government's communications, the aim being to secure the machine cipher against enemy attack. When in doubt, and perhaps somewhat naively, Colonel Roscher Lund and his Cipher Committee, as he wrote in his post-war London report, contacted 'the highest British cipher authorities' and showed them the Norwegian systems, for discussions and acceptance. Once wanting to establish a purely Norwegian telegraph circuit, Roscher-Lund got a firm no from the British who demanded that he used the Hagelin cipher system, known and accepted by them. 'German reports now tell us that the Germans broke the ciphers of a number of countries. However, no Norwegian cipher has been broken', he proudly concluded, not mentioning the ability he had given the British to read the Norwegian government's supposedly secret communications between London and its vital stations in Stockholm, Moscow, New York and Washington.[19]

Indeed I believe that this exchange of information and the mathematical study of the Hagelin machines, which had been introduced in Scandinavia before the war and which were extensively used on Allied communications throughout the war, is vital to the understanding of the post-war success claimed by European and US codebreakers. This success was to a large extent achieved against some of the many nations world-wide that later during the Cold War bought the Hagelin machines from Crypto AG.

Among the users of Hagelin equipment were many of the foreign diplomatic missions in Oslo, Stockholm, Helsinki and Copenhagen. The diplomatic traffic to and from these missions had, in deepest secrecy and as an emergency measure, been intercepted wholesale at least since 1939, an activity which continued after the war. Norwegian cryptanalysts were able to attack the ciphers of Italy, France, Belgium, Switzerland, Egypt and other countries after getting access to computer power in the mid-1950s.[20] The Danes claimed in 1960 to read the ciphers of 23 countries, including Japan.

In the light of later allegations against the Crypto company for giving NSA trapdoor facilities, one may wonder whether Boris Hagelin, to help his old Nordic friends, left more clues to his machines when he moved to Switzerland than so far has been acknowledged. Although final and written documentation is sparse, I believe that you cannot understand the success achieved by European codebreakers after the war without understanding the Hagelin history, a topic still awaiting thorough historical research.[21]

SCANDINAVIA, SIGINT AND THE COLD WAR 221

In Norway, Crypto got its main success in the field of communications security when two engineers in the early 1950s patented a random number generator, based on radioactive Cobalt. Connected to a teleprinter, the ETCRRM code machine produced a onetime cipher of high quality. Eventually 2,000 machines were sold and used on the main NATO communications in Europe, and also on the Hot Line between the Kremlin and the White House.[22] The code machine was considered secure until the discovery of the fatal dangers of electromagnetic radiation sent it straight to the scrapheap of scientific history.

POST-WAR EFFORTS

In April 1944 Roscher Lund had proposed the creation of a new Norwegian Secret Intelligence Service in which the collection of Sigint had to be pursued 'with great energy' and a highly qualified staff.[23] When his ideas were realised, one of the trusted men of his pre-war 'Black Chamber', Lieutenant Jon Krag Brynildsen, became head of the new Radio Control Office. Brynildsen, inspired by a visit to the Swedish FRA, proceeded in 1946 to establish small radio intercept stations at a former German Army camp in Oslo, at Madlamoen outside Stavanger on the south-western coast, and at Heimdal outside Trondheim in mid-Norway.

In July 1946, a small expedition, codenamed 'Torkel', was sent to the town of Vardø, some 40 miles from the Soviet border in the High North. During ten days the members of the expedition collected Soviet naval and air force signals, some of which indicated secret Soviet testing of captured German V2 rockets over the White Sea. A small cryptological department, headed by another of Roscher Lund's pre-war pupils, professor of mathematics at Oslo University, Erling Sverdrup, later replaced by another mathematician, Nils Stordahl, tried to break into the traffic.[24]

A year later, in March 1947, a detachment of nine radio telegraphists installed themselves in two ice-cold Nissen huts on an airfield outside the town of Kirkenes, even closer to the Soviet border. The first Sigint station, exclusively targeting the Soviet Union, had been provisionally established.[25]

As the Cold War developed, the head of the new Norwegian Defence Intelligence Staff, Colonel Vilhelm Evang, like his counterparts in Denmark, actively sought American and British financial and technical assistance. Alone, the Norwegians and Danes felt they had no chance to withstand an attack from the Soviets. As international tension rose throughout 1948, membership in NATO for both countries with their strong Western ties seemed a logical step. So acutely did the Norwegians feel the threat, even after the Atlantic Treaty had been signed, that five former senior Nazi German officers, led by Wehrmacht General Bernhard von Lossberg and Kriegsmarine Admiral Otto Ciliax, in July 1949 in deepest secrecy were brought to Oslo to advise the Norwegian High Command on the conduct of warfare under Arctic conditions.

General von Lossberg was secretly kept on the Norwegian payroll until the 1960s. He had in 1940 been one of the chief planners of Operation 'Weserübung', the surprise invasion of Denmark and Norway. Now, only nine years later, he was back, teaching the Norwegians how to beat back a feared Soviet cross-border attack, based on his experiences from the Murmansk front.[26]

'NEUTRAL' SWEDEN

Compared to the other Nordic nations, Sweden had had a relatively easy war by appeasing and to a certain extent collaborating with the Nazis in the early years. While Finland and Norway had to rebuild their war-torn and plundered countries, Sweden came out of 1939–45 with a strong economy, an undamaged industrial basis and modern intelligence services. Although FRA, the Defence Radio Institute, which employed 384 men and women in 1942, was reduced in size, the intercept stations were operational, and FRA continued its work on Soviet and European communications, even attacking the cipher systems of Japan and China in Asia and Brazil and Chile in South America.[27]

Up until Norway and Denmark chose the Western Alliance and became members of NATO in 1949, there was close contact between the Swedish, Danish and Norwegian cryptanalysts, based on the personal friendships established before and during the war. Work was being done to develop common ciphers for a possible Nordic Defence Union. When the German Geheimschreiber central in a bunker in Oslo was captured intact, the machines were taken apart and analysed. With certain improvements the Geheimschreibers thereafter were used on specific teleprinter channels, one of them being counter-intelligence landlines within Norway.

The FRA was the entity within the Swedish Ministry of Defence that handled all Swedish Sigint collection, processing, analysis and reporting. From its headquarters and processing centre outside Stockholm close to Bromma airport, the FRA operated five radio intercept stations, as well as mobile Elint collection platforms operated by the Defence Research Establishment (*Försvarets Forskingsanstalt*, FOA). The FRA also managed its own research and development program for new Sigint collection and processing equipment.

Navy Commander Torgil Thorén, trained as a torpedo officer, described as somewhat austere and harsh, but completely loyal to his staff, had been appointed head of the FRA when it was reorganised in 1942. He continued

SCANDINAVIA, SIGINT AND THE COLD WAR 223

as its leader until 1957, taking the FRA and most of the war-veterans with him into the post-war period, among them the highly intelligent leader of the Crypto Section, Åke Rossby.

As the Cold War developed, the FRA strengthened its Comint collection against Soviet military activities in the former Baltic republics. The FRA's listening posts, including a large station on the Baltic island of Gotland, intercepted all Soviet military radio traffic coming from the Leningrad and Baltic Military Districts, as well as from Poland, Czechoslovakia and East Germany.

In the early to mid-1950s, the FOA was flying DC-3 Dakota aircraft (the Swedish designation for the aircraft was TP-79) which were equipped with Elint collection equipment provided by the United States through the CIA station in Stockholm. The Swedish DC-3s spent most of their time flying Elint missions over the Baltic Sea collecting information on Soviet radar defences and movements of ships to and from Baltic ports. Of particular concern to the FRA were the Soviet long-range bombers, whose activities were closely monitored. The FRA was also copying and reading much of the diplomatic traffic coming in and out of Stockholm, including telegrams passing between Moscow and the Soviet embassy in Stockholm and KGB traffic coming from Leningrad and the Baltic states.[28]

THE AMERICAN CONNECTION

Despite claiming to be neutral, another picture of Sweden's political and military role during the Cold War has slowly emerged during the last decades. As the political scientist, Ola Tunander, has written, 'in a certain respect, Sweden has long been "a hidden member of NATO" or, more correctly, has been "plugged into NATO" with particularly close ties to the USA'.[29]

The first evidence of close intelligence sharing ties between neutral Sweden, the United States and Britain dates back to the end of World War II, and there are clear indications of a periodic exchange of Comint and Elint information, with the US agreeing to provide Sweden with Elint equipment for its DC-3 'Ferret' aircraft in return for the results of the flights.

Beginning in 1947, the United States and the UK solicited the assistance of FRA in collecting vitally needed Sigint on Soviet military activities in the Baltic region. The United States provided aerial cameras and other equipment to the Swedes, but reportedly refused to provide any sophisticated Sigint equipment. The CIA, however, was under no such restrictions and made such equipment available. According to a 26 January 1949 USAF intelligence summary, Sweden had provided the United States and Britain, through the British naval attaché in Stockholm, with Elint data which established the presence of a chain of 15 Soviet radars on the Baltic coast.

To plug the gap in the global Sigint encirclement of the Soviet Union, as it was disclosed during an interrogation of the convicted Soviet spy, Swedish Colonel Stig Wennerström, US defence personnel in 1949 provided state-of-the-art technology to the Swedes, flying the equipment into Swedish air bases at night in order to keep the arrangement fully secret. Wennerström also betrayed the American/Swedish collaboration to his masters in Moscow. Also in 1949, the Swedish Air Force extended its 'Ferret' flying missions on behalf of the Americans to the Kola Peninsula in the North, thereby violating Finnish airspace over Lapland, disregarding Finnish protests against these flights.

Available information indicates that Sweden supplied to the Americans Elint reports on Soviet radar locations and capabilities in the Baltic states during periodic meetings in Washington DC. In return, the US gave to the Swedes edited versions of their Elint summaries regarding Soviet radar activity in that same area.[30]

THE MASK FALLS

On 3 May 1973 two Swedish journalists, Peter Bratt and Jan Guillou, disclosed in the magazine *Folket i Bild/Kulturfront* the so-called 'IB-affären'. This article described, among other things, the ties between the US intelligence agencies and the Swedish foreign intelligence bureau, the *Informationsbyrån* (IB). The IB did not officially exist and, according to Bratt and Guillou, the IB came into action when the operational capabilities of the 'official' Swedish intelligence organisations had been exhausted.

The mysterious IB, closely tied to the ruling Labour Party, was formed in 1965. Its mission was both domestic and foreign, being authorised to collect Humint abroad, and spying on anything looking like a threat to national security at home through a network of informers.

The revelation of IB's existence became a watershed, not the least because the magazine story in detail exposed how supposedly neutral organisations in deepest secrecy seemed to cooperate closely with foreign intelligence agencies – like the CIA, MI6, Mossad and Shin Beth in Israel, BND in Germany, SDECE in France and IDB, BVD and MID in Holland.

Among the later revelations made by the authors, who were eventually arrested on 22 October 1973 and sentenced to one year in jail, was also the fact that Sweden and the US had traded Sigint data for many years. Pursuant to an informal agreement between the United States and the UK, the British Sigint agency, the Government Communications Headquarters (GCHQ), was given responsibility for handling liaison with the FRA. Nevertheless,

SCANDINAVIA, SIGINT AND THE COLD WAR 225

the United States maintained an independent liaison relationship with the Swedish intelligence services through the CIA's chief of station in Stockholm. It was revealed that the FRA had broken the diplomatic codes and ciphers of Brazil, Zaire, China, Iran, Turkey, Japan, Czechoslovakia and several other countries.

Later it came to light that the FRA had executed Sigint operations against the Soviet Union from within Finnish territorial waters, and even eavesdropped on the Finnish parliament in Helsinki. The Swedish embassy in the Finnish capital appears to have played an important role in the 1970s and 1980s as an intelligence gathering station. Well-informed US sources have disclosed that a Sigint unit of the FRA was established within the embassy, monitoring Soviet diplomatic and military communications emanating from the Leningrad district. The personnel was trained, and the equipment delivered by the National Security Agency.

In Stockholm itself, IB and Mossad agents broke into the Egyptian embassy to inspect the Egyptian codemachines and copy their codebooks. In exchange the IB received information about Soviet military hardware from the Mossad and the CIA.

The informal contacts between the FRA and the British GCHQ continued well into the 1960s, including a joint operation to spy on the Russian cruiser *Ordzhonikidze* that visited Stockholm in 1959. The FRA planned to operate against the ship, and the local MI6 station chief suggested that GCHQ and MI6 would help. The FRA accepted the offer, and the famous Peter 'Spycatcher' Wright flew to Stockholm. With the help of two FRA and two GCHQ technicians, a so-called 'Engulf' operation was conducted against the cruiser's cipher machines – apparently without success. A year later, Wright and other British intelligence officers persuaded the FRA to relinquish its 'neutrality' and to provide GCHQ with copies of Soviet military intelligence (GRU) messages exchanged between Stockholm and Moscow during 1941–45 as part of the still on-going Venona operation.[31]

JOINING THE GLOBAL US SYSTEM

Meanwhile, in Norway and Denmark, the Defence Intelligence Staff and 'the Villa' were finding their places in the world-wide US and British Sigint network. The two agencies had, on their own, started the collection and analysis of Sigint in 1946, but truly modern Signals Intelligence came into being some years later, in June 1952, during the Korean War, when two Americans paid a visit to Scandinavia. One of them, Louis Tordella, was travelling as Mr Jones, the other, Griffin Chiles, travelled as Mr Williams. They were both key figures in US Sigint and later leaders of the National Security Agency.

Since Norway as the only one of the Scandinavian countries has published an official study of its Defence Intelligence Staff from 1945 until 1970, we know most of what happened. At that time, the first permanent Norwegian Comint station, outside the small town of Vadsø and close to the Soviet border, had been operational for about one year with a staff of 20 radio telegraphists. In addition Norway had listening posts in Oslo, Stavanger and outside Trondheim on the western coast. Although much of the equipment was old and consisted of Adcock direction finders and other leftovers from the war, the radio telegraphists had proved very adept at understanding Soviet radio communications.[32]

Attempts at cryptanalysis gave few solutions, but traffic analysis had made great progress against the Soviet Navy, Air Force and land forces in the North. The telegraphists had solved the very complex Red Navy system for distributing callsigns and could therefore follow the movements of Soviet warships, from hour to hour.

The Air Force analysts were greatly helped by the rigidity of the Soviet command system, whereby every pilot and commander had to file complete flight plans for any airborne mission – and usually did so in clear text, the night before the mission took place. By carefully analysing and filing away every bit of data, the Norwegian Defence Intelligence Staff had, in cooperation with the Swedish FRA, acquired a huge knowledge of Soviet call signs and patterns of communication, and were able to locate and identify most of the Soviet military units in the north-western area of the USSR. They had also broken the coded Soviet grid system for navigation, a system that had caused the Americans a lot of trouble.[33] According to legend, Tordella and Chiles returned to Washington very much impressed with what they had learned in Scandinavia and immediately suggested extensive co-operation.

As for Norway, it was clearly established that if the Norwegians would modernise and expand their Comint stations, they would find a willing customer in the new National Security Agency, which needed a continuous supply of intercepts for early warning purposes. Basically the Americans, with the help of the Nordic countries, wanted to reconstruct and follow *all* communications in the north-western part of the Soviet Union. In exchange Norway would get US financial support and access to the latest in US high technology.

The controversial chief of Norway's Defence Intelligence Staff from 1947 to 1966, Colonel Vilhelm Evang, a revolutionary Marxist before the war who had served 30 days in jail for anti-military agitation, was an intellectual, a graduate of the University of Oslo. He had worked with Roscher Lund in London during the war, and with the Swedes in the late 1940s. He immediately understood the importance of the deal being offered. With access

SCANDINAVIA, SIGINT AND THE COLD WAR 227

to American capital and knowledge, the Intelligence Staff could finally reach the professional level he thought necessary to counter the Soviet threat, and he therefore persuaded the Labour government to go ahead.[34]

INTO UKUSA

After hurried preparations the first early warning intercepts were delivered to the Oslo CIA station on 16 September 1952, being followed by a steadily increasing stream of more or less raw intercepts as more people were hired and the network of Comint stations expanded.

For the next 40 years the cooperation blossomed, unknown to the general public and unknown to most of the politicians in the Norwegian parliament. The belief in 'the Bomber gap' led to renewed pressure from Washington to triple the capacity of Norwegian Comint from 100 to 300 Russian-speaking telegraphists, until the cooperation finally was formalised on 10 December 1954.

Louis Tordella and Frank Rowlett came to Oslo with a delegation from the NSA and the CIA, to sign the top secret bilateral 'Communications Intelligence Agreement' between the United States and Norway, known as the NORUSA pact, thereby probably also giving Norway 'Third Party' status in the world-wide UKUSA Comint partnership.

With the signing of the formal agreement, Norwegian Comint saw a huge expansion during the next ten years, the stations being continually modernised, and the number of staff reaching more than 800, working round-the-clock to monitor Soviet communications. Very soon the daily product reached enormous proportions. As early as 1957, the Intelligence Staff couriers brought, every day, to the NSA/CIA cell at the American embassy in Oslo for transmission across the Atlantic, over 1,000 intercepted telegrams, 350 pages of radio journals, 500 direction-findings, four or five big spools of magnetic tape, containing intercepted voice communications, in addition to daily summaries of various kinds. The materials were not completely raw, as it was Norwegian policy to do the preliminary traffic analysis themselves to secure a rough overview of what the Americans got – partially to be able to control the quality of what they later got back of analysed material.

A CHANGE OF PRIORITIES

By the late 1950s, codebreaking attempts had more or less been given up, and usually only about one to ten per cent of intercepts could be read. This situation further deteriorated in 1965, when the Norwegian Security Service was established as a separate organisation, and was given control over Norwegian ciphers. The Norwegian Intelligence Staff then became one of

228   SECRETS OF SIGNALS INTELLIGENCE DURING THE COLD WAR

the few – if any – foreign intelligence organisations in the Western world which did not work on cryptanalysis – a tragic loss, according to many veterans, as Norway now became totally dependent on the United States in codebreaking matters.

The areas covered by Norwegian Comint expanded from the Kola region, the Barents and White Seas to include the vast areas of northern Russia east to Dickson and south to Leningrad and Moscow, including some of the satellite states of the USSR. While the Northern Fleet and the Soviet bomber force had been at the top of the priority list in the 1950s, the early 1960s saw a change. At a meeting in Oslo in 1962, Louis Tordella, now Assistant Director of the NSA, presented a completely new list of targets with the new Soviet nuclear submarine fleet, the Soviet Strategic Rocket Forces and nuclear bomb testing at the top. Below that on the list came Soviet Air Forces, Soviet Air Defences, the Northern Fleet, the general activity in north-western Russia and some civilian targets, including shipping, fishing vessels and civilian air traffic.[35]

THE EXPANSION OF 'THE VILLA'

It is widely believed, but not yet officially confirmed, that the Danish government at about that same time entered into a similar 'DANUSA' bilateral Comint agreement, giving Denmark and 'The Villa' 'Third Party' status within the global UKUSA pact.

The secret cooperation between the US and Danish intelligence communities was certainly close. In the summer of 1954, the Danish intelligence chief, Colonel Hans Lunding, informed the US air attaché in Copenhagen that he was willing to assist the United States in conducting special reconnaissance flights across Danish territory, flights which could not be cleared through normal diplomatic channels.

These so-called 'Fluorescent' flights passed across Zealand and penetrated deep into the Baltic before returning. Lunding ordered his Comint stations to monitor all Danish air traffic control channels to prevent reporting of the US spy flights. He even assisted the 'Fluorescent' flights with Danish decoys. As the US 'Ferret' plane crossed into the Baltic, a Royal Danish Air Force plane would take off and fly to the south of Zealand, thereby attracting the attention of the East German radar and radio stations. The monitored signals were then exchanged between the Danes and the Americans. Four Danish airfields were earmarked as emergency landing fields, and the enterprising Colonel even promised that a security guard would be on alert to seal off and patrol the runway to prevent any information reaching 'unauthorised' personnel, which probably included several Danish politicians.

SCANDINAVIA, SIGINT AND THE COLD WAR   229

Historically, Denmark had enjoyed good relations with Poland and had, since the beginning of the century, given home to a small but well organised Polish immigrant community. Colonel Lunding was an equestrian of international distinction, having won a bronze medal at the Berlin Olympics in 1936. He had likewise spent a lot of time in Poland before 1939 and established friendly relations with some Polish Army dragoons.

As the Cold War intensified, Poland was seen as a key Eastern Bloc intelligence target, situated as it was between the Soviet Union and East Germany. Without secure roads and railway lines through Poland, the Soviets would not be able to conduct ground war operations against the West. Just one hundred kilometres to the north of the Polish coast, the Danish island of Bornholm lay as a perfect listening post, with a ringside view over military and political developments on the mainland. The attempts by the CIA and MI6 to infiltrate into Poland had been a disaster.[36] When a representative of Danish intelligence in 1954 met John Bross, head of CIA's Eastern Bloc Department, he was first of all interested in what information about Poland the Danes could provide. 'Could they find only one source among a thousand, willing to give information of value, it would mean better results than all previous efforts', Bross said, according to the Danish report.[37]

When Denmark at about that time entered the UKUSA pact, it is therefore fair to assume that 'The Villa' was given Poland as its main target.

Danish Sigint even reached beyond Europe. When Christmas-Møller, who for some years served as an intelligence analyst in Copenhagen, in the 1960s was briefed about 'The Villa' by Commander Mørk, he was given a sheaf of intercepts, describing negotiations in which South Africa invited Rhodesia and Portugal to join a military alliance. When the secret report, based on these intercepts, was shown to the Danish Foreign Office, they wanted more – especially intercepts revealing the Japanese position on textiles in the tough trade negotiations going on at that time.

It was on this occasion that Commander Mørk said that 'The Villa' had grown to more than 1,000 employees and were able to read, whole or in part, the ciphers of 23 foreign countries.

A NEW FINNISH SIGINT SERVICE

Ten years apparently passed between the end of World War II and Finland's first careful attempt to re-build its Sigint services. When the Finns eventually started, in 1955, they had to start from scratch under the threatening shadow of the Soviet Union, and with many of their best codebreakers in permanent exile, Paasonen retiring to Portugal, and Hallama ending as a carnation grower in Spain.

Afraid of provoking the Soviets, Finland has up until quite recently shrouded their Sigint efforts in the outmost secrecy. Of the sparse information available, the first Comint intercept station, targeting Soviet Russia, seems to have been opened in October 1955 at an Air Force base near the town of Jyväskylä in central Finland, under the covername, Viestikoelaitos (VKL), the Signals Experimental Institute. The driving force is said to have been Army General Lauri Sutela, head of Finnish Military Intelligence and in the 1970s Commander-in-Chief of the Armed Forces. Despite misgivings among the few senior Finnish politicians who knew about Sutela's initiative, the service was secretly expanded and became a unit under the General Staff in 1960. Intercept and direction-finding stations were set up, mostly in abandoned and empty buildings throughout Finland, while centres for codebreaking and analysis were established in the cities of Tampere, Rovaniemi and Kuopio.[38]

In October 1967 Elint and Telint became a separate department of the VKL, headed by Engineer Corps Major Lasse Lehtonen, who also worked with the Electro-technical Research Institute in Espoo outside Helsinki. Some ten years later, VKL had the first Finnish 'Ferret' flights airborne, using old DC-3 Dakotas, codenamed 'Leena' and 'Ursula'. The Dakotas were in the 1980s replaced with modern Fokker Friendship aircraft.

In 1988 the VKL was re-organised as a department within the Finnish Air Force with Colonel Keijo Kepsu as its commander. Its headquarters was then and now discreetly placed within the perimeter of the Tikkakoski air base, some 25 kilometres to the north of Jyväskylä. While for example Norway and Sweden up until recently exchanged Comint summaries on a daily basis, similar exchanges between Finland and the other Nordic countries have never been reported. It is however widely believed that Finnish Sigint, at least in recent years, has developed contacts with the US intelligence community.

EARLY WARNING

Elint and Telint in Scandinavia, outside Finland, evolved in parallel to Comint, with the US Air Force as the driving force. The USAF wanted early warning of a possible Soviet bomber attack from air bases in northern Russia, and they needed precise intelligence about Soviet radar and air defences for their own 'route planning' for the B-52 bombers which would sweep over the Arctic regions towards Moscow and reduce central Soviet Russia to 'a smoking radiating ruin' within hours, should war break out.

In Norway the first attempts to intercept radar and other electronic emissions took place from mobile stations in 1954, and led, in the next 15 years, to the establishment of highly advanced Elint stations – some of them

SCANDINAVIA, SIGINT AND THE COLD WAR   231

placed on various mountain tops close to the border, and with a free view to a number of installations on the Soviet side.[39]

These worked in conjunction with the huge combined Elint station and search-and-tracking radar, codenamed 'South Sea', just outside the small coastal town of Vardø, 40 miles from the Soviet border. When the first signs of Soviet missile testing in the Barents Sea were detected in 1961, the focus was redirected from the ground and into space. From now on, the base complex was expanded and co-ordinated with Comint to follow the flight of Soviet test missiles, including cruise missiles and SLBMs testfired from submarines, and ICBMs testfired from Plesetsk, 150 miles to the south of Archangel.

From 1957, the Soviet nuclear bomb tests at Novaya Semlya and Semipalatinsk, and the Chinese tests at Lop Nor, were monitored from stations in central Lapland, which provided the Western Allies with information crucial to the implementation of the 1963 Test Ban Treaty.

When the new Telint station, codenamed 'Cod Hook', at the township of Fauske, near the Polar circle, came into being in 1965, the re-orientation towards Soviet nuclear missiles and space activities was complete. From this station Soviet satellites in a polar orbit could be tracked, as could rocket launches from Plesetsk.

Since 'Cod Hook' lay within the footprint of Soviet satellites downloading over Moscow Control Centre, the station intercepted telemetry and other emissions of high intelligence value about space-based Soviet spying, both with regard to photoreconnaissance and to Elint. This expanded the area Scandinavian Sigint could cover beyond the northern hemisphere and into the Middle East, the Indian Ocean and the South Atlantic.

During the Gulf War in 1991, Norwegian Sigint technicians were decorated by NSA because they provided the US Air Force with vital intelligence from the Iraqi battlefields.[40]

The bilateral agreement between Norway, NSA and CIA was reconfirmed on 1 August 1970 in a pact called NORUSA II, this one including both Comint and Elint. At that time about 800 people were employed in the around-the-clock Comint effort and about 200 in Elint and Telint. Some 80 per cent of the cost was paid for by the US.

SOSUS IN THE NORWEGIAN AND BARENTS SEAS

Finally, a separate development took place underwater, when the Norwegian Sound Surveillance System (SOSUS) was established in the late 1950s on an experimental basis. The first underwater listening cable was placed on the seabed to the west of the Lofoten Islands to detect Soviet submarines passing between northern Norway and the island of Jan Mayen.

In the 1960s and 1970s, in a secret cooperation with the US Navy, a number of other cables were laid, gradually moving to the east of North Cape as the Soviet submarines got more and more quiet and sought other transit routes from their Kola bases to their patrol stations. From the Norwegian land station in the heart of Lapland, Soviet nuclear submarines were fingerprinted and tracked, in close cooperation with SOSUS stations in the United Kingdom and Iceland, and with airborne reconnaissance.

The intelligence produced was a result of a very complex process, involving both Comint, Elint and Acoustint. To track a Soviet submarine, all radio traffic, on various nets, concerned with submarines entering and leaving harbour was monitored, including what could be gleaned from the Soviet 'Glonass' navigational satellites. These intercepts were crosschecked against information from US spy satellites, from SOSUS and from sonar buoys dropped from Orion and Nimrod aircraft. It is known from the recently published book *Blind Man's Bluff* that US spy submarines penetrated into Soviet waters and placed bugs on underwater Soviet communication cables in the Barents Sea.[41]

It is fair to assume that, in the end, the picture emerging from all these different intelligence sources gave the Western Allies a very good insight into the movement of Soviet strategic submarines – even up to recent days.

When the Soviet submarine *Komsomolets* started burning in the Norwegian Sea in 1989, the fire was first discovered by a US spy satellite, which found a hot spot in the ocean. After that, Norwegian Comint took over, picking up and relaying to Washington the voice communications between the aircraft flying over the scene and the home base in Severomorsk. Finally, when the submarine went down, SOSUS could follow and tape the underwater sounds created by that tragic event; first *Komsomolets* sinking, then about half a minute of silence until the hull imploded with a cracking sound that made echoes through the ocean from coast to coast, none of which, of course, was ever reported in the press.[42]

In 1965, the 'South Sea' Sigint station detected its first Soviet submarine-launched ballistic missile, a SS-N-6 'Sawfly' rocket which could carry a thermonuclear warhead from a submerged 'Yankee' submarine 1,300 miles towards enemy territory. It marked the start of the development of the Barents and White Seas into becoming the main Soviet testing ground for new weaponry – both ballistic and cruise missiles.

When the complete history of Sigint during the Cold War is written, I think one will find that the continuous supply of precise and updated information about the Soviet Rocket Forces as each new generation of missiles were developed and tested, will prove to have been the one major contribution from Scandinavia – of crucial importance to the USA, and to NATO.

CROWDED SKIES OVER THE BALTIC

The Swedish FRA operated in the 1980s an extensive network of listening posts situated along Sweden's coastline. FRA managed its main station at Lovön near Stockholm, which was equipped with a circularly-disposed antenna array (CDAA) antenna, and where FRA's analysis was done. There is also evidence to suggest that FRA operated another listening post at Umeå in northern Sweden, as well as another two important sites on Gotland in the Baltic. In addition to these stations there was also a FRA unit situated in Blekinge. (The FRA only performs Comint collecting and processing. The responsibility for collecting and processing Elint is the responsibility of the Defence Logistics Establishment.)

In Denmark, 'The Villa' continued to expand with financial and technical help from the United States. In the 1980s, from the Aflandshage headquarters outside Copenhagen, six listening posts were operated throughout Denmark, copying Polish, Soviet and East German traffic. These listening posts were located at Aflandshage itself, which worked on diplomatic and commercial traffic, the three HF intercept and direction finding stations at Hjørring, Løgumkloster and Almindingen on Bornholm, and the big Elint intercept sites at Gedser on the southern tip of the island of Falster, just across from the East German port of Rostock, and a second identical site at Dueodde on the southern coast of Bornholm, just off the Polish coast.

The land stations were not alone. The first issue in 1978 of the Royal Swedish Air Force Bulletin disclosed the flood of spy planes from both Eastern Bloc and Western countries, which were crowding the skies over the Baltic Sea. The magazine published maps, showing the 'highways' used by the spy planes, and described the various types that had been observed by Swedish intelligence. The story told the readers that the Cold War was being fought on Sweden's and Denmark's doorsteps – and that the intelligence activity easily could turn into a political nightmare.

For example, in late 1976, an American Sigint RC-135 aircraft almost collided with an Aeroflot commercial airliner just to the south of Gotland. The Swedish government protested and pointed the US embassy in Stockholm towards two other near misses in September that year, also involving an American RC-135.

The number of Sigint flights had increased dramatically, making the Baltic and Gulf of Bothnia an intelligence-gathering playground. The main target of the West was the new Surface-to-Air Missiles (SAMs) being deployed by the Soviets in the Baltic border regions, and in East Germany, especially the SA-5 'Gammon' missile batteries near the port of Rostock, later the SA-10s.

Similarly, the Soviets had shifted many spy planes to airfields further west, at one time reportedly operating more than 66 aircraft as Elint and photoreconnaissance platforms.

Sweden had their own reasons for worrying. A few years earlier, on 13 June 1952, one of their DC-3 Dakota Elint aircraft was shot down over the Baltic Sea by Russian fighters while on a spy mission. Despite claiming that the Dakota had crashed close to the Swedish island of Gotska Sandön, the Swedish government knew better, and later has had to accept that the crash site was near to the Latvian coastline. A Convair TP-47 Catalina flying boat, which had been sent out to look for survivors, was also attacked by Soviet MiG-15 jet fighters near the island of Huumand. Swedish radio intercept operators, who recorded the conversations of the Soviet fighter pilots, monitored the entire attack. Despite the loss, the Swedish Air Force continued to fly 'Ferret' missions off the Soviet Baltic coastline.

Rumours that the crew of the DC 3 survived and was picked up by the Soviet Navy and detained for life in the Gulag, have never died. Some are of the opinion that crew members were still alive in the 1990s, and that one of them may have been an American intelligence officer.[43]

ELINT AIRCRAFT

In 1953, after the loss of the Dakota, Sweden purchased a British TP-82 Vickers Varsity Hastings transport aircraft for the nominal purpose of conducting high-altitude radio trials. In fact, the heavily modified Varsity performed Elint collection missions over the Baltic until 1973. The nickname of the plane was the 'Flying Pig'. It now sits rusting on the grass at the Swedish Air Force museum at Linköping.

In 1960, Sweden purchased from the RAF two Canberra B.2s high altitude reconnaissance aircraft for the same purpose, and, in 1971, added two second-hand Aerospatiale SE 210 Caravelle passenger aircraft from SAS to replace the Varsity and the Canberras which shortly thereafter were pensioned off.

In March 1974 the planes had received Sigint modifications and each of the two aircraft had been modified with a plethora of antennas. It was rumoured that the planes carried a crew of eight to ten Sigint technicians in the rear of the aircraft. The missions were flown in Swedish as well as international airspace at altitudes up to 14 km to obtain as large a 'footprint' to monitor as possible. Missions usually lasted four to six hours and attracted from time to time the attention of foreign fighters.

The two Caravelles, stationed at Malmslatt Air Base, near Linköping, would fly for some 25 years and perform exactly the same task as the British RAF Nimrods, the German Breguet Atlantic Sigint gatherers and the

American RC-135s, which frequently could be encountered on Barents and Baltic Sea patrols. In the 1990s their role has been taken over by a new fleet of Gulfstream IV high-altitude Sigint aircraft.

The Swedish Navy operates its own spy ship, the 201-foot long *Orion* (A201), which was commissioned in June 1984. With a displacement of 1,400 tons and a maximum speed of 15 knots, the *Orion* carries a crew of 35, conducting Sigint collection in the Baltic Sea. In November 1985, a Soviet minesweeper in international waters in the Baltic rammed the *Orion* when she got too close to a Soviet naval exercise. According to Swedish sources the *Orion* was built with extensive support from the National Security Agency. The Swedish military negotiated directly with the director of NSA, Admiral Bobby Inman, and the final go-ahead for building the ship came immediately after US Secretary of Defense, Casper Weinberger, had visited Stockholm in 1981.

Despite outwardly keeping its distance from the enormous intelligence efforts directed against the Soviet Union and her Warsaw Pact allies, Sweden's true and secret role was quite different, hidden from the Swedish public. As Ola Tunander sums it up:

> According to a Swedish senior intelligence officer, the Swedish Signals Intelligence Agency (FRA) was responsible 'within NATO' for covering the Soviet Baltic Republics; it was also vital for adapting American bombers and later cruise missiles to the Soviet air defences in the Baltic, so that they could break through to Leningrad and Moscow. In the 1980s, a Swedish Signals Intelligence vessel, the *Orion*, was built to cover the Soviet Baltic area and adapt US cruise missiles and B-1 bombers to the low-level radars of the Soviet air defence system, SA-10.
>
> Sweden had the same responsibility for the Soviet Baltic Republics (primarily because of the Swedish island of Gotland in the Baltic Sea) as 'other' NATO states had for their neighbouring areas of the Warsaw Pact: Norway, with its northernmost border to the USSR, was responsible for covering the Kola Peninsula and the naval and air defence bases around Murmansk; Denmark was responsible for Poland (because of the island of Bornholm); and West Germany was responsible for East Germany...[44]

May be this new interpretation of the political and military realities explains the reasons behind the angry American reaction against the late Swedish Prime Minister Olof Palme's criticism of the Vietnam War: He did not tell the Swedish public about his government's contribution to the war. The Americans had got much of the intelligence about Soviet air defences, radars and surface-to-air missiles from Swedish Sigint in the Baltic. It was precisely this intelligence that helped American bombers penetrate the airspace above Hanoi, which explains why many US intelligence officials regarded Palme as a hypocrite.

CONCLUSIONS

As a function of geography, Scandinavian Sigint was superbly placed to cover the Soviet war machine. All available evidence indicate that most, if not all, resources were committed against the Eastern Bloc military target.

The combined intelligence collected from the Sigint land stations, from the Norwegian spy vessel *Marjata* in the Barents Sea, from the Swedish *Orion* in the Baltic Sea, from continuous 'Ferret' flying missions along the Soviet perimeter, and from numerous Humint missions into the ports of Murmansk, Archangel and Leningrad by agents on-board merchant ships, gave the United States and the Western Allies an enormous amount of intelligence about the Soviet war machine.

We have not, as yet, any reports that finally confirm the breaking of postwar Soviet military ciphers. On the contrary, despite sending literally tons of intercepted communications to the NSA, no report has ever come back, for example to Norway, that indicates that the massive codebreaking efforts were successful. What we have is some incidents where communications were read – not because of successful codebreaking, but because the Soviets made mistakes in the uses of their communication procedures.

We know that 'black-bag' jobs were done by breaking into foreign embassies and stealing codes and ciphers, and we know that cipher clerks were regularly targets of blackmail and other kinds of pressure and temptations. Despite that, no reports have reached the public domain that confirm that codebreaking in later years did produce great results against the Communist bloc. The Defence Intelligence Staff and the Security Service in Norway did for years monitor the radio communications between all East European embassies in Oslo, and their home stations. In the 1970s, when hunting for a spy in the Norwegian Foreign Office, it was seen that the radio traffic between Oslo and Moscow Central peaked once every month, and it was assumed – rightly as it turned out – that the Soviets were sending home important information after meeting a spy in Oslo. Despite an enormous effort, with the help of the NSA, not a single word of the traffic could be read.

The agent, a female secretary, was only arrested in 1978 through police surveillance. She had then been a spy for nearly 30 years. The case was filed away as Operation 'Impossible'.

Nevertheless, it is reasonably clear that the combined Scandinavian Sigint effort gave the United States and the Western Allies an almost complete insight into the order of battle and state of readiness in the Soviet

Baltic Republics, Poland and north-western Russia, and, crucial to defence planning, a fundamental overview of the capabilities of Soviet armaments.

When, in the future, all is revealed, one may conclude that US nuclear attack submarines, through Scandinavian Sigint and cable-tapping operations, were placed in a position to be able to wipe out the Soviet strategic submarines, thereby coming close to robbing the Soviets of their second strike capability.

Scandinavian Sigint also helped ensure that vital elements of the Soviet missile arsenal were rendered obsolete, and would have been quite useless against American naval targets in case of war. We do not yet know how this brilliant product of Scandinavian Sigint influenced the new and offensive US Maritime Strategy and other important policy decisions in the later stages of the Cold War – but the influence may prove to have been great.

THE DOWNFALL

In 1992, the great downfall of Scandinavian Sigint occurred, when the US Congress drastically reduced US payments for signals intelligence operations directed against the former Soviet Union.

Norway and Denmark were hit very hard. In Norway alone, 350 people immediately lost their jobs, when four out of six Comint stations were closed.

The Soviet war machine had to a large extent disintegrated. There was no longer any need for huge hour-to-hour Sigint operations, as the level of activity within the Soviet forces diminished, and the threat appeared to vaporise. Today, in Norway, only the Comint and Sigint stations in the Soviet border area are operational. They have just gone through expensive modernisations, making the intercept process more or less automated, and integrating them with American and Canadian stations to cover the whole Arctic area. The 'South Sea' Sigint station is being equipped with a huge radar which can look deep into space, while the new 5,000-ton spy ship *Marjata* continues to prowl the waters of the Barents Sea and elsewhere, spying on Russian submarines and missile tests. Seventy people at SOSUS headquarters listen to the faint sounds of the few Russian nuclear submarines still operational, while the 'Cod Hook' station tracks Russian satellites.[45]

The changes and the cutbacks have been dramatic – likewise in Sweden and Denmark, where many old Comint stations have been closed.

For more than 40 years the threat from the Soviets was the main *raison d'être* for Scandinavian Sigint. We do not know if it would have been able to fulfil its primary mission: to provide the West with a strategic warning of Soviet war preparations – as the threat never materialised.

The paradox is, of course, that none of the Sigint agencies – either in Scandinavia, nor in the US, or in the rest of Europe – foresaw the dramatic changes which appeared in Eastern Europe, and which culminated in the fall of the Wall, despite the enormous amounts of sophisticated equipment, and the billions of dollars spent.

LOOKING FOR A NEW MISSION

Today, Scandinavian Sigint operators are desperately looking for a new mission. They try to find it in the Balkans where a Scandinavian Sigint cell still is operational in Bosnia as an element of the Stabilisation Force. They are trying to find it in other areas – whether it be fighting terrorism, organised crime, or other threats against society.

Curiously, we have not one single reported and confirmed example of Scandinavian Sigint being used to cover civilian targets – except one instance where Norway or Sweden allegedly intercepted the communications of the Quebec separatists in Europe, on behalf of the Canadian government.[46] This may be explained as the effect of excellent cover-ups – or because the Soviet Union was the all-dominating priority, which I think, to a large extent, is the truth.

This again may soon change as it is evident that the Scandinavian agencies are about to involve themselves in intercepting satellite communications. As late as 1994, the NSA encouraged the Norwegian Defence Intelligence Staff to monitor civilian traffic to combat possible terrorist threats against the Lillehammer Olympic Games.[47] The plans included a Satcom intercept station with 125 employees, but the Norwegian government said no. The 100 million-dollar investment and the political risk was deemed to be too high. It is widely believed among intelligence officials that, at the same time, both Sweden and Denmark said yes. They entered the Satcom intercept business with new and sophisticated stations, which give them the possibility to intercept a huge variety of military and civilian communications, to conduct economic spying, and even enter the famous Echelon partnership.[48]

This may signal a new era – and indeed a new life. During the Cold War, Scandinavian Sigint became a one-eyed monster, dedicated to the overthrow of Soviet Communism. If the Sigint organisations can adapt to a new and completely different world and lift the pre-World War II heritage, where I started this account, they may again become providers of independent information for the decision-makers in the Nordic capitals.

One thing is a least for certain: in a world full of conflicts, and with only one remaining superpower, the value of qualified and independent intelligence for small nations is higher than ever.

NOTES

1. The 'base policy' was announced by the Norwegian government on 1 February 1949 in response to a Soviet note asking whether Norway, as a consequence of its membership in NATO, would allow air and naval bases to be established on Norwegian territory. The answer was negative with some rather subtle qualifications. No foreign troops would be stationed as long as Norway was not attacked or under threat of attack. Denmark later announced similar reservations at the same time 'allowing' the Americans to station nuclear weapons on Greenland. For discussions about the base policy, see for example Mats R. Berdahl, *The United States, Norway and the Cold War* (London: Macmillan 1997) p.6ff, or Rolf Tamnes, *The United States and the Cold War in the High North* (Oslo: Ad Notan 1991) pp.60ff and 79ff. For a recent study of Sweden's problematic 'neutrality', see Ola Tunander, 'The Uneasy Imbrication of Nation-State and NATO. The Case of Sweden', *Cooperation and Conflict, Nordic Journal of International Studies* 34 (1999).
2. A parliamentary commission, appointed to investigate claims that the Norwegian Secret Services illegally had spied on Norwegian citizens, got access to all archives and officers of the three services in question: the Defence Intelligence Staff, the Defence Security Service and Norwegian Counterintelligence, which is a police unit. Despite interviewing more than 600 former and serving officers during 1994–96 and getting, in the Western world, unprecedented access to Intelligence records, the commission found very little to support the allegations. Although the commission left some questions open, its main criticism was directed against Norwegian Counterintelligence Service which had been too diligent in following the orders they got from the different post-war governments. They had spent too much time and resources on spying on the Communists and other left-wingers for too long. The commission was named after its chairman, supreme court judge Kjetil Lund. The Lund Commission's 600-page report was published on 28 March 1996 as *Parliamentary document Nr. 15* (1995/96). In another move to stem the flood of more or less fantastic accusations against the Defence Intelligence Staff, historians Olav Riste and Arnfinn Moland were engaged to write the staff's official history. The study was called *'Strengt Hemmelig': Norsk etterretningsteneste 1945–1970* [Top Secret] (Oslo: Universitetsforlaget 1997) and appeared in English as *The Norwegian Intelligence Service, 1945–1970* (London and Portland, OR: Frank Cass 1999).
3. Yves Gyldén, *Chifferbyråernas insatser i världskriget till lands* (Stockholm 1931). For more information about Gyldén: David Kahn, *The Codebreakers* (NY and London: Scribner 1966) p.422ff. A more recent study was published by Bengt Beckman as *Svenska kryptobedrifter* [Swedish achievements in cryptology] (Stockholm 1996). Beckman came to the Swedish Sigint Service, *Försvarsväsendets Radioanstalt* (FRA) or National Defense Radio Institute, as a young recruit in 1946 and retired as head of cryptanalysis in 1991. Gyldén was one of the founding fathers of Swedish Sigint before World War II, but could not stand his fellow codebreaker Arne Beurling, later a professor at the Institute for Advanced Study at Princeton and by many considered the most brilliant of the Scandinavian cryptanalysts. Gyldén was also president of the Swedish Rugby Union, and fought his most famous battle when tensions between him and the great Beurling reached such proportions in 1940 that it ended in a bloody fistfight outside the headquarters of the Defense Radio Institute in Stockholm. See Beckman (*supra*) p.124.
4. Alf R. Jacobsen and Egil Mørk, *Svartkammeret: Den innerste hemmeligheten* [The Black Chamber] (Oslo: Cappelen 1989). I wrote the book with Egil Mørk, who, at only 15 years of age, in 1936 became the youngest member of Roscher Lund's Cryptology Club. Colonel Roscher Lund died from cancer in 1976, spending his last years working on transportation matters in the Norwegian Department of Trade without getting the public recognition he deserved.
5. For more information about Damm, Hagelin and the Cryptograph company: Kahn (note 3) p.415ff. and Beckman (note 3) p.29ff.
6. For a good, brief account of the exciting history of Finnish Sigint; see David Kahn, 'Finland's codebreaking in World War II' in *In the Name of Intelligence. Essays in Honor of Walter Pforzheimer* (Washington DC 1994).

7. Beckman (note 3) p.55ff.
8. When a member of Roscher Lund's private network of informers broke into the German consulate in Bergen and stole the content of the wastepaper baskets, it gave him and his group the ability to break the code used on messages between the German Abwehr agents in Bergen and their headquarters in Hamburg. On that occasion Roscher Lund apparently went to Stockholm and got help from his friend, Yves Gyldén, although no written documentation exists. The cooperation seems to have been close between Stockholm and Helsinki, although Arne Beurling in 1940, at the famous Operakällaren bar in Stockholm, castigated Erkki Pale for Hallamaa's apparent unwillingness to share Finnish results against Soviet systems with the Swedes. The young Beurling had visited the Finnish codebreakers in January 1940, during the Winter War. The temperature stood at 40 degrees Celsius below zero when he was invited into the sauna followed by a roll in the snow. 'It's the first time I've experienced a temperature difference of more than one hundred degrees', said Beurling who enjoyed physical hardship.
9. Tore Pryser, 'From Petsamo to Venona', *Scandinavian Journal of History* (1999).
10. See Nigel West, *Venona: The Greatest Secret of the Cold War* (London: HarperCollins 1999) p.1ff.
11. Jacobsen and Mørk (note 4) p.38ff.
12. In *The Destruction of Convoy PQ17* (London: Cassell 1985). David Irving credits a Luftwaffe unit in Northern Norway with intercepting and breaking British naval ciphers. However, Beckman (note 3) writes (p.187f.): 'When the Germans attacked Soviet Russia, the Swedish codebreakers got unexpected help. The Germans were intercepting and breaking Soviet traffic in Northern Norway and Northern Finland and sent their results to Berlin on the Geheimschreiber teleprinter lines through Sweden. The Swedes [*who read the Geheimschreiber, my comment*] now could check their own codebreaking results against what the Germans had achieved...When a convoy approached Northern Russia, it was overflown by Soviet aircraft who reported the number of ships, their speed and positions. Later, when Soviet naval vessels were sent out to escort the convoys, the radio traffic increased. Swedish Comint intercepted and read the Soviet telegrams and so did the German/Finnish radio intelligence service in Northern Finland, unfortunately...We sat ringside and could follow the progress of the convoys...We knew what the Germans knew, and we knew how they planned to destroy the merchant ships...' According to Beckman a young Finnish intelligence officer, listening to Soviet Air Force traffic, in June/July 1942 intercepted and broke a telegram, giving a complete description of the famous PQ17 convoy which a few days later was destroyed by German bombers and U-boats. Some time later the Finns broke another Soviet telegram, describing the next convoy, PQ18. The decoded telegram was given to the Germans and transmitted through Stockholm to Berlin. Thirteen of 40 ships were sunk. The Swedish cryptanalysts, reading both the German/Finnish and the Soviet traffic, and knowing what might happen, found the experience nerve-racking, according to Beckman. The Swedish chief of Combined Intelligence, Col. Bjørnstjerna, gave some of the intercepts to the British Naval Attaché in Stockholm, Capt. Henry Denham. When it was discovered, the Colonel was fired. See Henry Denham, *Inside the Nazi Ring* (London: John Murray 1984) p.79ff.; F.H. Hinsley, *British Intelligence in the Second World War 2*, Vol. II (London: HMSO 1981) p.210ff, and Beckham (note 3) p.184.
13. Beckman (note 3) p.194ff.
14. Pryser (note 9). Mainly for reasons of Cold War politics, I suspect, few seem to differentiate between the Winter War (Nov. 1939–March 1940) when the Soviet Union attacked Finland, and Operation 'Barbarossa' in June 1941, when the Finns joined the German attack against the Soviets in the North. A large number of German and Finnish codebreakers worked side-by-side behind the Murmansk front in Northern Norway and Northern Finland. They had to flee in the autumn of 1944 when the Russians broke through the German lines and advanced into Norway, a little known chapter of WW2. The codebreakers escaped in two groups. The 'Stella Polaris' group came directly to Sweden when Finland capitulated in Sept. 1944. The German *Meldekopf Rovaniemi* (later *Meldekopf Nordland*) who had worked with two Finnish Sigint companies, plus its own German troops, in Northern Finland, retreated to the south of Norway. When the German Army in Norway finally surrendered in May 1945, the

SCANDINAVIA, SIGINT AND THE COLD WAR   241

remaining 36 codebreakers crossed into Sweden from the town of Lillehammer with their crypto material in boxes and suitcases. What happened after the group was flown to Wiesbaden and taken into custody by the OSS, is not known. It is not unreasonable to think that some of their material ended in the hands of the US National Security Agency, and therefore may help explain the remaining questions surrounding 'Stella Polaris'. See the discussions in West's Venona book (note 10) p.8ff., and in C.G. McKay's, 'Debris from Stella Polaris: A Footnote to the CIA-NSA Account of Venona', *Intelligence and National Security* 14/2 (1999) pp.198–201.

15. In recent years there has been a growing willingness among historians in the Nordic countries to penetrate the myths of WW2 and the rhetoric of the Cold War and look more soberly on the historical events. Two excellent Danish studies are Henrik Dethlefsen and Henrik Lundbak (ed.) *Fra mellemkrigstid til efterkrigstid* (Copenhagen 1998) and Claus Bryld and Anette Warring, *Besættelsestiden som kollektiv erindring* (Copenhagen 1998). See also note 16.
16. Wilhelm Christmas-Møller, *Obersten og kommandøren* [The Colonel and the Commander] (Copenhagen 1995) p.29ff.
17. Interview with former British Naval Intelligence officer, 6 Nov. 1998.
18. Christmas-Møller (note 16) p.145ff.
19. Col. Roscher Lund's report on the work of the London Intelligence Staff, p.11.
20. In 1954 Norway signed a contract for the very first commercially produced computer in Britain. The Mercury computer was delivered from the Ferranti works in Manchester in 1957 and paid for by secret Intelligence funds. The money, about 200,000 US dollars, a considerable sum at that time, apparently represented the US/British investment in the Stay Behind networks in Norway. Nothing was entered into the books, and the machine was smuggled to Norway in the diplomatic post. Nominally owned by the Norwegian Defence Research Institute, the computer was used to solve their mathematical problems, and also, to try to break Soviet ciphers – without success. See, Jacobsen and Mørk (note 4) p.199ff., and Olav Njølstad and Olav Wicken, *Kunnskap som våpen* [Knowledge as a weapon. The history of the Defense Research Institute, 1945–1975] (Oslo: TANO Aschehoug 1997) pp.395ff.
21. For the allegations against Crypto AG, see Scott Shane and Tom Bowman, 'No Such Agency. Rigging the Game', *Baltimore Sun*, 10 Dec. 1995 and Wayne Madsen, 'Crypto AG: The NSA's Trojan Whore?', *Covert Action Quarterly* No. 63 (Winter 1998) pp.36–42.
22. See 'From the memoirs of a Norwegian cryptanalyst,' paper delivered at the Eurocrypt 1993 conference by Professor of Mathematics, Ernst Selmer, who worked in Roscher Lund's cipher office in London during WW2. Selmer writes:

> I was on a course in the Hagelin factory in Stockholm, and had just taken an electrical machine into parts and pieces, when Hagelin senior passed by. Addressed to me, he said that 'it is easy to take it apart, more complicated to reassemble it, and much more difficult to make it work afterwards'...But the next day, I could show him a working machine...(We) suspected that the Hagelin cipher might be broken, at least from corresponding plaintext and cipher, that is, from pure key. We therefore shuffled different parts of each message, and used a strange mixture of, often abbreviated, Norwegian and English...I continued (to work on breaking the Hagelin machine cipher) after the war, and it was quite clear that only manual calculations would be too time-consuming. In 1952, I still did not have an electronic computer at hand, but now I did at least know very much about computer design [*Selmer had worked on computers with von Neumann in the United States, my comment*]... I used this to draw a special purpose relay computer, earmarked for breaking the Hagelin cipher. It was built...and was used for several years breaking the diplomatic cipher of some foreign countries.

> Selmer's work as a consultant for the Defence Intelligence Staff was considered to be so secret that he had to get a special permission from the taxman to leave his crypto income out from his tax returns, which in Norway in those days was harder to get than Top Secret clearance.

23. Roscher Lund's plan, 28 April 1944. See Jacobsen and Mørk (note 4) p.157ff.
24. Erling Sverdrup interviews, 1986.

25. Riste and Moland (note 2) p.24ff.
26. Private information and Riste and Moland (note 2) p.50.
27. Beckman (note 3) p.230.
28. Dr Cees Wiebes, 'Sssh! It's Swedish SIGINT. A short history of the FRA', private paper, Amsterdam 1999.
29. Tunander (note 1).
30. Wiebes (note 28).
31. Ibid. pp.11. See also, Peter Bratt, *IB och hotet mot vår säkerhet* (Stockholm 1973) and Peter Wright, *Spycatcher* (NY: Viking 1988). The warship had of course featured in the Commander Crabb incident at Portsmouth in 1956.
32. Riste and Moland (note 2) p.131ff.
33. Confidential interviews with Norwegian intelligence officers, 1996–99.
34. Because of his revolutionary youth, Col. Vilhelm Evang was regarded by many Americans as a communist and a security threat. One of the men who followed Evang with suspicion was James Angleton, head of CIA Counterintelligence. When the paranoid defector Anatolij Golitzin in 1961 told Angleton about a female spy, who had served at the Norwegian embassy in Moscow, they pointed their fingers at an unmarried Norwegian woman who belonged to Evang's staff. She had served in Moscow 1956–59 and occasionally done services for the CIA. Encouraged by Angleton, the woman was arrested in 1965 and spent three months in jail before she was released. She was innocent. The real spy was another woman who had served in Moscow before her, from 1947 to 1956.

  Angleton and Golitzyn, supported by Evang's enemies in Norway, went after the wrong woman to get rid of 'the red colonel' at the top of the Norwegian Defence Intelligence Staff. The results were tragic. Both Evang and the head of Norwegian Counterintelligence were sacked. The woman, who had suffered a wrongful arrest, never recovered from the shock. The other woman kept on working as a Soviet agent for another 12 years. See Tom Mangold, *Cold Warrior. James Jesus Angleton: The CIA's Master Spy* (London: Simon & Schuster 1991) p.114 ff. and Alf R. Jacobsen *Iskyss* (Oslo: M. Aschehoug 1991) p.239.
35. Riste and Moland (note 2) p.150ff.
36. Stephen Dorril, *MI6. Inside the Covert World of Her Majesty's Secret Intelligence Service* (London: Free Press 2000) p.249ff.
37. Christmas-Møller (note 16) p.133.
38. Private information and *Helsingin Sanomat*, 22 Nov. 1998.
39. Riste and Moland (note 2) p.156ff.
40. Confidential interviews.
41. Chris Drew and Sherry Sontag, *Blind Man's Bluff* (NY: Public Affairs 1998) p.209ff.
42. Confidential interviews.
43. Wiebes (note 28) p.8.
44. Tunander (note 1).
45. A heated debate has been going on for several months about the new 'Have Star' radar at the old 'South Sea' site in Norway. Recently transferred from Vandenberg Air Force Base in the United States, the critics claim that the radar is part of a planned US anti-ballistic missile defence system.
46. Mike Frost, *Spyworld* (Toronto: Doubleday 1994) p.239. Despite vigorous efforts, no one so far has been able to confirm Frost's story in Scandinavia. The Quebec separatists, at that time, had their European headquarters in Paris. Later, they opened an office in Stockholm. The telegrams mentioned by Frost may have been intercepts of traffic in and out of this office.
47. Confidential interviews.
48. The Danish Satcom intercept site is believed to be at Aflandshage, the old 'Villa' headquarters just to the south of Copenhagen. See the articles in the Danish newspaper *Ekstrabladet*, 17 Sept. 1999 and 18 Sept. 1999.

8
Dutch Sigint during the Cold War, 1945–94

CEES WIEBES

On 28 November 1973 the Intelligence Co-ordinator dispatched a memorandum to the Netherlands Prime Minister and the Foreign Minister. In this Top Secret minute the co-ordinator stated: 'Dear Prime Minister and foreign secretary. I think that I do not have to convince you that the WKC is the most valuable asset we have to collect an intelligence product that is valuable for all interested parties.' Indeed, there was no need to persuade these two in those days, Prime Minister Joop den Uyl and Foreign Secretary Max van der Stoel, of the value of the signals intelligence product produced by the WKC.[1] The two ministers needed no convincing because from the start of the oil crisis in October 1973 they received many and excellent communication intercepts from the *Wiskundig Centrum* (Mathematical Centre, WKC). It is the Dutch equal of the American National Security Agency (NSA) or British Government Communications Headquarters (GCHQ). The name of *Wiskundig Centrum* (Mathematical Centre, WKC) was changed in 1982 into *Technisch Informatie Verwerkings Centrum* (TIVC, Technical Information Collection Centre). This Sigint organisation played an important role as regards interception and decoding of satellite- and other communication traffic. During the October war in 1973 and subsequent oil embargo against the Netherlands several times every day a whole bundle of intercepts (the so-called Red Edition containing European and Middle Eastern diplomatic intercepts) was dispatched by a special courier to the residence of the Dutch Cabinet.

Writing a contribution about the history of signals intelligence in The Netherlands was not an easy undertaking. As the good book says: Sigint is intelligence derived from a combination of Communications Intelligence (Comint) and Electronic Intelligence (Elint). However, I have to postulate that Elint is done in the Netherlands solely on very small scale by the Navy Intelligence service. Accordingly, in Holland we are mainly dealing with

Comint. Nevertheless, here I will write about Sigint and not distinguish all the time between Comint and Elint.

As previous contributors in this work already have pointed out: getting hold of archival materials and in particular as regards Comint and Elint related matters, is extremely difficult and in some countries even impossible. For this reason, I am very grateful that the Netherlands Ministry of Defence, the Ministry of Foreign Affairs and the Cabinet Office declassified hundreds of formerly top secret official documents to me dealing with this signals intelligence service. This service, nowadays named the *Afdeling Verbindingsinlichtingen* (AVI), is more or less the roof over three Sigint pillars being the Sigint units of the Navy, Army and Air Force. However, in the rest of this contribution I will stick to its former name TIVC.

As regards the period of the declassified documents which were made available to me: they range from 1946 to 1994 and deal not only with the internal history but also with the private discussions in the Cabinet as regards the financing and future tasking of the TIVC. Of course, not everything was declassified, as you will understand. More than nine pages solely dealing with the successes and achievements remained top secret.[2] I am also grateful and indebted to various 'old intelligence hands' who worked in the past at the TIVC or in the Dutch Directorate of Navy Intelligence (DNI) whom were willing to share their knowledge with me about the operations of the TIVC.[3]

Nevertheless, seen in this light, it will be clear to the readers that this analysis has to be regarded as a real first attempt to reconstruct the Sigint history of the WKC/TIVC during the Cold War. This is a hard job enough already in view of the many sensitive issues involved. Accordingly it is certainly and absolutely not the full, complete and final history. On the contrary, many blank spots still have to be filled in, in particular the Sigint operations against the former communist enemy. Giving this aspect and the limited space, I will therefore not pay much attention to the role of the Code Coordination Bureau, the various Sigint activities of the Netherlands Army (898th Radio Battalion) and Air Force (1st LVG) and other organisations involved in the Netherlands with Sigint related matters.

THE PRESENT SITUATION

The foundation of the Netherlands intelligence community was laid between 1940 and 1945. Already during World War II the Netherlands government-in-exile realised that it had bitterly failed as regards its pre-war intelligence capabilities. In London therefore the new structure of a future Netherlands intelligence community would be developed. The basic elements of such a future community were worked out and soon after the liberation various new services were established. As regards national security a service was established soon after the liberation of the Netherlands in May 1945 and it mainly dealt with tracking down collaborationists, members of the German intelligence and security services like Abwehr and Sicherheitsdienst and possible units of the Nazi Stay Behind organisation, the Werewolf. This service functioned until 1949.

The Cold War had changed the scene and a manifestly strong Communist Party led to the establishment of a National Security Service (*Binnenlandse Veiligheidsdienst*, BVD) which is comparable to MI5 in Britain. The BVD came under the Minister of Interior despite efforts by the Justice Department to get the responsibility for the security service. However, the memories of World War II were still very fresh. In order not to create the general impression that the BVD would be a kind of 'new Gestapo' it was decided to make a separation between the counter-intelligence and executive functions. The former was given to the BVD. Investigation, arrest and detention became a responsibility for the police.

The BVD was formally established in 1949 by confidential royal decree. This defined its functions as collecting information about all persons who tended to, were or had been involved in activities dangerous to the Netherlands or to friendly countries; gathering information about politically extremist movements; the furtherance of security measures in all vital and vulnerable government and private institutions and industries; and the maintenance of liaison with friendly foreign security and intelligence services.[4] The BVD is a service with roughly 500 employees.

From 1945 onwards there were also three intelligence and security services of the armed forces being the Army, Navy and Air Force Intelligence services. However, plans to integrate these three services into one Defence Intelligence Agency (*Militaire Inlichtingendienst*, MID) never materialised because the three military services were refusing (as in many Western nations) to relinquish their sovereignty and each service went its own way. The MID was formally created in 1987 when the three intelligence services of the armed forces were integrated into this new organisation. The prime task of the MID is to gather intelligence as regards the strength, activities and intentions of foreign armed forces which in the past meant mostly Soviet and Warsaw Pact nations. The MID has also a close liaison with the Intelligence Division of NATO. Altogether in 2000 more than 700 officials were working at the MID.[5] Their present activities are also dealing with peacekeeping and/or peace-enforcement operations in which Dutch military forces were or are involved. Operations have been executed in Cambodia, Haiti, Cyprus, Namibia and Bosnia.

FIGURE 8.1
THE HQ OF THE DUTCH DEFENCE INTELLIGENCE AGENCY

Another important element of the intelligence community has been its foreign intelligence service. In 1946 the Dutch government decided to establish such a service which was later called the *Inlichtingendienst Buitenland* (IDB).[6] In a confidential royal degree its task was the gathering of intelligence concerning developments abroad which could be advantageous to the Kingdom of the Netherlands. The service was dismantled in 1994 by the Dutch Cabinet.[7]

THE BEGINNING

Late in 1946 it was decided at the request of the Ministry of Foreign Affairs to establish a service, which would deal with decrypting foreign codes and cipher traffic. In January 1947 the Cabinet formally gave the green light and the first head of the TIVC would become Colonel J. A. Verkuyl, the head of the Code Coordination Bureau. This unit was initially responsible for codebreaking but later in particular involving the manufacturing of codes, namely the art of cryptography, and keeping up to date with the know-how as regards cryptanalysis, being the process and science of converting encrypted or encoded messages into plain text.[8] The TIVC was from the start a part of Dutch Naval Intelligence (*Marine Inlichtingendienst*, MARID), called Marid VI.

The primary reasons why the TIVC became a naval outfit were actually threefold. First of all, it was much easier to camouflage the yearly TIVC budget in the overall budget of the Netherlands ministry of the Royal Navy. Second, the Dutch Navy had ample space to house the TIVC, which other departments did not have. Third, the Navy had already various Sigint experts, which would sooner or later come back from the Netherlands East Indies (NEI) where they were still on active military duty. In particular they were working in this former Netherlands colony as cryptanalysts.[9]

In the beginning the TIVC had no interception capabilities and the material in the beginning consisted mainly of coded messages which arrived by safe hand. In 1948 a group of Navy telegraph operators were hired and two intercept facilities were founded: one in Amsterdam, which existed from mid to late 1948. Hereafter came Hellevoetsluis, which is situated south-west of Rotterdam, and this site operated from late 1948 until October 1952.

From the start the activities of the TIVC were greatly helped because Dutch intelligence officers had recruited an American expert who worked at the NSA. This spy was the cryptologist Joseph Sidney Petersen who worked at Arlington Hall as head of the Russian section.[10] He forwarded useful and valuable information as regards codebreaking activities and methods. For example, he provided to his Dutch caseofficers and handlers a top secret

American analysis of the inner workings of the Hagelin Cryptograph Type B-211 which machine was used in those years by many countries. He also informed his Dutch counterparts that the Americans were reading the Dutch diplomatic traffic.

American officials regarded the Petersen case as a very serious matter. No one less than James Jesus Angleton, working at the counter-intelligence unit of the CIA, flew to Holland. He was wearing a belt in which a onetime pad was hidden for his personnel secret communications with Langley. This immediately led to a row with the CIA station chief in The Hague, Justin O'Donnell. He gave Angleton a choice: either to communicate to Langley via O'Donnell or 'get the hell out of here'. After these threats Angleton caved in.

In talks with officials of the Netherlands Ministry of Foreign Affairs Angleton demanded to speak to Petersen's Dutch contacts. This was flatly refused. He then asked for files and correspondence dealing with Petersen. He was told that there were only two letters, which he was not allowed to read. This infuriated Angleton. However, the Ministry allowed him to put forward a list of questions.

From Angleton's list of questions transpires that he was mainly interested in the question if Petersen had given any secret intelligence to his Dutch runners as regards foreign ciphers and codes, foreign cipher systems and in particular secret material on American efforts regarding North Korean codes and ciphers. Had Petersen for instance given away a very sensitive document entitled 'Routing of North Korean Political Security Traffic'?[11] Dutch officials denied this. Angleton was also interested if the Dutch had forwarded American intelligence documents to third countries. Again this was denied. Interestingly, Angleton also repeatedly wanted to know if Petersen had told the Dutch anything about the American relationship with Boris Hagelin, the constructor of cipher machines. The answer was again a denial. In order to calm down Angleton Dutch officials told him that everything Petersen had given them had been destroyed.[12] Angleton left Holland empty handed and in a very angry mood: the Dutch had kept him completely in the dark as regards the documents Petersen had given them.

After a trial and conviction of Petersen the story hit the Dutch press. In particular the Communist daily *De Waarheid* paid quite some attention to the spy case to the dismay of many American officials. The daily alleged that the CIA had set up networks in Holland. However, the first secretary of the American embassy, Oliver M. Marcy, sent a calm report to the State Department.

> If the communist press had significant circulation in The Netherlands, which it does not, there might have been occasion for some concern

at this interpretation. However, communist assertions are viewed with some scepticism in this country, and public interest in the Petersen affair had already flagged by the time the communists came up with their idea. It therefore seems unlikely that many Dutchmen have been persuaded that in the Petersen case the finger of suspicion should be pointed at the United States.[13]

Also the British embassy reacted relieved. The Petersen case produced only 'a mild stir'.[14]

Nevertheless, Dutch intelligence officials are certain that the Petersen case produced quite a chill in the American-Dutch intelligence relationship. It certainly did not lead to a situation like in Britain, Norway and many other countries whereby the NSA would finance a large part if not most of the yearly budget of the Sigint service. The Dutch had to do it on their own and in the end the outcome was not that disadvantageous.

In 1948 roughly 50 officials were working at the TIVC. They booked their first successes in breaking various Indonesian codes delivered to them by safe hand. The machinery they used for this effort was hired from IBM.[15] The TIVC was originally housed at the Navy Barracks at Kattenburg in Amsterdam and was a support unit of the Commander-in-Chief of the Royal Netherlands Navy. The TIVC which eventually had world-wide capabilities was not only intercepting maritime military networks for the Intelligence Bureau of the Royal Dutch Navy but also started to intercept diplomatic traffic by radio and international postal and telegraph communication networks used by companies and individuals.

From the start there been a secret arrangement with the main postal and telegraph office in the centre of Amsterdam where all open but also coded and ciphered telegrams were brought to in order to be dispatched by telegraph. A similar arrangement was executed in The Hague. At these main postal and telegraph offices copies were made of most if not all of the important telegrams and then delivered to the cryptanalysts of the TIVC. In this manner extremely valuable economic intelligence was gathered because in this way many commercial secrets and other trade information flowed into the hands of intelligence officials. Very often the economic traffic (cipher or encoded) by telegram and later telex posed few problems for the codebreakers. The codes were often simple to solve and decipher.[16]

The TIVC (unlike the Army and Air Force Sigint units) never had any analytic and distribution capabilities for non-military traffic. For this reason the *Inlichtingendienst Buitenland* (Netherlands Foreign Intelligence Service) was used as conduit to the various ministries and in particular to the Foreign Office and the Departments of Trade and Agriculture. For example, Sigint was responsible for the fact that the Ministry of Trade came to understand why Czech industrial goods were not delivered to the Netherlands but were diverted to the Soviet Union and Poland. This happened after the so-called *coup d'état* in Prague in February/March 1948. When a new trade agreement had to be negotiated with the Czech government the Dutch were in a far better position because of the intelligence derived from these intercepts.[17]

Except for the traffic to and from foreign embassies in The Hague the TIVC also intercepted many commercial messages dealing with quotations and tenders, with which the interests of Dutch companies and firms were involved. A big take of this kind of Sigint traffic went to the Department of Trade and in particular the unit dealing with foreign trade and commercial relations and the unit Industrial Projects.[18] These units were used as a conduit to forward these commercial intercepts to multinationals and large firms like Philips, Shell and the aeroplane manufacturer Fokker.

To the Netherlands Department of Agriculture intercepted and deciphered traffic was passed along dealing with for example the results of harvests and crops in foreign nations. The Dutch ministry was for instance in particular very keen on learning more about the yearly harvests of potatoes in Malta. While the Netherlands was and still is a big producer and exporter of potatoes and products made from potatoes themselves the Ministry of Agriculture could learn in this way more about the harvest results of one of their main competitors on the European agricultural market.

The independence of Indonesia in late 1949 resulted immediately in the return of many specialists and their interception equipment from the Netherlands East Indies. Some of these specialists stayed in the region and were transferred to the last colonial Dutch possession in Asia, the island of New Guinea. There they were later involved in the defence of the island against Indonesian military incursions and attacks, a matter to be dealt with extensively in the next contribution to this work, written by Wies Platje. In the meantime the plans for expanding the intercept capabilities in Holland were right on track. More staff were hired and receivers, direction-finders and other equipment were ordered in the United States

In 1950 the director of the TIVC, Verkuyl, who also recruited and ran the American spy Petersen at the NSA was forced to resign. The circumstances remain clouded, even to this day. Some claim that he was fired because there was a lack of management and that he was not capable of steering the WKC internally.[19] Verkuyl took revenge because he remained head of the Code Coordination Bureau (CCB). He and his successor were apparently angry with the new Director of the TIVC, which led to the situation where the CCB refused to transfer one of the older Hagelins in their possession to the TIVC. This very vulnerable cipher machine was still used in the 1950s and 1960s by many nations, in particular those which recently had become independent in Africa and Asia. Also many multinationals and big firms still used the Hagelin. Within the TIVC they did not possess this certain type and for this reason they could not do anything with many intercepted ciphered Hagelin messages.[20]

LIAISON WITH THE AMERICANS

The successor of Colonel Verkuyl was J. Spanjaard, a retired lieutenant colonel of the Royal Netherlands Indies Army. He would give a great impetus to the expansion of the TIVC. In the background also a conflict with the Netherlands National Security Service played a role. It turned out that the director of the BVD, Louis Einthoven, had made an agreement with Verkuyl but also with the CIA station chief in The Hague, Justin O'Donnell. Einthoven could lay his hands on a copy of an original Czech codebook, which he had obtained from a Czech defector. He subsequently had asked Verkuyl to decipher the diplomatic telegrams from the Czech embassy in The Hague. This arrangement was executed without informing the director of the Directorate of Navy Intelligence.

This conflict also shed an interesting light on how the BVD and TIVC viewed cooperation with the American services. The Netherlands Directorate of Navy Intelligence Service wanted to keep a certain distance from this American intelligence service. Although there was cooperation it remained afraid of disclosing its capabilities. However, the BVD embraced the CIA wholeheartedly. There had already been talks between the BVD and the CIA in Brussels about the shipment of five direction finders to the BVD for communication intercept purposes. The BVD was in particular interested in tracking down secret Communist agent transmitters.[21] In return the BVD would deliver raw intercepts from the Soviet embassy in The Hague.

For some time Verkuyl was the liaison in this arrangement and he had forwarded on a regular basis this raw intelligence to the CIA station chief O'Donnell in The Hague. However, Spanjaard had in October 1951 stopped the weekly shipment of intercepts to the American embassy because the CIA never replied or did not return valuable intelligence on a quid pro quo basis. The American intelligence agency apparently was not willing to share its secrets with the Dutch services and the exact content of these telegrams (whether or not they were deciphered) was never made known to the Dutch services.[22]

Einthoven of the Dutch BVD talked this over on 6 March 1951 at CIA headquarters in Langley, Virginia. The Agency had made clear that a response from the forerunner of the NSA could not be expected.[23] The

American services would guard their secrets about whether they had had any success with these Dutch intercepts of Soviet diplomatic, military and even perhaps KGB and GRU traffic. However, this episode shows that the TIVC was in this early period not the sole service trying to lay its hands on intercepts. The BVD was also active in the field of signals intelligence. For example, as the Dutch intelligence historian Frans Kluiters revealed, it had its own eavesdropping site inside a building opposite the Polish embassy in The Hague. This BVD listening post was probably involved in monitoring Polish diplomatic and military traffic coming from the embassy.[24] Actually, this was almost a mirror of the situation in Warsaw. In the 1970s the Poles did the same eavesdropping with respect to the Dutch embassy in Warsaw. In those days the Dutch embassy represented Israel in Poland and was heavily involved in organising the emigration of Polish citizens who wanted to settle in Israel.

Spanjaard vigorously began expanding the TIVC. He proposed to the Cabinet various plans and ideas as regard the future status of Marid VI. The staff had to be enlarged from 92 to 165 and he introduced various units within the TIVC. There existed already from 1947 a cryptanalytic unit but he also introduced new ones such as a traffic analysis unit and also installed a documentation unit. This had to keep track of all latest developments in the domain of Sigint. Of course, Spanjaard was helped by the circumstance that the Netherlands government had signed the North Atlantic Treaty in 1949. This meant that the Dutch services now also had international obligations. Dutch Naval Intelligence for example had obligations vis-à-vis SACLANT while the Air Force and Army Intelligence closely cooperated with SACEUR. This led also to a geographical division: the latter were mostly continental and Warsaw Pact oriented. They had focus on identification, the actualisation of the order of battle and early warning.

However, the TIVC as a marine intelligence organisation had worldwide military tasks and obligations but it also had to produce political and commercial intelligence products for several ministries like the Cabinet Office and Foreign Office. For this sole reason the TIVC was in need of a greater variety of technical intercept equipment and a larger staff compared to the two other military Sigint services. The TIVC had to keep track of the Soviet, Polish and East German maritime build-up in the Baltic, Atlantic Ocean and North Sea. Although these fleets were not yet a direct maritime threat the TIVC continued to monitor all developments and especially hull-identification of Soviet naval vessels. Later the TIVC was also obliged to deal with out-of-area matters as in the Indian Ocean and the Pacific.

THE FINANCIAL PROBLEMS

With respect to the civilian tasks: the TIVC for example was instructed to monitor foreign postal radio connections for diplomatic and economic traffic, diplomatic radio communication networks of foreign governments and to study material and intercepts which was delivered by safe hand via the BVD, IDB or other sources. This led to a situation which hampered the steady grow of the TIVC and which would bother the various directors almost constantly over the last 50 years. The main problem would always remain who would finance the TIVC? Would that be the Department of Defence or should other ministries also take a part of the growing financial burden? It was a problem, which was actually never solved in a satisfactory manner.[25]

In 1950 the budget of the Department of the Navy was scaled down and it turned out that the Ministry of Defence was no longer able to pay for all the costs of the TIVC. A proposal was tabled in the Cabinet to make the various recipients of the TIVC product pay for the intercepts they received on a daily or weekly basis. Every day a so-called Green Edition was produced containing the most important diplomatic intercepts. However, all departments refused and the Department of Defence was forced to pay all costs of the TIVC. The debate over which would finance the TIVC started from this year: an issue to which I will return to later. One cost saving solution proposed was the idea of hiring conscripts. However, this was rejected due to the fact that the security clearance for the very delicate work at the TIVC was extremely high. Another idea was to separate the TIVC from the Navy and make it an extension of the Foreign Intelligence Service. However, this plan also never materialised.[26]

Despite all these financial problems the search for good intercept sites continued. One was established in October 1952 in the north of Amsterdam at Zeeburg. This site replaced Hellevoetsluis. However, in view of the 150,000-volt power lines crossing the station it soon turned out that Zeeburg had limited intercept capabilities. The TIVC had no access to the intercept stations of the Army Intelligence in Gorkum and Air Force Intelligence in Zeist. From those sites these units could monitor the communication radio networks of the Soviet military forces in Poland, East Germany and Czechoslovakia. However, sites belonging to the Dutch Postal and Telegraph Service (PTT) were used like the ones in Goes en Ockenburg. From there the Eastern European and Middle East diplomatic traffic was monitored. Goes also monitored the traffic of East European embassies.[27]

Later also the PTT sites at Kerkrade and Waalsdorp were used for the interception of diplomatic radio networks.[28] Kerkrade for example monitored the Yugoslav transmitter YTC in Belgrade, which transmitted primarily messages of economic importance. The transmitter YTC also produced diplomatic traffic from Yugoslav Foreign Secretary, K. Popovic, which he sent to the Yugoslav ambassadors abroad.[29] The Army was not happy with the TIVC use of Goes and Kerkrade. It accordingly could not operate these sites and this hindered its own eavesdropping activities.

Until 1950 there was no formal cooperation between the Army and TIVC. This is not so strange as it appears. The Navy was primarily interested in world-wide maritime targets and was involved in intercepting diplomatic, political and economic traffic. The Army and Air Force were mainly interested in the military developments in the German Democratic Republic and Poland. They mainly targeted Soviet and Warsaw Pact armed forces. In the framework of assigned tasks within NATO Army and Air Force intelligence were soon expanding their intercept capabilities. Despite the different targeting, the TIVC tried to improve the cooperation and a formal merger was proposed but on conditions as formulated by the Navy. This was rejected by the two other military intelligence services because they probably feared that their signals requirements were to not be taken care of sufficiently in such a new-fashioned TIVC.[30] Nevertheless, the later diplomatic and military conflict with Indonesia over New Guinea would force the Dutch Army and TIVC to cooperate more closely. But it has to be said: cooperation with the Army Sigint service would never reach an ideal situation in the Netherlands.

What certainly improved in the early 1950s was the governmental steering of the overall collection effort of Sigint. Various official bodies were established such as the National Telecommunications Board (NTR), the Radio Intelligence Committee (RIC), the Radio Intelligence Bureau (RIB) and the Crypt-Analytic Interception Committee (CIC), These official bodies would steer the overall Sigint gathering effort better and more smoothly. In these committees representatives from the various departments and the foreign intelligence service would have a seat in order to give explicit guidance to the TIVC and the Sigint units of the Army and Air Force. More important was the direct involvement of the Coordinator (more or less the Netherlands intelligence tsar), which brought the Cabinet Office closer in contact with the work of the TIVC.

In the meantime the collection effort of the TIVC continued of course and a so-called Green Edition of sometimes 30–50 pages was produced on a daily basis. It contained mainly diplomatic intercepts about political and economic matters and was sent off six times a week. In the beginning there was also a special Blue Edition which contained economic Sigint but this was later integrated into the Green Edition.[31] A few automatic recorders were bought which enlarged the Sigint take, although the TIVC in this phase did not work with computers such as the NSA and GCHQ had at their disposal. However, various requests for additional radio receivers and extra direction finders were consequently rejected between 1953 and 1957. Accordingly, the cryptological section was sometimes forced to even devise its own equipment and machinery mainly because of financial reasons.

For this reason Spanjaard kept on hammering for more financial resources. For example, after the Hungarian Uprising in 1956 he dispatched a damning memorandum to the Director of Naval Intelligence. He complained in this memo that because of a lack of staff and rather insufficient and not up-to-date radio equipment the TIVC had not been able to intercept enough Hungarian and Soviet diplomatic radio traffic before, during and after the uprising. Valuable Sigint was accordingly never intercepted. Spanjaard claimed that the result was 'practically nothing'.[32] For Spanjaard this experience of the TIVC not being able to keep pace with an international political and military crisis was enough reason to renew his earlier attempts to separate the TIVC from the Directorate of Navy Intelligence in 1957. There was absolutely no future for his organisation in this service.[33]

Interestingly, Spanjaard never contemplated merging or working more closely with the two other military Sigint services. And this was exactly the RIC and CIC were meant for: closer cooperation in the domain of codebreaking activities and traffic analysis. However, it never materialised in the Netherlands and in the day-to-day situation the cooperation of the TIVC with foreign Sigint services was much better compared with the collaboration between the Dutch Sigint services of the Army and Air Force.[34]

These attempts did nevertheless produce some results and shook up the government in The Hague. Extra financial support was made available and in 1958 another five Hell recorders, two Marconi direction finders and ten Collins receivers were bought in the United States. Also some new staff was hired. In this year also the first computer 'Zebra' was introduced. Actually those who worked with the machine described it to me as being more or less an advanced electronic calculator. However, this Dutch computer 'Zebra' which had the size of a small container had a serious problem. It went completely out of control when it became too hot. The night shift of the TIVC was sometimes not particularly in the mood to work all night (like many people on long and sometimes boring work). So, what did they staff do? Often they closed all windows and doors. Accordingly the temperature in the room went up and the 'Zebra' computer quickly collapsed because it ran too hot. The night shift then had nothing to do and was of course subsequently sent home.

It was not until 1963 that the 'Zebra' computer was moved to an air-conditioned area where there was also ample room for a second computer. Later the TIVC would buy several IBM Tempus computers. Soon the machinery was expanded with computers from Bell (in 1952 and in 1965), DEC (1974), the German company Siemens and also the British Racal.[35] Both were well known names and products in the world of Sigint.

Also special printers to absorb the heavy load of daily Sigint intercepts were required. However, these printers were quite expensive in those days: roughly $35,000 each. The TIVC could not pay for this but this time the NSA was willing to assist. The Americans eventually offered seven printers (almost $300,000) as a gift to the TIVC. By nature the TIVC was confronted with the question: had the NSA not build in a so-called 'backdoor' which would enable the NSA to spy on the systems of the TIVC. To counter this danger the TIVC placed all seven printers in a cage of Faraday in order to be 100 per cent sure that the NSA would not be able to eavesdrop on the Dutch Sigint service.[36]

With so much new equipment: the search for good intercept sites continued constantly which resulted in 1962 in the closure of Zeeburg. Soon a new locality south-east of Amsterdam was discovered and it turned out to be an ideal location. There was absolutely no interference with the reception of traffic and it was situated quite close to Amsterdam. The nearest town had no plans to build near this piece of real estate. After the mayor of the nearby town was convinced (he did not like the view of the huge antennas) the TIVC had a brand new site.[37] The construction of this intercept site started in 1961 and in 1962 the station was formerly opened. It was Eemnes, which still exists.

The 'Granger' antenna, which is still at Eemnes was bought from US Navy Intelligence. Actually, it was some sort of clandestine operation because the American intelligence service was not allowed by law to sell this antenna abroad. Nevertheless, the Americans decided to sell it to the Dutch service at a huge discount. The antenna 'disappeared' in the USA and the 'Granger' would finally end up in Eemnes. Nobody ever asked any questions about how it came to be there in the first place or who paid for the antenna. It was partly the American taxpayer. It was at the same time also a clear message that the Netherlands Naval Intelligence Service preferred to have close ties with their American naval counterparts and not so much with the CIA.[38] In 1972 there were in Eemnes three logperiodic antennas and about five V-antennas. Eemnes was and still is a Sigint station dealing with intercepting diplomatic and maritime HF traffic.

SIGINT AND INDONESIA

The expansion of the TIVC in staff and technical capabilities continued despite the fact the government refused to take care of proper finances for a real growth of the organisation. In the 1950s the TIVC was probably most successful in decoding Indonesian diplomatic and military traffic. The intercepts were partly done from New Guinea and were very rewarding. In part the Dutch were helped by the simple fact that the Indonesian government and military used the Hagelin cipher machine for their secure communications.

When in 1954 a rebellion broke out on various islands in Indonesia, the Indonesian government in Djakarta simply forgot that the rebels captured during the revolt many of these Hagelins. The Indonesian authorities did not change or alter the settings of their Hagelins and the negligence in the end produced quite staggering results for the Dutch codebreakers who managed to get hold of some of these Hagelins. Only the keys sometimes produced some minor problems. In view of the British and American involvement and support of these Indonesian rebels one can take it for granted that the NSA and GCHQ also obtained similar results. Actually, documentary proof of such results by GCHQ can be found in the latest release by the NSA to the National Archives in Washington DC. In the release of roughly more than 1.3 million pages of documents a researcher can find several GCHQ intercepts of Indonesian political and military traffic.[39]

These developments and codebreaking activities led to many Dutch successes in this part of Asia. To give a few examples: in January 1959 more than 1,000 enciphered telegrams were intercepted. By July 1959, this number had risen to almost 2,200. At the height of the New Guinea crisis, this number had gone up in January 1962 to almost 10,000 per month. This sheer volume, a computer and excellent staff enabled the TIVC in Amsterdam to break the Indonesian code in five weeks.[40]

This huge number enabled the TIVC to read almost all Indonesian diplomatic and military traffic. Apart from these cryptanalytic accomplishments the TIVC was also quite successful because the Netherlands Foreign Intelligence Service had recruited several high-ranking spies within the Indonesian government. For example, within the Indonesian delegation, which negotiated at Geneva in 1956 with the Netherlands government about the transfer of the last colonial possession to Indonesia, the Dutch service had recruited several agents. This major intelligence undertaking was called Operation 'Virgil' and the recruitment would even include the Indonesian foreign minister and other highly placed officials in the Indonesian diplomatic and military community in Djakarta.

These spies but also other recruited agents working at Indonesian embassies and the Foreign Ministry in Djakarta itself proved to be very valuable. They provided the clear text of many diplomatic telegrams, which the Indonesian delegation during their complex negotiations with their Dutch diplomatic counterparts exchanged with Djakarta. These clear texts of course enabled the TIVC to decipher the still coded Indonesian messages.

It is probably safe to say that Indonesia as regards intentions and capabilities did not harbour many secrets for the government.[41] A fascinating topic and quite successful intelligence story will be dealt with extensively by co-author Wies Platje in the next contribution of this study. It is also a fine example of how Sigint and Humint can be complementary.

In this respect it is all the more peculiar that the government did not pour more money into the TIVC. For in the 1950s and the start of the 1960s the organisation had proven to be capable of delivering an excellent Sigint product to the Dutch Cabinet and its diplomatic negotiators. But this did not happen. On the contrary: simple typing and other administrative work at the TIVC had to be done by trained cryptanalysts while their real intelligence work sometimes came to a standstill. Many complaints by the Director of the TIVC at the highest level at the Foreign Office were met with sympathy but not with a substantial improvement in the overall situation.[42] This was very remarkable because in particular Foreign Minister Joseph Luns, who later became Secretary-General of NATO, was very pleased with the materials he received from Sigint.

The conflict with Indonesia did however improve the contacts with the Dutch foreign intelligence service, the IDB. The heads of TIVC and IDB conferred several times and an agreement was reached. Greater coordination was worked out and the Director of TIVC even agreed to share his archives and documentation with the IDB. Interestingly enough is that again as in the past this arrangement was executed without informing the Director of Dutch Navy Intelligence, who was nominally responsible for the TIVC.

Spanjaard, as I mentioned before, a retired lieutenant colonel in the Royal Netherlands Indies Army, also conducted secret negotiations with the Dutch Army in order to improve the status of his TIVC but with no great accomplishments. The formal cooperation with the BVD did not increase either and remained also sparse.[43] One insider who attended some of those meetings dealing with expanded collaboration claimed that there were simply no mutual fields of interest. The anxious interest by the TIVC in cooperating with the Army and Air Force was, according to some sources, mainly aimed in furthering the interests of the TIVC. It was exclusively interested in using the intercept stations of both military Sigint units.[44]

THE INTELLIGENCE LIAISON

There was also an attempt to start international cooperation and Comint exchange. This was tried in the late 1950s with the British GCHQ but it led to absolutely nothing. For example: the TIVC handed over to the British Sigint service the Indonesian Air Force code, which the Dutch had broken.

It of course expected to get something in return. However, GCHQ produced nothing valuable: predominantly quite outdated material. When after various complaints the British service did not produce more substantive intelligence products, the Dutch Sigint service subsequently completely broke off the exchange of documents.[45] In later years this cooperation was resumed.

Apart from the successes against Indonesian targets more accomplishments were achieved. The traditional Dutch political distrust of the French national security policy in Western Europe led to another operation. TIVC intercepted, on behalf of the Dutch Foreign Office, the diplomatic traffic of the French embassy in The Hague to and from the Quay d'Orsay in Paris for quite a long time. More than one official worked full-time at the TIVC on these French ciphers and codes and there was even a French chamber at the TIVC. Some sources even suggest that this was done over a period of at least 20 years and perhaps even more. The French codes were apparently not too difficult to decipher because the French embassies used the same codes over a very long period of time. These intercepts probably had already started in 1946 and continued until the late 1960s.[46]

It was a similar operation, as was actually executed in more European capitals such as London, to those Peter Wright revealed in his book *Spycatcher*.[47] Whether the TIVC or another Dutch intelligence or security service gave any help or professional advice to GCHQ in this clandestine eavesdropping intelligence operation is not known. However, the fact indicates that these two Western European countries were particularly suspicious of the French political intentions, in particular during the various British bids to join the European Community (EC). The fact that Britain's entry into the EC was twice stopped by a French veto irritated the Dutch government considerably and especially Netherlands Foreign Minister Luns was often vexed about the French stance.[48]

Spanjaard died in 1960 without having been able to get more substantial funding for the TIVC. Nevertheless, he is regarded by many intelligence professionals as the man who built up the service from scratch and who laid in very difficult times a firm and sound basis for the TIVC. Only between 1960 and 1963 was more financial support given to the TIVC, which enabled the organisation to buy additional intercept equipment, Siemens printers, direction-finders and about 15 Racal receivers. Also 15 new telegraphers formerly belonging to the Netherlands pilotage were transferred to the TIVC.[49]

At the time Spanjaard died there was more or less still no contacts between the TIVC and the Sigint units of the Netherlands Army and Air Force. The three military Sigint services each followed their own path.

Again, this is not so strange in view of the fact that areas of interest and potential and actual targets were quite different. Nonetheless, cooperation in the field of Sigint was restored in 1969. In this year the working relationship between these three elements doing the same kind of work was re-established although quite carefully. The association of the TIVC with its Dutch Air Force counterpart was also quite cautious. Although it officially began in 1953 not much was achieved in the late 1950s.

For this reason it was quite logical that the Dutch Air Force Sigint service worked more closely together with its German counterpart than with the TIVC. In 1967 for example an agreement was drafted which established a regular albeit unofficial Netherlands-German exchange of data, stemming from Sigint. In 1970 the cooperation with the German Fernmelderegiment 70 was officially approved and formalised. And Dutch Army Intelligence worked together with the German *Bundesnachrichtendienst* (BND) and Army. This intelligence cooperation concentrated for example on tracking down numerous burst transmitters in the German Democratic Republic.[50]

LATIN AMERICA

However, the intercept activities of the TIVC were not limited to Asia, Western and Eastern Europe. At the beginning of the 1960s the TIVC also played an (albeit minor) role during the Cuban Missile Crisis in October 1962. An interesting detail in this respect emerged from the official British archives. In a document dealing with a NATO Experts Working Group on Latin America, which was held at the beginning of November 1962, officials from the French intelligence community claimed that they had received an indication in August 1962 of the presence of surface-to-surface missiles in Cuba. However, officials from the Dutch intelligence community claimed that they had already received similar evidence in September. From the lengthy roundtable discussion it transpired that no other NATO member had received any forewarnings of the upcoming crisis.[51]

However, this valuable information came only partly from Sigint because the Netherlands Military Intelligence Agency had a Humint source in Cuba who checked the Sigint evidence. Soon the Humint reports came back and it turned out that the Sigint was solid: another fine example of Sigint and Humint sources working together. The source of the Netherlands Military Intelligence Agency also reported the arrival of Soviet bomber and transport aircraft in Havana. Nevertheless, probably this time it was the Air Force Sigint unit, which had eavesdropped on Soviet military traffic in the German Democratic Republic. It discovered the movement and subsequent transfer of Soviet military advisers and equipment to Cuba.

However, there was probably also another Dutch service involved. The BVD monitored all conversations by phone to and from the Cuban embassy in The Hague. It is reasonable to assume that likewise the Cuban diplomatic telex and telegraph connections were monitored. One day a Cuban diplomat did read out loudly on the phone in Spanish the plaintext of a coded diplomatic telegram. On the other end of the line another Cuban diplomat listened and compared also in Spanish the plaintext with the cipher text. An easier instance of an open goal could not be envisaged.

The Cuban cipher was broken in this simple manner. The Cuban conversation on the phone was clearly one of the biggest violations of communication security ever heard of. The BVD forwarded this material quickly to the CIA station chief in The Hague, the late Dr Cleveland Cram, who was posted to the Netherlands from 1966 to 1970. The BVD did not tell Cram how they were able to break the Cuban diplomatic code. A few weeks later the NSA via CIA channels told the BVD jokingly that they really had to stop these kind of activities because it turned out that they had a 100 per cent score. All NSA personnel 'would lose their job' if the BVD continued with these kind of excellent codebreaking activities. The BVD had also immediately forwarded the results of their codebreaking operation to the TIVC but there this material was hardly appreciated and for unknown reasons not much enthusiasm was shown.[52]

In this respect presumably the Dutch listening post in the Netherlands Antilles (West Indies) did not play an active role in this phase. The main reason for this was the following. There were indeed intercept facilities in the Netherlands Antilles in the Caribbean. From a site on the island of Curaçao in particular a country like Venezuela but also other Latin American and Caribbean nations were monitored constantly. This site had been operational since World War II but became more actively involved in intercepting traffic after 1945. In the beginning the Venezuelan intercepts were forwarded to Amsterdam for decryption but this did not turn out to be a success. The intercepting activities were not executed on a frequent basis and soon the intercepts did not produce a breakthrough in the sense that the military codes and ciphers of Venezuela were broken, a success the forerunner had already accomplished during World War II. After 1956, most personnel were withdrawn from Curaçao in order to help out in the conflict with Indonesia.

However, in 1957 it was the British GCHQ at Cheltenham, which asked the TIVC if they were willing to establish a permanent listening site at the island of Curaçao. The British intelligence service informed the Dutch that there was a hiatus in their Sigint network, which operated within the UKUSA agreement. In this arrangement which dates from 1947 the United States, United Kingdom, Canada, Australia and New Zealand entered into a treaty that created a network of security and Sigint cooperation that continues to this day. Nearly 30 years passed before any participating government acknowledged the existence of this agreement.[53]

As many readers will know by now: UKUSA nowadays plays an important role in the Echelon and other global eavesdropping networks. The British request for Sigint collaboration was a unique chance for the Dutch to try for closer cooperation with the UKUSA countries. However, lack of funds and staff made the Dutch decline the request by GCHQ. In retrospect the drafters of the official internal TIVC history acknowledged that an excellent opportunity had been missed to begin some sort of collaboration with the UKUSA nations. It was observed that this would undoubtedly have been very valuable in international crisis situations.[54]

After the conflict with New Guinea had ended the Director of Dutch Naval Intelligence decided to dispatch a small staff to the island of Curaçao in order to step up the eavesdropping activities. In January 1963, a Sigint team of seven persons was transferred to the Caribbean island. It was to target Venezuela. In June 1963 a second team was sent and after finishing their assignment it was reported that from Curaçao the military traffic of all three Venezuelan armed forces could be easily intercepted with excellent results. Venezuela was worthwhile monitoring because of its sizeable armed forces and closeness to the last Dutch colonial possessions in the Caribbean. Also the aspirations of Venezuela with respect to the continental shelf were of importance because this area was quite rich in oil. Operations were mounted from the listening post Saint Joris, which worked in 1982 with nine Racal monitor-receivers and six Racal search-receivers.[55]

Helped by the TIVC the Dutch government was well informed about the plans of the government in Venezuela. The TIVC was thus able to keep the Cabinet well informed about the military order of battle in this region of Latin America.[56] According to various well-informed sources and released documents the military traffic of other Latin American countries apparently also posed no problem for the Dutch codebreakers. Plans were made to station a permanent interception group of one officer and a staff of 18 on Curaçao. A budget of almost 600,000 Dutch guilders was also proposed. The under-secretary of the Navy approved the plan and it was decided that all efforts would concentrate on Venezuela.[57] In addition it should be mentioned that the CIA also ran its own Sigint station on the Dutch island of Curaçao.

Furthermore, the Dutch Army began in 1963 its Sigint operations from Surinam soil. The Comint operations were in particular aimed at issues related to the political developments in British Guyana. Between 1963 and 1966 various political questions produced increased Comint activities from Surinam soil like the activities of Soviet spy trawlers off the coast of Surinam and the independence struggle in British Guyana. In this British colony there were still British troops stationed until the country became independent on 23 May 1965. Netherlands and British armed forces cooperated in this period.

From 1967 onwards the tension between Surinam (still a Dutch colony) and British Guyana heightened because of a territorial border dispute. Consequently the Sigint activities from Surinam directed against Guyana were stepped up. Additional Sigint equipment, analysts, translators and technicians were flow in. Guyana remained the main target until 1975.

Sigint was hereafter also used at length to monitor the smuggling activities in the jungle of Surinam. Successes were soon accomplished and a international smugglers syndicate was rolled up because their codes were broken by the Netherlands military eavesdroppers in Surinam. The cooperation with other Sigint services (particularly the US) expanded and this led among others on the basis of Dutch Sigint to the arrest of a spy from a Warsaw Pact country who worked on Cape Canaveral (presently the Kennedy Space Centre) in Florida. However, the frequent attacks on Soviet Navy codes and ciphers did not produce that much success. The Sigint unit of the Netherlands Army in this colony operated until the independence of Surinam and finally returned in 1975 to the Netherlands.[58]

FINANCIAL PROBLEMS STILL UNRESOLVED

However, the history of the TIVC, which until now has been portrayed as reasonably successful, had also a darker side. It is a shadow, which has been hanging over the service from the start. It is the simple question: who is going to pay for its activities, very expensive equipment and its experienced, well educated and trained staff? The financial budget for the TIVC grew from more than 172,000 guilders in 1947 to almost 400,000 guilders in 1948. In 1950, the Minister for the Navy wrote to the Prime Minister and Minister of the Treasury that his department was no longer able to pay all the costs involved. He wanted a cost sharing arrangement because the main bulk of the TIVC output went to the Foreign and Cabinet Offices. A basic element in these discussions was of course always: should one department charge another for the product it delivered. Probably to this day this is still an unresolved issue in the Netherlands intelligence community. As a Sigint analyst told me privately: 'It is a cumbersome topic nobody likes to talk about. It is a dead issue.'[59]

The Ministry of Foreign Affairs as well as the Department of Trade claimed constantly that other intelligence and security services like the BVD or IDB also delivered their intelligence products but never charged any money for it. Although all departments realised the importance of the Sigint product: there was no willingness to pay for it. And although the

Treasury Department promised to increase the overall yearly budget of the Royal Dutch Navy with an extra 500,000 guilders this pledge never materialised. In the meantime, in 1952 the annual budget had grown to a mere 500,000 Dutch guilders per year.

In 1956 the debate about finances was reopened but to no avail. The Prime Minister's Office, the Ministry of Foreign Affairs and Department of Trade refused to contribute 100,000 guilders each. Finally, in 1957 the Cabinet decided that the National Telecommunications Board would fund a part of the overall budget of the TIVC. A steady albeit marginal increase of the budget would from now on take place on a yearly basis. The Directors of the TIVC were of course not pleased, but there was no alternative. The various recipients of the TIVC eavesdropping intelligence product simply refused to pay for it. In 1965 the proposed budget had risen to 2.6 million guilders a year. The biggest chunk went to salaries: 1.9 million, which was four times as much compared to 1950/52. This rise of the annual budget was no surprise in view of all earlier achievements and successes in Europe and Asia and the new operations in the Netherlands Antilles.[60]

NEW TARGETS

In the 1960s, after the transfer of New Guinea to Indonesia, the TIVC had to concentrate their efforts on new targets although Indonesia and its military remained an important object of study. For example, between 1965 and 1969, a part of the intercept capacity was tasked to monitor all Indonesian diplomatic and military traffic. After two weeks the traffic and transmitting schedules and times of all Indonesian embassies were mapped out. In view of the huge earlier successes in breaking the Indonesian diplomatic and military traffic either via codebreaking or the use of high-ranking spies it is beyond doubt that again much of the Indonesian traffic was read. One should not forget that Operation 'Virgil' was still producing excellent results.[61]

Of course, the year 1965 was important in view of the political and military developments in Indonesia when in the night of 30 September and 1 October 1965, an abortive coup d'état took place when the Communist Party tried to remove the government of Sukarno. The attempt failed but sealed the fate of President Sukarno who in 1967 was replaced by General Suharto. It is also safe to assume that the Dutch codebreaking results were shared with the Americans and British although by that time the NSA and GCHQ had probably no problem at all in deciphering the Indonesian diplomatic and military traffic.

In the same period the TIVC enlarged the targeting of the Middle East and by 1965 it started to monitor all radio teletype traffic dealing with the movements of Egyptian ships. This information was shared with Israel and was probably used by the Israelis in 1967 during the war in the Middle East. Other targets were the Turkish diplomatic and military traffic, which was intercepted and decoded successfully. Also traffic from Somalia, which began to follow a Moscow-oriented political course, was successfully read.

As said before, in these years also the cooperation of the TIVC with the Sigint units of the Netherlands Army and Air Force did not improve considerably because there was no real need to. Actually, in the mid-1960s the Army Signals Intelligence Service was doing well and was capable of monitoring the movements of troops on the night of 20 and 21 August 1968 when the Warsaw Pact invaded Czechoslovakia. These reports were the first and the intelligence was immediately forwarded to the HQ Army Intelligence. However, this service for unknown reasons decided not to pass on this crucial information to the other intelligence services, which later of course led to a huge internal discord.[62]

Another problem was the interception of the traffic by telex. The Royal Mail (Dutch PTT) and BVD did this kind of interception by ministerial authorisation. The cover name for this was Operation 'Codeword 231'. In the first part of the 1970s the PTT used two telexes for the registration of intercepted messages. That was not enough, according to various intelligence and security service sources. They wanted to expand this number from 2 to roughly 20. At the beginning of the 1970s, Prime Minister Barend Biesheuvel had authorised the interception of this traffic by telex. However, in 1976 the PTT demanded a new authorisation to continue this intelligence work from the newly elected Social Democratic government led by Joop den Uyl. The whole affair was discussed in the ministerial intelligence committee of the Cabinet in 1976. The intelligence coordinator argued in favour of expansion. The services wanted two intercept telexes for Warsaw Pact traffic and three telexes for the Middle East and North African countries to choose from Morocco, Algeria, Tunisia, Libya, Iraq and Saudi Arabia. They also wanted two telexes for European countries such as Portugal, Spain, Italy and Turkey, and three telexes for Africa and the rest of the world.

Another problem, which was raised in this special ministerial intelligence committee of the Cabinet, was the request for cooperation by the American National Security Agency (NSA) and the West German Bundesnachrichtendienst (BND). There was already some sort of collaboration with the BND but not on the official level. Both services had discreetly inquired if the Dutch Sigint services were not willing to officially exchange intercepts. Various Dutch ministers in this special intelligence committee had some doubts because if they concurred with this request the Netherlands were admitting in that case that they were quite active in the field of Sigint. This was of course quite a silly argument because the ministers involved apparently still had the impression that the foreign services had until that day not noticed these intelligence activities. However, in the NATO framework it was widely known that the Dutch Sigint capabilities and cryptanalytical know-how plus equipment were indeed very good. The coordinator for this reason held a strong plea for an expansion and this had to begin with sharing of intelligence. A chance that the whole intelligence sharing operation would be compromised could be overlooked in his view.

The coordinator informed the ministers involved that security measures and a good compartmentalisation would prevent this. All intercepted messages in which Dutch citizens were mentioned by their real name were delivered to the BVD. The intercepts, in which foreigners were mentioned, were distributed to the BVD, IDB and Ministry of Foreign Affairs. The TIVC in principle solely documented topics and issues: not persons. All diplomatic and political-economic intercepts would be delivered via the IDB to the Foreign Office and BVD. Purely economic intercepts would go to the Ministry of Economic Affairs. The special committee of the Cabinet gave thereupon the definite green light.[63]

THE OIL CRISIS OF 1973

The 1970s was an era in which more and more traffic was sent through communication satellites. Accordingly, apart from Project Codeword 231 an enormous operation began. This time with the cover name: Project Codeword 631. At Burum in the province of Friesland the Dutch Royal Mail (PTT) established a satellite ground station for transmitting and receiving via regular satellite communication traffic sent by phone, telex and telegraph. However, unknown to the outside world was the simultaneous establishment of a relay station, which forwarded a part of the telegraph traffic passing through Burum to a TIVC unit where the intercepts were stored in a computer. In this machine messages were sorted out and later transferred by landline to the TIVC in Amsterdam. However, employees of the PTT and in particular its director-general, were utterly unhappy with these eavesdropping activities. They wanted to get rid of it. The Radio Intelligence Committee (RIC) proposed to the Special Intelligence Committee of the Dutch Cabinet that the PTT would no longer execute this operation on behalf of the TIVC. The Sigint organisation would have to perform this in the future. This was also common practice in Britain, the United States and other Western European nations.

The PTT was very pleased with this proposal by the RIC. In 1980 the PTT proposed to give a satellite dish and receiver with which the TIVC

FIGURE 8.2
SECRET TARGETS FOR DUTCH SIGINT, JANUARY 1981

GEHEIM

29/HR/81
d.d. 19/1/81

Aan: H-WKC
Van: H-IDB
Betreft: Onderwerpen van belangstelling

A. Europa

1. Oostbloklanden
   a. Reacties van Oosteuropese landen op de ontwikkelingen in Polen (van partij/regering, vakbonden, bevolking).
   b. Economische steunmaatregelen voor Polen en de consequenties daarvan voor de eigen economie. (Hoe worden eventuele tekorten opgevangen?)
   c. Buitenlandse handel van Oosteuropese landen (onderling en met westerse landen).

2. Albanië
   a. Economische en politieke contacten met Joegoslavië.
   b. Wijzigingen in de buitenlandse betrekkingen en handelscontacten met vooral kleinere Westeuropese landen.

3. Bulgarije
   a. Relaties met Joegoslavië (Macedonische kwestie).
   b. Relaties met het Midden-Oosten (met name met Libië op het gebied van kernenergie en militaire samenwerking met Irak, Algerije).
   c. Wapenleveranties aan Zuid-Afrika?

4. C.S.S.R.
   a. Dissidente groeperingen.
   b. Rol bij de produktie van kerncentrales in de Comecon.

5. D.D.R.
   a. Verhouding partij/kerken.
   b. Duits-Duitse betrekkingen.
   c. Betrekkingen met Afrikaanse landen en Cuba.
   d. Activiteiten in het Midden-Oosten.

GEHEIM

FIGURE 8.2

BIJLAGE 2

Belangstelling voor TIVC berichten.

I. Volgens IDB vraagstelling (vide 86/CF/86)

1. Handelspolitieke info in de meest uitgebreide zin over SU en Comecon partners.

2. Handels- en betalingspolitiek van: ▬▬▬

3. Financiële transacties van ▬▬▬ met: ▬▬▬

4. Mogelijkheden voor Nederlandse bedrijfsleven tot participeren in de uitvoering van "grote militaire produktieprojecten", ▬▬▬

5. Tenderinfo t.b.v. bedrijven met zgn. "monopolistische positie" in de categorie: telefoon
    luchtvaart
    vliegtuigbouw
    militaire scheepsbouw
    civiele scheepsbouw
    militaire electronische apparatuur

II. Aanvullingen volgens opgave directie Operaties

1. **Aannemingsprojecten** (havens, civiele scheepsbouw, consultancy, vliegvelden, pijpleidingen, waterbouwkunde en irrigatie, leverantie constructiemateriaal) in Midden Oosten ▬▬▬
   Bedrijven: ▬▬▬

2. **Petrochemie** (olie-gas en verwerking) en methanolprojecten (ureum, kunstmest) in M.Oosten ▬▬▬
   Bedrijven: ▬▬▬

1

VERTROUWELIJK

FIGURE 8.2

VERTROUWELIJK

3. **Chemie/medische apparatuur en uitrustingen** (pulse generators, pacemakers, leverantie chemicaliën) van ▓▓▓▓▓▓▓▓▓▓▓▓▓▓▓

4. **Agrarische** projecten en handel in agrarische produkten (zuivel, pootaardappelen, bio-industrie) in M. Oosten ▓▓▓▓▓▓▓ en Afrika ▓▓▓.

5. **Electronica, communicatie- en computerapparatuur** ▓▓▓▓▓▓▓▓▓▓

6. **Financiële- en handelstransacties** (kredietfaciliteiten, bankberichten) over ▓▓▓▓▓▓▓▓▓▓▓▓▓▓▓▓▓▓▓▓▓▓▓▓▓▓.

7. **Militaire projecten** en handel in mil. goederen (explosieven, beveiligingssystemen) naar ▓▓▓▓▓▓▓▓

8. **Narcoticahandel** m.b.t. ▓▓▓▓▓▓▓▓▓.

9. **Terroristische activiteiten** m.b.t. M.Oosten, Oost Europa, Car. gebied, Azië.

10. **"Personeelswerving"** voor M. Oosten▓▓▓▓▓

11. **Nieuwe en toeristische projecten** ▓▓▓▓▓▓▓

could intercept the traffic which was transmitted by satellite. The Cabinet readily agreed with this idea. The IDB also became more actively involved and a direct hotline was created using a secure landline between the TIVC in Amsterdam and the headquarters of the IDB in The Hague. Raw intercepts were passed along to the IDB and it started to weave this material into its finished intelligence products. At the same time the IDB also passed along to the TIVC specific requirements regarding the targets they wanted Sigint on.[64]

However, the PTT wanted to sever its ties with the TIVC completely and was very willing to give the Sigint service a huge financial alimony payment if they were willing to agree to a divorce. The TIVC would receive enough money to build its own satellite intercept site at Zoutkamp in the northern Netherlands in the province of Groningen, a site close to Burum with excellent intercept conditions. This intercept station, also known as Project Codeword 632, would begin commence its intercept operations in

1983. The Cabinet Office and the Ministries of Defence, Trade and Transportation paid the roughly 4 million guilders needed to build this site. It was planned to build this site up to a full Satellite Interception System (Project Codename 'Delta'). The funds for the first phase (roughly 700,000 Dutch guilders) were available for this operation. The money for the second and third phases for the years 1987–89 was estimated in 1985 to be roughly 6 million.[65]

It is claimed that this satellite interception site was able to search with a computerised dictionary through roughly 20,000 messages every 24 hours,[66] going through and scanning the intercepted traffic by using the keyword search method. These keywords could of course vary and were actually changed on a daily basis. Nevertheless, the keywords dealt mainly with political, military but also economic, technical, non-proliferation, narcotic and scientific related matters. For example, using these entries and keywords information was sought dealing with nuclear, biological and chemical proliferation and missile technology.[67]

FIGURE 8.3
DUTCH SATELLITE INTERCEPT STATION

Not only great results were achieved in the 1960s but also in the 1970s. According to reliable sources, the TIVC was exploiting in September/October 1973 the whole communication network of the Egyptian Navy as well as the Egyptian Air Force. This information was undoubtedly shared with Israel. According to a former employee of the Mossad, the IDB even warned Israel, on the basis of intercepts, about the upcoming Arab offensive a few days before war broke out. However, these and other warnings were tossed aside by Israel. Was this on purpose or is that too far fetched? What is interesting is that Israeli Prime Minister Golda Meir in November 1973 spoke at length to the Leaders Conference of the Socialistic International. Dutch Prime Minister Den Uyl was present at this meeting. He later told one of his trusted personal national security advisers that Golda Meir had informed secretly him that she had been aware of the upcoming attack 15 hours before it actually took place. The government of Israel had considered launching a pre-emptive strike against Egypt and Syria but decided against such a military move.[68]

A primary recipient of intelligence derived from Sigint was thus in those days Israel. During the October 1973 war the Dutch services tipped off their Israeli counterparts about telephone and telex cables on the seabed of the Mediterranean between Egypt and Libya. An elite Israeli unit succeeded in blowing up this cable, which forced the Egyptians and Libyans to rely on radio links. The relationship with the Mossad was actually very intimate because the IDB also gave the Israeli services raw intercepts from certain Arab embassies in The Hague. Mossad was in particular looking for traffic to and from representatives of the Palestinian Liberation Organisation (PLO). And in a black bag operation, the IDB also opened an Arabic diplomatic mailbag. They made copies of the original diplomatic telegrams and these materials were given to the Mossad. This enabled this service to compare the raw Sigint with the original telegrams and considerably enhanced the chances of deciphering the Arabic codes. However, repeated requests by Israel to forward intercepts of traffic of French diplomatic and military authorities were constantly refused. On the other hand, the Mossad was given intercepts of traffic between German, Italian and French companies and various Arab capitals in the Middle East. These intercepts dealt with the export and sale of weapons and the technology needed to produce weapons. Special Mossad couriers flew over from Paris to collect this important intelligence.[69]

The TIVC was in these years very successful against Middle Eastern targets and broke the diplomatic codes of many Arab nations. Based on intercepts, the TIVC at a very early stage could inform the Dutch Cabinet that a decision to impose an oil embargo against the Netherlands and in particular Rotterdam had been taken already in July 1973. This was confirmed by a report from the Dutch national security service, BVD, which was sent in November 1973 to the Dutch intelligence tsar and coordinator, Admiral Frits Kruimink. The leaders of the major political Dutch parties were discreetly informed of this fact in a closed session of the Parliamentary Intelligence Committee.

The intelligence services did have a pretty good picture of the oil supply situation because the TIVC could also read the coded messages of Royal Dutch Shell, British Petroleum, Mobil Oil and various other oil companies. All multinationals in the oil sector used very primitive codes which were very easy to decipher. Actually, this situation did not pertain solely to the Netherlands. Also the NSA received in those days much intelligence derived from intercepts of communications from the major American oil companies. The President-director of Shell, G. A. Wagner, was warned by the Dutch intelligence community of this lapse in security several times. However, not much changed. Wagner was told over and over again that not only other Western services but also the Soviets and other East European services were able to read the codes of Shell because they were so easy to break. Shell was asked several times to do something about this and to improve the codes. Nevertheless, the situation would remain the same and enabled the TIVC to deliver the confirmation that there were ample oil stocks for six months available in Rotterdam harbour and in fact the Netherlands was flooded with oil.[70]

Accordingly, the TIVC played an important role during the oil crisis. In those days the TIVC devoted 15 per cent of its resources on the Middle East but because of sufficient analyses and research this 15 per cent capacity could be expanded quite quickly. The TIVC did for example break the Tunisian diplomatic code and the reports of the Tunisian embassy in Damascus, Syria, one of the nations at war with Israel, was an important source of information for the Dutch government.[71] Everyday a special Middle East edition full of intercepts was distributed by the TIVC to the members of the Dutch Cabinet. The Dutch government was well informed about the political and military developments in the Middle East because in a war situation most of the time no ciphers or codes were used by many parties. This pertained to Israel, the Arab nations but also for the Soviets. Much was 'read' like for example non-encrypted Soviet military traffic to and from Yugoslavia and because of this the Russian linguists at TIVC discovered much about the Soviet deliveries of arms via Yugoslavia to the Arab nations. From Sigint much was also learned about the Arab operations and Soviet strategy and the location of various military units.[72]

Apart from this, Middle Eastern and Soviet, the diplomatic traffic about the crisis produced by various European countries was also intercepted and read by the Dutch Sigint services. In later years, Dutch Sigint material, which was forwarded to Israel contained excellent information about the Iraqi nuclear reactor at Osirak. The problem was not so much the location of the reactor: this was already known by the Israelis from aerial reconnaissance and satellite pictures. The difficulty was more how much concrete (with or without steel?) had Iraq used in the construction of the reactor. The vital information was needed by the Israeli Air Force in order to determine which bombs to use. This information was collected in 1981 by the TIVC from the Italian companies who built the reactor because those firms used open communications, which could be intercepted.[73]

RECURRENT FINANCIAL PROBLEMS

Nevertheless, despite all successes during the oil embargo of 1973–74 the problem of which ministry would finance the TIVC was still not solved in this period. Only 30 per cent of the output of the TIVC went in those days to the Dutch Navy although insiders maintain that it was often very difficult to achieve this percentage.[74] The remainder went to the Cabinet Office and Ministries of Foreign Affairs, Trade, Treasury and other departments in The Hague. The numerous attempts by the Royal Dutch Navy to have the real customers pay for the intelligence product would still continue to fall upon deaf ears. A yearly subsidy by the National Telecommunication Board in the 1970s to the Ministry of Defence of 500,000 guilders was only a meagre contribution. On the other hand various departments also explicitly forbade scaling down the activities of the TIVC because the quality and results would diminish quite fast. It was indeed a vicious circle from which an escape did not seem possible.

This problem had also been discussed in October 1976 in the special intelligence committee of the Cabinet. Minister of Foreign Affairs, Max van der Stoel, who profited enormously from the Sigint take during the oil crisis, was of course in favour of the continuation of the Sigint efforts of the TIVC. However, Prime Minister Den Uyl was less enthusiastic. He was concerned about the enormous costs and the payments made to the hundreds of employees. On a yearly basis roughly 30 million guilders would go to the TIVC. Actually, the total costs of the TIVC were estimated in 1985 to amount between 50 and 100 million guilders per year.[75] In 1992 an article in a newspaper even estimated that the TIVC was costing roughly 150 million guilders on a yearly basis.[76]

However, these financial figures are probably far too high, and deal in all likelihood with investments and not with operational costs. For example, it was claimed in an official publication that in 1977 the yearly TIVC budget was about 10 million guilders, which is significantly less than earlier estimates in the press.[77] However, the overall cost is probably much higher because this figure includes only the salaries, operational costs and finances involved in acquiring equipment.[78] Financial figures for later years remain obscure because the real figures are hidden in the overall budget of the Ministry of Defence. Hopefully new research and future Freedom of Information Act requests will shed some more light on this financial aspect.

Apart from this, the doubts of the Prime Minister were reinforced by the fact that the Sigint reporting was rather raw, sometimes unbalanced because the TIVC did not produce any finished analytic intelligence products based on intercepts. The coordinator could only agree with the Prime Minister. However, this was not quite fair because the TIVC was only responsible for collecting Sigint and not responsible for the overall analysis of the Sigint. This had to be done by the IDB. Nevertheless, the coordinator himself had compared the costs against the benefits and had likewise concluded that this was a negative balance sheet. However, he assured the Prime Minister that he had ordered an increased cooperation between the various Sigint services. More analysis had to be done and less raw intelligence had to be delivered to the intelligence customers.[79]

PROJECT 'MORE' AND SIGINT LIAISON

These fresh instructions by the intelligence coordinator were presumably the outcome of the October 1973 crisis, during which the Army Sigint unit had also produced excellent results. In 1976, roughly 125 staff was attached to the TIVC, 360 to the Army Sigint unit and about 100 to the Air Force Sigint unit. The number for the Army Sigint Unit is of course higher because many military personnel were involved in tactical and strategic military Sigint work. The staff of the TIVC did not grow spectacularly. In 1990 the number of staff of TIVC had increased to a mere 135. Nonetheless, increased cooperation would only really start in 1987 when the three Armed Forces services merged into a single Defence Intelligence Agency. In 1988/89 the HF intercept station at Eemnes became somewhat outdated and it was proposed to replace it with a brand new station which would cost the Dutch taxpayer an amount of 40 million Dutch guilders over a period of eight years. This operation was named internally Project Codeword 'MORE'.[80] Actually, the word MORE was the abbreviation for 'Modernisation Radiostation Eemnes'. This project was completed around 1991/92.

Finally, with whom was this intelligence internationally shared? The intelligence derived from Sigint was primarily shared with the United States and Britain. In the beginning the TIVC did not share its intelligence with the CIA. Rather everything went to the US Office of Naval Intelligence (ONI). This was done because the naval intelligence unit of the CIA was not to be considered as the most ideal partner in the field of intelligence liaison. This unit often produced low quality information contrary to what ONI provided to Dutch Navy Intelligence. However, as a former Dutch Naval Intelligence official told me, this situation pertained to the period 1950–80.

After 1980 apparently much finished intelligence based on intercepts from the TIVC was customarily shared with the CIA. This happened several times a week. In particular when on the basis of intercepts the Dutch discovered that some sources or agents of the CIA in foreign countries were in acute danger. This kind of life saving information was also shared with for example the Mossad and in this manner an assassination of two Mossad agents in France by a Palestinian terrorist group was prevented. It was thwarted because the intelligence derived from Sigint was given urgently to the Israelis.

The relationship of the TIVC with the CIA became much better when William O. (Bill) Studeman went from ONI to the NSA and later became Deputy Director of Central Intelligence (DDCI) at the CIA. From that moment on, the access of the Dutch service to the top of the CIA was very easy. There was also formal cooperation with NSA and Agency officials who visited the Netherlands on a frequent basis. One of the most recent visits was that of Deputy Director Barbara McNamara late in 1999 only a month before she was transferred to London. During the visit of McNamara the topic of increased American-Dutch Sigint cooperation regarding the Netherlands Antilles was discussed. Apparently an agreement was drafted on this subject. Also joint operations were executed. In 1979 a frigate from the Royal Netherlands Navy visited the Barents Sea and the vessel carried American Comint operators on board trained for eavesdropping on Soviet communications and military traffic in the high north.[81]

The relationship with the British services never reached the same level of intensity and depth compared to the US. For example, during the Falklands War (1982) the TIVC was intercepting Argentine traffic from the eavesdropping site Eemnes. The Argentine military forces had increased the power of their military radio networks because they wanted to be sure that all Argentine military forces would receive the messages. For this reason it was no problem at all to follow the Argentine traffic from Eemnes. However, the Dutch were of course not the only ones intercepting this traffic. Also other intelligence services like the German BND were reading this Argentine traffic. Nevertheless, one day, information based on deciphered Argentine intercepts dealing with a downed British pilot was quickly forwarded to the Chief of Station of the British foreign intelligence service, MI6, in The Hague. With a reddish face he had to confess that GCHQ had already the same information.[82]

Actually, it turned out that GCHQ was pretty unprepared for an Argentine invasion of the Falklands. Although GCHQ was able to decrypt the Argentine diplomatic and military traffic the Sigint service was unable to analyse the enormous amount of intercepts that suddenly poured in. There were also not sufficient linguists available to handle the extra workload. Also MI6 was unprepared. For example, the handbook *Latin America* of MI6 was updated for the last time in 1972! It appears that the British intelligence community had practically nothing on the Argentine military forces and the Falklands.[83]

Also within the framework of NATO important information derived from Sigint was shared with the Western allies. For example: on the basis of traffic analyses Dutch Sigint was able in the early 1980s to predict a buildup for a future Soviet invasion of Poland, which luckily never took place. Nonetheless, these messages produced long and hectic nights at the desks of the various Dutch intelligence and security services and at NATO headquarters in Brussels and Mons. The main problem remained of course the decipherment of Soviet, Warsaw Pact and Chinese diplomatic and military traffic. This was too difficult for the TIVC, which was the sole service doing the cryptanalytic work in Holland. The Sigint units of the Army and Air Force were not involved in any codebreaking activities. Less difficult in the 1980s was the traffic of many countries in the Middle East including Afghanistan, Pakistan and Iran.

That the TIVC did read the diplomatic traffic of the latter country was revealed in an article in the Dutch weekly *Vrij Nederland* on 9 September 1978. In the article it was disclosed that the Sigint service was reading the diplomatic traffic between the embassy of Iran in The Hague and Tehran. It was also exposed that the police cooperated with the embassy of Iran in tracking down Iranian students and dissidents living in the Netherlands. It was primarily students who protested on a frequent basis against the harsh policy of suppression by the Shah of Iran. With this capacity the TIVC probably could also monitor and read the other encrypted Iranian diplomatic exchanges. *Vrij Nederland* printed the original intercepted coded message as well as the translation. This article led to an enormous row inside the TIVC and a molehunt began led by the military police. Although suspicions centred on a certain person working at the TIVC who probably leaked this very sensitive information to the press, an arrest and subsequent trial never materialised.[84]

SIGINT AND NATO

In later years the exchange of purely military intelligence took place within completely new frameworks such as the NATO Advisory Committee for Special Intelligence (NACSI) and the Signals Intelligence Data System, which was abbreviated as SYGDASIS.[85] Originally it was some sort of computerised back-up system in case one of the nations lost its national Sigint capacity. Within this framework the Sigint cooperation took place in the so-called multinational SYGDASYS committee. From the start the sharing of entirely military intelligence took place on a quid pro quo basis. All nations involved poured into SYGDASYS as much military Sigint and other information as possible. Each nation could present its requirements which, however, meant that this also at the same time showed its strengths and weaknesses.

More important was that the enormous overlap in targeting slowly disappeared. SYGDASYS played an important role during the Gulf crisis of 1990–91. However, the US wanted to involve more countries in SYGDASYS, which led to some panic among NATO allies. They did not trust some of the Arab friends of the United States and of course, a kind of mutiny also started in Israel.

SYGDASYS also had a regional database system, which specialised on certain areas. The members of the SYGDASIS committee had to report to the Sigint Senior Meetings (SSMs) in which all heads of Sigint organisations were represented like NSA, BND, GCHQ, DGSE, the Italian SISMI, the Norwegian Military Intelligence Service, the Danish Defence Intelligence Service, the Belgian Military Intelligence Service, etc. The chairman of the SYGDASIS committee did this reporting. The Director of the MID (often accompanied by the Director of the TIVC) was always in a strong position in these meetings of the Sigint Senior Meetings because he could trade off strategic and tactical Sigint. The exchange of acoustic intelligence was done in a separate framework.

Sweden was probably discreetly informed of the decisions in the Sigint Senior Meetings by the American services and/or other participants. However, the Swedes were always quite sensitive in these kinds of contacts. For example, they always travelled in plain clothes and always wanted the lowest profile possible. All the same, Swedish Sigint officers from the *Försvarsväsendets Radioanstalt* (FRA), or National Defence Radio Institute visited the Netherlands for discussions with their Dutch counterparts on a regular basis.[86]

However, the diplomatic traffic of several NATO allies with embassies in The Hague, which was intercepted and deciphered by the TIVC over a longer period of time, was not shared. In conversations and articles the Belgian diplomatic traffic is an example often mentioned. The Dutch Ministry of Foreign Affairs always had certain doubts as regards the future political orientation of Brussels. From intercepts of the Belgian diplomatic traffic to and from The Hague and other capitals much more was discovered about the Belgian intentions. Various other countries apart from Belgium were also monitored like Italy and Turkey.[87]

Because the traffic of the Soviet, Chinese and Warsaw Pact embassies was difficult to decipher, it was all the more interesting to read the traffic of various foreign Western European embassies and consulates behind the Iron Curtain and in China. The TIVC concentrated on diplomatic missions of nations, whose traffic it could read. For instance, the BVD was interested in reading the diplomatic reporting of the Italian embassy in Moscow. The visits of prominent Italian Communists and PCI members to Moscow often turned out to be great ideological and political disappointments. These frustrations by Italian Communists were quite frequently uttered to Italian diplomats at the Italian embassy in Moscow who in turn forwarded this to the Italian Foreign Ministry in Rome. The TIVC intercepted and read this Italian diplomatic traffic, giving the BVD an excellent insight into the latest developments in the European Communist world. Of course, officials at the Netherlands Ministry of Foreign Affairs were also pleased because the diplomatic reporting of foreign officials in The Hague often landed on their desk.

However, some officials were shocked when they were confronted with these intercepts. A Dutch diplomat even informed his Turkish colleague that the Turkish diplomatic traffic was being intercepted, deciphered and translated by the TIVC. This stupid move of course infuriated the staff of the TIVC. Nevertheless, he was not the only one who leaked this information to the Turkish government. From a Turkish intercept, the TIVC discovered pretty soon after the incident in The Hague that an Italian source had told the same story to the Turks. Apparently, the Italian Sigint service was also able to read this traffic. However, the Turks apparently did not believe these stories because to the amazement of the Dutch TIVC officials, they did not change their cipher and code systems, which enabled the TIVC to continue to read their diplomatic traffic.[88]

Finally, a touchy subject: economic intelligence derived from Sigint. In a briefing on 7 March 2000 before the Washington Foreign Press Center former CIA director James Woolsey vigorously denied that the United States was engaged in industrial espionage in the sense of collecting or even sorting intelligence that it collects overseas for the benefit of or to give to American corporations. He took this remarkable step in view of the 'tumult' in Europe as regards the revelations about Echelon and economic espionage.[89]

However, the Dutch Sigint services certainly have not behaved in the past like 'angels' and probably will not do so in the future. Via intercepts the TIVC collected quite a lot of intelligence for Dutch industry. This material was usually passed on to the Department of Trade and the IDB. In particular intelligence about foreign tenders and contracts was of importance to companies and firms in the Netherlands. Also developments in the international tourist industry were monitored via Sigint. In addition, Sigint on large construction projects like airports, harbours, irrigation, etc. was also looked for. Also various Dutch multinationals like Philips and Shell received this kind of economic intelligence. The TIVC also intercepted German contract information dealing with the construction of frigates for foreign governments. This enabled Dutch shipyards, for example, to present bids which were more attractive than the German ones. In the end the Dutch wharf could sign the contract. On the other hand, the German services like the BND sometimes also outsmarted the Netherlands by intercepting Dutch tenders enabling the Germans finally to get the contract.

There was also a tricky aspect. The IDB exchanged economic intelligence derived from Sigint. However, this sometimes caused a kind of incestuous relationship between this service and Dutch economic life. The something in return mentality within the intelligence world was also applied to Dutch multinationals and other companies. In exchange for economic intelligence, industry was often asked to 'do something' on behalf of the Dutch intelligence service. Often there was no need to ask for it because the companies understood all too well that if they gave nothing in return the next time they would no longer receive intelligence.[90] For example, it is rumoured that Philips during the construction of enormous telephone switchboard systems built in a backdoor for the Dutch intelligence community whereby eavesdropping equipment was installed at the same time.[91] Intelligence was also gathered on behalf of the Ministry of Agriculture. The TIVC for example forwarded information about the agricultural output of various European countries.[92]

THE GULF WAR

Via intercepts the TIVC sometimes also monitored the export or attempts to export strategic materials, whose export was forbidden. This did not always mean that the shipment was halted. On the contrary, as in Britain, a modest Iraqgate took also place in Holland and the services tracked the flow of deliveries to countries like Iraq in order to establish who was delivering what to whom. On the basis of these exports and various intercepts, it was often possible to determine how far science and technology had progressed in certain nations, like Algeria, Libya, Iran and Iraq.[93]

In the 1990s it was again the TIVC which produced excellent results during the Iraqi invasion of Kuwait and the subsequent war in the Persian Gulf. Exceptionally valuable Comint and Elint was gathered about the Iraqi Air Force, Air Defence systems and Navy. This intelligence enabled Dutch Navy ships taking part in the naval operations in the Gulf to operate in a safe manner. In particular, intelligence about the location of the Iraqi Air Force was priceless. Also the interception of diplomatic traffic during the war produced valuable information regarding the political and military plans of Iraq. In addition, I do have to mention that not only the operations by the TIVC were a great importance. This was also true for the special operations by the IDB which 'scored' very high ratings from the CIA and MI6. These secret activities were publicly praised in parliament by the Prime Minister, Ruud Lubbers.[94]

Another target was the former Dutch colony in Latin America, Surinam. In particular it became a target after the rise to power of Desi Bouterse in 1980. The brutal murder of his opponents in 1982 and the involvement of Bouterse in cocaine trafficking increased the attention of the Army Intelligence unit of the Netherlands Military Intelligence Service. This time the TIVC hardly played any role at all. The main reason was that the Surinamese diplomats stationed in the Netherlands and Western Europe did not use ciphers and codes. They communicated with the Surinam capital Paramaribo in plain language. Accordingly, the diplomatic pouch and couriers carried many of the confidential and secret messages. It soon turned out that the diplomatic reports by the Surinam embassy in The Hague therefore never produced many problems. The messages were very easy to read because they were not encrypted. The contents of the Surinam diplomatic traffic was accordingly seldom very valuable from a political or military viewpoint. Nevertheless, some of the intercepts found there way into a special Yellow Edition with some Surinam and other Latin American intercepts, which were of higher value.[95]

THE FUTURE

Needless to say that the activities of the TIVC continue until today although the organisation is hampered by budget cuts and is like many other nations in search of new targets. It is rumoured in the United States that the TIVC might be invited to join the Echelon network as run by the UKUSA countries.[96] Perhaps this step is still far away but at the same time it is also some sort of world-wide recognition of the quality of the Sigint product which the TIVC produces with a small budget and a modest staff. The Dutch Sigint service is presently internationally ranked belonging to the best five. Perhaps the chill in the American-Dutch Sigint relationship caused by the Petersen spy case was fundamental to this achievement. The Dutch service had to do it all alone without American financial support and it did so. At the same time this brought the TIVC in a somewhat unique position compared to other European Sigint services: the Dutch always remained independent because the Sigint service was not walking on an American financial leash.

The year 1999 went down in the MID history as the year of the reorganisation. This was a final step in a process of integration and centralisation that started in 1987. The Minister of Defence approved the reorganisation plan by late 1999. The plan improves internal alignment and transparency in the primary process of the MID. The plan also postulates to transfer the MID from a primary supply-oriented service into a demand and service-oriented organisation.[97]

In the Netherlands itself the newly fully-integrated Netherlands Military Intelligence Agency (MID) will cherish its excellent Sigint capability. The TIVC no longer exists. Also the Sigint units of the Netherlands Air Force (1st LVG) and Army (898th Vbdbat) have been fully integrated into this new AVI organisation, a process that took place from 1 January 1996 onwards. There is presently an Operational Signals Intelligence Division (OVIC) at Eibergen in the east of Gelderland dealing mainly with intercepting military communications. OVIC is the outcome of the fusion between the Sigint units of the Army and Air Force. Apart from OVIC, there is since 1 January 1998 a Strategic Signals Intelligence Division (SVIC) in Amsterdam, which is dealing primarily with the interception of diplomatic and maritime HF and satellite traffic and cryptanalytic activities. Both OVIC and SVIC have been integrated into the *Afdeling Verbindingsinlichtingen* (AVI) where presently the analysis of the intercepts is done by a separate unit.[98]

SVIC presently mainly reports via AVI to the Intelligence Department of the Netherlands Military Intelligence Agency, the Cabinet Office and the Ministries of Foreign Affairs, Trade and the BVD. OVIC is to report primarily to the intelligence units of the various components of the Netherlands Armed Forces like the Army, Navy, Air Force and the Military Police.[99]

However, there has already been in the past a tendency within the Dutch Navy to scale down its Sigint activities. Former Navy Admiral Nico Buys once privately claimed: 'I can buy a frigate for the money you guys spend on Sigint.' However, he apparently forgot that Sigint probably saved and will save in the future the lifespan of the same frigate in future United Nations peacekeeping and enforcement operations like the war with Iraq. Nevertheless, the present situation tends to suggest that there will be an additional merger. SVIC in Amsterdam will be closed in the future and its staff will be transferred to the MID headquarters in The Hague. The two main intercept stations, Eibergen and Zoutkamp for satellite communications, will continue to function for the time being but Eemnes will be abandoned in the future and its mission moved to Eibergen.

Nevertheless, overlooking more than 50 years of history it can be safely stated that during the Cold War the Netherlands Sigint organisations of the Army, Navy and Air Force played an important role. Their work kept Dutch policy-makers and military up to date with current diplomatic, political, military and economic events, trends and developments. However, its should be said that the main successes of the TIVC were not so much achieved against the Soviet bear and its East European allies and China although small Sigint achievements were made. Important accomplishments in this field were mainly achieved by the Comint activities of the Army and Air Force, which is another story that one day has to be told. The accomplishments of TIVC lay mainly in the domain of the national interests of the Netherlands such as those during the conflict with Indonesia between 1946 and 1962.

Also in later years valuable intelligence derived from Comint about this former colony would flow to the Dutch policy-makers and the military. Needless to say that also in conflicts in Europe and the Middle East the Dutch service played an important role of keeping the government up to date regarding the latest developments. This was particularly the case during the oil crisis of 1973. In this respect it is all the more peculiar that the consumers of the intelligence product produced by the Netherlands Sigint organisations never have been willing to pay more for this product. The Green, Red and Yellow Editions were often highly valued, but when a higher budget was asked for all ministers involved said most of the time: no way.

NOTES

1. See for an analysis of the impact of the oil crisis on the Netherlands: Duco Hellema, Cees Wiebes and Toby Witte, *Doelwit Rotterdam. Nederland en de oliecrisis 1973–1974* (Den Haag: SDU 1998) *passim*.
2. Netherlands Military Intelligence Agency, MARID VI/Wiskundig Centrum 1947–1965, Top Secret, 31 Oct. 1965, S.Marid 6110/000780/65/zg (hereafter *History WKC/TIVC*) pp.80–9.
3. I am also indebted for the comments on an earlier draft made by various former Dutch intelligence officials. They prefer to remain anonymous. I also would like to thank Matthew Aid for his stimulating comments.
4. Bob de Graaff, 'From Security Threat to Protection of Vital Interest: Changing Perceptions in the Dutch Security Service, 1945–91', *Conflict Quarterly* 12/2 (Spring 1992) p.13.
5. F.A.C. Kluiters, *De Nederlandse inlichtingen en veiligheidsdiensten* (Den Haag: SDU 1993) pp.253–5 and *Jaarverslag MID* (Den Haag 1998) p.34.
6. Prior to 1972 it was called *Buitenlandse Inlichtingendienst* (BID) before it changed its name in 1972.
7. For the history of the IDB: Bob de Graaff and Cees Wiebes, *Villa: Maarheeze. De geschiedenis van de Inlichtingendienst Buitenland, 1945–1994* (Den Haag: SDU 1998) *passim*.
8. F.A.C. Kluiters, *De Nederlandse inlichtingen- en veiligheidsdiensten* (Den Haag 1993) pp.153–6 and Norman Polmar and Thomas B. Allen, *Spybook, The Encyclopedia of Espionage* (NY: Random House 1998) pp.141–5.
9. Confidential interview.
10. James Bamford, *The Puzzle Palace. Inside the National Security Agency, America's Most Secret Intelligence Organization* (NY: Houghton Mifflin 1985) pp.173–7.
11. D. Lawrence, 'Today in Washington', *New York Herald Tribune*, 16 Nov. 1954.
12. Archives of the Netherlands Ministry of Foreign Affairs, Code 921.0 B.K., Van Tuyl to Van Roijen, No. 7173, 12 Oct. 1954.
13. National Archives, Washington DC, RG 59, box 3745, 756.5211/10-2954, Marcy to State Department, 29 Oct. 1954.

DUTCH SIGINT 1945-94   283

14. Public Record Office, London, FO 371/109482, Chancery The Hague to Foreign Office, CN 10345/3, 3 Nov. 1954.
15. *History WKC/TIVC* (note 2) pp.1–4.
16. A confidential interview. Also 'PTT werkte jarenlang als spion', *Utrechts Nieusblad*, 13 Nov. 1995.
17. F.A.C. Kluiters, *De Nederlandse inlichtingen- en veiligheidsdiensten. Supplement* (Den Haag: SDU 1995) p.82.
18. Archives of the BVD, File BID/IDB (Part 2), No. 2.279.266-a0, Memo DOB/EH to PHD, 19 Oct. 1962.
19. Private information.
20. *History WKC/TIVC* (note 2) p.93.
21. Dick Engelen, *Inlichtingendienst Buitenland. Een institutioneel onderzoek naar de Buitenlandse Inlichtingendienst/Inlichtingendienst Buitenland 1946–1996* (Den Haag: SDU 1996) p.78.
22. Confidential interview.
23. Archives of the Netherlands Ministry of Foreign Affairs (hereafter NMFA), *Code 921.0 B.K.*, Memo by Van Tuyll to Einthoven, Nov. 1954.
24. Kluiters, *Supplement* (note 17) p.81. Also Willem Heyboon, 'PTT-dienst onderschepte jarenlang telegrammen, *Het Parool*, 13 Nov. 1995.
25. *History WKC/TIVC* (note 2) pp.5–18.
26. Ibid. pp.19–20.
27. Kluiters, *Supplement* (note 17) p.84.
28. A confidential interview.
29. Kluiters, *Supplement* (note 17) p.84.
30. *History WKC/TIVC* (note 2) Appendix 16.
31. Archives of the Netherlands Cabinet Office (hereafter CAB), Folder No. 3, FOIA No. 99M008634, Memo to Director IDB, Secret, 16 June 1988.
32. *History WKC/TIVC* (note 2), Appendix 37, Memo Spanjaard to Director Marid, 19 Nov. 1956.
33. *History WKC/TIVC* (note 2), Appendix 24, Memo, Top Secret, 1 Feb. 1957.
34. A confidential interview.
35. Private information.
36. A confidential interview..
37. *History WKC/TIVC* (note 2) p.30.
38. A confidential interview.
39. National Archives, Washington DC, Record Group RG 457, NSA Records, Box 171, Indonesia General, GCHQ Intercepts, various dates.
40. *History WKC/TIVC* (note 2) pp.43–4.
41. De Graaff and Wiebes, *Villa Maarheeze* (note 7) pp.121–243.
42. *History WKC/TIVC* (note 2) pp.31–2.
43. Ibid. p.92.
44. Private information.
45. A confidential interview and *History WKC/TIVC* (note 2) p.44.
46. Various confidential interviews. See also: Kluiters, *Supplement* (note 17) p.122.
47. Peter Wright with Paul Greengrass, *Spycatcher. The Candid Autobiography of a Senior Intelligence Officer* (NY: Viking Press 1987) pp.138–41.
48. Duco Hellema, *Buitenlandse politiek van Nederland* (Utrecht: SDU 1995) pp.211–12.
49. *History WKC/TIVC* (note 2) p.34.
50. A confidential interview. See also: Kluiters, *Supplement* (note 17) pp.108–9.
51. Public Record Office (PRO), London, FO 371/162405, Minute by A.D. Parsons on Cuba, Secret, 9 Nov. 1962. Thanks to Jim Hershberg for pointing to this document. See also Kluiters, *Supplement* (note 17) p.121.
52. A confidential interview.
53. Jeffrey T. Richelson and Desmond Ball, *The Ties that Bind* (Boston: Unwin Hyman 1990) passim.
54. *History WKC/TIVC* (note 2) p.73.
55. Kluiters, *Supplement* (note 17) p.120.

56. M.W. Jensen and G. Platje, *De MARID. De Marine Inlichtingendienst van binnenuit belicht* (Den Haag: SDU 1997) p.298 e.v.
57. *History WKC/TIVC* (note 2) pp.73–4.
58. P. Yska, 'Radio-Interceptie bij de Troepenmacht in Suriname', *Intercom* 28/4 (Dec. 1999) pp.19–27.
59. Confidential interview.
60. *History WKC/TIVC* (note 2) pp.75–9.
61. De Graaff and Wiebes, *Villa Maarheeze* (note 7) pp.232–42.
62. Various confidential interviews. Also Kluiters, *Supplement* (note 17) p.121.
63. De Graaff and Wiebes, *Villa Maarheeze* (note 7) pp.280–1.
64. CAB, Inv. File No. 2, FOIA No. 98M008634, Director IDB to Director WKC, No. 29/HR/81, 19 Jan. 1981.
65. CAB, Inv. No. 116, FOIA No. 98M008820, Memo Interception, Top Secret, 1 May 1985.
66. Kluiters, *Supplement* (note 17) p.112.
67. CAB, File 1, FOIA No. 99M008634, Director IDB to Director TIVC, No. 453/CF/88, 8 Nov. 1988.
68. Hellema, Wiebes and Witte, *Doelwit Rotterdam* (note 1) pp.24–32.
69. De Graaff and Wiebes, *Villa Maarheeze* (note 7) pp.280–1.
70. Hellema, Wiebes and Witte, *Doelwit Rotterdam* (note 1) pp.161–96.
71. NMFA, Code 6, File III, 613.211.45, Box 38, Folder 440, Kruimink to Van der Stoel, No. 20, Top Secret, 22 Nov. 1973.
72. Various confidential interviews.
73. A confidential interview.
74. A confidential interview.
75. Robert Schouten, 'Koude Oorlog is ook oorlog', *Haagsche Courant*, 30 March 1985.
76. J. de Haas and Ch. Sanders, 'Zien onze spionnen spoken?', *De Telegraaf*, 1 Feb. 1992.
77. Dick Engelen, *De Militaire Inlichtingendienst 1914–2000* (Den Haag: SDU 2000) pp.112–14.
78. CAB, Inv. No. 116, FOIA No. 98M008820, Memo Interception, Top Secret, 1 May 1985.
79. De Graaff and Wiebes, *Villa Maarheeze* (note 7) p.273.
80. CAB, Folder No. 3, FOIA No. 99M008634, Memo Interception, Top Secret, 26 Jan. 1987.
81. A confidential interview.
82. A confidential interview.
83. Various confidential interviews.
84. Aukje Holtrop en Rudie van Meurs, 'De geheime Wassenaarse dossiers van de Sjah', *Vrij Nederland* 39, 9 Sept. 1978.
85. Mike Frost and Michael Gratton, *Spyworld. Inside the Canadian and American Intelligence Establishments* (Toronto: Doubleday 1994) p.23.
86. A confidential interview.
87. Zie ook R. Schouten, 'Afluisteren op grote schaal', *Haagsche Courant*, 30 March 1985, idem, 'De ondergang van de IDB', HP/De Tijd, 6 May 1992.
88. A confidential interview.
89. Transcript of briefing by former CIA Director James Woolsey at the Foreign Press Center in Washington DC, 7 March 2000. See: http://cryptome.org/echelon-cia.htm
90. De Graaff and Wiebes, *Villa Maarheeze* (note 7) pp.295–6.
91. M. Thomassen, 'Philips in diskrediet', *Algemeen Dagblad*, 24 April 1982.
92. A confidential interview.
93. Vgl. ook M. Koolhoven, 'Geheime agenten', *De Telegraaf*, 19 April 1997.
94. De Graaff and Wiebes, *Villa Maarheeze* (note 7) p.402.
95. Ibid. pp.363–4. Also a confidential interview.
96. A confidential interview.
97. *Annual Report Nederlandse Military Intelligence Service 1999* (The Hague: MID 2000) p.24.
98. See also *Tweede Kamer der Staten-Generaal, Vergaderjaar 1995–1996*, Vragen gesteld door kamerlid Hoekema, No. 1240, p.2521, 2 May 1996.
99. Engelen, *De Militaire Inlichtingendienst* (note 77) p.147.

9

Dutch Sigint and the Conflict with Indonesia

WIES PLATJE

When I left the former Netherlands New Guinea in October 1962, it was impossible to foresee that 38 years later an invitation would arrive to contribute to a book about Sigint in Western Europe during the Cold War. I was a junior analyst in the intelligence section of the Commanding officer of the Netherlands Forces in Hollandia. My long flight home marked the end of a period of hard work, tropical heat, round the clock intelligence shifts and an endless stream of intercepted Indonesian messages that had to be processed.

For the time being no more Sigint for me. Although, I realised that the effort of the Netherlands Navy Intelligence Service and especially the product of the radio interception section on the island of Biak, played a crucial role in successful Dutch military operations in, around and in the skies above the Netherlands' last foothold in the Far East. Before coming to the core of this contribution: 'Dutch Sigint and the Conflict with Indonesia', I would like to give a brief historical résumé of Dutch communication intelligence activities and the development of cryptology and cryptanalysis in the Netherlands. In this review, emphasis will be put on Sigint operations in South East Asia.

REVIEW

In the early history of Europe, characterised by frequent and long lasting wars, efforts have been made to decipher secret messages of the enemy. The nineteenth century introduction of the telegraph, and later, the twentieth century use of wireless radio systems for military and diplomatic communications, presented new possibilities to fulfil this requirement. The new techniques provided a unique fast way to communicate, but rendered the enemy the possibility to intercept the message. The use of codes to encipher the message was the answer.

Enciphering of sensitive information became common practice, while the same time much effort was put in the development of knowledge necessary to decipher foreign code systems. It is obvious that also in the Netherlands, with its world-wide international commercial interests, a neutral position between powerful and aggressive neighbours and overseas colonies, efforts were made to intercept foreign communication lines and to develop cryptological expertise.

WORLD WAR I

The outbreak of World War I forced the Netherlands government to restructure and to enlarge the intelligence organisation. In order to maintain its safe and neutral position amid the powers at war, a military intelligence service with the capability to monitor foreign communication lines and to decipher secret messages became urgently needed. So, when in 1914 the Dutch armed forces were brought to a state of 'war readiness' the Army increased the organisation of the military intelligence service (GS 3) with a new section: GS 4. The task of this new section included censorship, counter-espionage, counter smuggling and cryptology. Herewith, Dutch 'Black Room' operations in modern history became a fact.

In this period the 'Godfather' of Dutch cryptology, Henri Koot, plays a very important role. He was born in the Netherlands-Indies in 1883 as son of a European father and a Chinese mother he went to the Royal Military Academy in the town of Breda in 1901 and returned to the Dutch colony in 1904 as lieutenant in the Royal Netherlands-Indies Army. Until 1911 he was involved in military actions against rebels on the islands of Ceram and Ambon. In 1911 he started a study at the Higher Military Staff College in The Hague. He would never return to his native soil. Europe faced the outbreak of World War I and Koot, a man with a mystical mind and a non-Western world of thought and gifted with an amazing talent for mathematical and rhythmical analysis, was assigned to the new Army section GS 4.

At the outbreak of war there was hardly any cryptological expertise, a great challenge for Koot who started to study intercepted radio messages looking for mistakes and repetitions. With his natural feeling for solving puzzles and his statistical insight he laid the foundation of the revival of an almost forgotten science in the Netherlands. His work was aimed at improvement of national code systems and to create more possibilities to explore foreign communications by deciphering encrypted messages.

The importance of the Army sections GS 3 and GS 4 was clearly recognised by the government. The number of intelligence personnel increased significantly. At the end of the war 12 officers and a proportional number of enlisted personnel were involved in 'Black Room' operations and another important decision was made: the sections were not demobilised after the war. In 1919 Captain Henri Koot became head of the enciphering service of the General Staff, subsection GS 3c. One year later he became head of the Cryptological Bureau, an inter-governmental service of the Department of Foreign Affairs, manned with army and navy personnel. In this function he put much effort in the training of new communication and intelligence personnel of both the Army and the Navy. Lessons given by Koot were hard to keep up with for his students because of the complex and abstract character of cryptology and the high pace he displayed. Koot knew this but he never changed his methods, because he was convinced that by doing so he would lose a selection criterion for natural talents.

For budgetary reasons Major Koot was discharged from his function as head of the Cryptological Bureau in 1933. This measure caused a deep personal crisis. He remained head of subsection GS 3c. In 1939 Koot was convinced that the Netherlands military communication and security organisation was insufficiently prepared for war and much was done by him to improve this situation. During the war he became a key figure in the Resistance, among others using the cover of the International Red Cross. In 1944 he became commanding officer of the Internal Armed Forces, a function in which he became involved with ceasefire and surrender negotiations with the defeated German Army. Because of the fact that in 1945 his counterpart, a German general refused to negotiate with a lower ranked officer, Koot was promoted to Major General. In 1947 he retired from the service and died in 1958. Nationally Koot is well known for his outstanding activities during World War II, but it is also fairly well known that he was one of the world's most brilliant cryptologists.[1]

JAPAN

In the Netherlands, the need for communication interception capabilities and cryptological expertise intensified when in the early 1930s a growing number of prominent Japanese politicians and both military and civilian authorities began to speak openly of territorial expansion as a plausible solution for the economical problems Japan was faced with. The Dutch authorities became more and more concerned about a possible Japanese attempt to seize oilfields in the Netherlands-Indies or other targets of interest for the Japanese economy. In that time it was assessed by the central intelligence service in the Netherlands-Indies, the 'Service of East-Asian Affairs' (Doaz) that, if such an action should be carried out, the limited Dutch military forces, facing the geographical extensiveness of the

Netherlands-Indies, would probably fail in a successful response. Also, the Dutch military commanders realised that in fact little was known about Japanese military developments and capabilities, in particular intelligence on shipbuilding programmes, tactical doctrines and military training standards in the Japanese Imperial Navy was urgently required.

The structural military weakness and the initial Dutch policy of strict neutrality in the Pacific region consequently resulted in the development of a national expertise in the field of cryptology and codebreaking. One of the first and important steps was the installation of a navy radio monitor station in 1933. This station was situated in a building of the Staff of the Royal Netherlands Navy in Batavia.[2] The purpose of the unit was to monitor the naval communications of France, the United Kingdom, the United States and Japan. It is evident that Japan soon became target number one. In the mean time some Netherlands Navy officers started a study in the Japanese language both in Holland and in Japan, while others attended cryptology courses given by the already mentioned Koot.

THE FIRST SUCCESS

In 1934 the Netherlands Navy restructured its organisation in the Netherlands and in the Netherlands-Indies. As a result of this an intelligence section was established on the staff of the Commander of the Royal Netherlands Navy in Batavia. Lieutenant-Commander J. F. W. Nuboer became the first head of the section. He had a background as radio officer and was trained in the use of code systems and in cryptanalysis.[3] Appointing Nuboer to this job proved to be very successful.

From 1935 onwards, Nuboer managed to provide a very reliable and current order of battle of the Japanese Navy. Step by step, he and his assistants managed to break Japanese navy codes, a very remarkable result in a time when one had to rely on 'one-human-brain-processors' only. Not only an accurate naval order of battle was produced, but also a good insight into naval ship movements by intercepting daily weather reports containing the name of the ship and its position.[4]

ROOM 14

The cryptological bureau of the Netherlands-Indies government was established in Bandoeng together with Section 7 of the General Staff of the Royal Netherlands-Indies Army, responsible for deciphering intercepted diplomatic messages. This section was located in 'Room 14', a very well protected and secret part of the General Staff building in Bandoeng. In the history of 'Room 14' in particular the year 1932 stands out, when the

DUTCH SIGINT AND THE CONFLICT WITH INDONESIA 289

FIGURE 9.1

Captain Henri Koot, Royal Netherlands-Indies Army.

Japanese diplomatic codes were broken as a result of a systematic approach and stiff perseverance.[5]

In July 1937, when Japan de facto declared war against China, the Japanese Navy communication security intensified and wartime codes were applied. From that time it became almost impossible to decipher the intercepted messages. In the fall of 1938 a new Japanese code system was introduced. It consisted of five numbers coded by a table of 10,000 numbers. This was too much for Dutch naval intelligence.

In the meantime an officer without any feeling for or experience in cryptology and codebreaking relieved Commander Nuboer. Consequently Japanese naval traffic was no longer intercepted. Until the attack on Pearl Harbor, intelligence on Japanese naval movements was mainly based on surveillance by submarines and aircraft reconnaissance. Navy analysts now joined their army colleagues in 'Room 14' in Bandoeng. The joint efforts became successful.

The professional standard and dedication of personnel enabled the Dutch authorities on almost every occasion to anticipate promptly and adequately Japanese diplomatic and subversive actions.[6] When in 1939, Europe faced the outbreak of World War II, 'Room 14' had to deal with another problem. Many German messages were intercepted and from these messages the Netherlands authorities learned that German diplomats in Batavia had frequent contacts with Japanese secret agents and with Indonesian nationalists.

PEARL HARBOR

From the early 1930s onwards Dutch codebreakers were able to decipher Japanese code systems without support from the United States and the United Kingdom. Long before 1941, when the United States provided London with knowledge about the Purple, Red and other Japanese systems, Dutch analysts had already deciphered these systems independently. In that time both the United States and the Netherlands maintained a neutral policy against Japan and a structural exploitation of intelligence did not take place.

When in January 1941 Captain William R. Purnell, Chief of Staff of the US Pacific Fleet, had a secret meeting in Batavia with the Dutch Commander-in-Chief Vice-Admiral C. E. L. Helfrich, both parties maintained silence about the progress of breaking Japanese codes. In 1941, the Dutch authorities assessed that a Japanese attack on the Netherlands-Indies would not take place as long as the US Pacific Fleet in Pearl Harbor posed a threat to the Imperial Navy. The American author Robert D. Haslach, writes in his book *Nishi No Kaze Hare* interesting details about the development of Dutch-US intelligence exchange and the fact that this intelligence assessment, although released to the US, until now not has been found in the US National Archive. *Nishi No Kaze Hare* means 'Westerly winds, clear sight' and this phrase was used in a Top Secret Message from Tokyo to the Japanese diplomatic station in Batavia. It indicated that an attack on the Netherlands-Indies was imminent and that Japanese international diplomacy was about to collapse. The same message indicated the probability of military actions against the United States.

Currently there is a very interesting international exchange of views, facts and opinions between historians about the question: was President Roosevelt informed about the Japanese plans to attack Pearl Harbor? On 1 December, seven days before the attack, the Japanese navy codes were changed, not only the highest code 'JN-25' but also the list of call signs. This was a serious problem for traffic analysis, but within two days, 200 of the most important callsigns could be re-identified by FRUPAC, the US Fleet Radio Unit Pacific.[7]

Until the occupation of the Netherlands Indies and the inevitable surrender, 'Room 14' remained successful in providing tactical intelligence on Japanese operations. The benefit of this was decreased gradually by a fast growing and bitter shortage of war material and equipment. Especially

DUTCH SIGINT AND THE CONFLICT WITH INDONESIA   291

the lack of sufficient air power increased the speed of the military defeat significantly. The Japanese forces occupied Java on 8 March 1942, almost 100 days after Pearl Harbor. The Royal Netherlands Navy had lost 2 cruisers, 7 destroyers, 8 submarines, 2 gunboats, 7 minelayers, 20 minesweepers and several auxiliary vessels. The remains of the fleet, including 1 cruiser, 3 destroyers, 4 gunboats and 14 submarines, escaped and joined the Allied naval forces in the ongoing war against Japan.[8]

THE JAPANESE OCCUPATION

Only a few members of Dutch intelligence personnel were evacuated. Two men were detached to the British communication intelligence service in Ceylon. One of them was the navy officer L. Brouwer, who studied the Japanese language and cryptology before the war. Early in 1942 Lieutenant-Commander Brouwer joined the new Netherlands Forces Intelligence Service (Nefis) in Australia and became head of Section III, responsible for special intelligence and special operations.

At the end of the war Brouwer was tasked to design a plan for an intergovernmental communication intelligence organisation. Herewith the foundation was laid for the WKC in Amsterdam. Brouwer resumed his career as a navy officer outside the intelligence community and became Commander-in-Chief of the Royal Netherlands Navy in 1959 as a vice-admiral. A third cryptologist, Verkuyl, an officer in the Royal Netherlands-Indies Army, went to the United States where he was attached to the US Sigint organisation SIS. In 1947 Colonel J. A. Verkuyl became the first head of Marid 6, the new communication intelligence centre in Amsterdam, later called the WKC.

THE NETHERLANDS FORCES INTELLIGENCE SERVICE

Early in 1943, General Douglas MacArthur complained about the lack of sufficient intelligence on the Netherlands-Indies. He wanted the Netherlands military command to set up its own intelligence service to fill this gap. The Netherlands Commander-in-Chief, Admiral Helfrich acted quickly and in May 1943, a joint Dutch intelligence organisation was formed in Australia, called the Netherlands Forces Intelligence Service (Nefis).

First head of this service was Commander G. B. Salm. After the war Rear-Admiral Salm became the first head of the newly formed Netherlands Navy Intelligence Service (Marid). The Allied command in the Pacific tasked Nefis to operate in the occupied Netherlands-Indies (not including the island of Sumatra). In 1943 Nefis consisted of three sections: Section 1 'general intelligence', Section 2 'internal safety and security' and Section 3 'special intelligence' and 'special operations'. Nefis Section 3 executed 'hit and run' operations and 'intelligence parties' deep inside occupied territory. Nefis 3 was part of the Allied Intelligence Bureau (AIB), formed in July 1942 to coordinate various national intelligence operations. The AIB was organised in 4 sections: Section A 'British Secret Service (SOE and SOA)', Section B 'Australian section of the British Special Operations Unit', Section C 'Field intelligence' and Section D 'Propaganda'.[9] Until the end of the war Nefis did not carry out Sigint operations, but there are interesting examples of Japanese Sigint activities against Dutch intelligence commandos operating in New Guinea.

One of the most successful intelligence parties was Operation 'Prawn'. A Dutch submarine, ten miles east of the Japanese army headquarters near Sorong, dropped the Nefis party consisting of seven secret agents. The objective was to report daily by short-wave radio communication:

- patrol flights of Japanese aircraft along the north coast
- Movements of Japanese naval vessels along the coast
- Dislocation of Japanese forces in the area
- Suitable-landing areas for imminent landing operations of Allied forces and weather reports in support of Allied air attacks.

The intelligence reports were very successful. Allied Liberator bombers often attacked reported Japanese naval reinforcement and supply vessels in the area.

Of course this did not remain unnoticed by Japanese intelligence. They activated Elint operations resulting in locating the general dropping area by means of direction finding of intercepted radio communications. Japanese patrols were deployed to eliminate the Dutch agents. A warning by local people enabled the party to escape by submarine safely after three months of fruitful intelligence collection.[10] One of the Nefis agents who took part in Operation 'Prawn' was J. F. (Fred) Bastiaans, a former officer in the Royal Netherlands-Indies Army. After the war he left the military service and became head of the traffic analysis section of the WKC in Amsterdam.

A NEW INTELLIGENCE CHALLENGE

When Japan capitulated in August 1945, 54 Nefis employees were transferred to Batavia. There they faced a new intelligence challenge: a confrontation with a very determined and dedicated adversary, the new Republic of Indonesia proclaimed by Achmed Sukarno. The nationalists

DUTCH SIGINT AND THE CONFLICT WITH INDONESIA 293

rapidly formed military units. They managed to build, almost from scratch, a well-organised high-frequency radio communication network.

In order to re-establish Dutch authority, intelligence on enemy plans, strength and dislocation was urgently required. To enable Nefis to carry out this requirement a new 'Section 8' was formed. The task of the new section was radio-interception, radio traffic analysis and cryptanalysis. Nefis 8, with its wartime-trained arsenal of Sigint experts, realised the good possibilities for Sigint operations but at the start the new section faced many problems. First it had to carry out a broad survey in the entire short-wave frequency spectrum, looking for manually-operated Indonesian Morse code transmitters. Information on working frequencies, callsigns, used codes and transmitting schedules was not available.

The high-frequency spectrum had to be searched patiently, but success came soon: one by one Indonesian military radio stations were found and localised and messages were intercepted. It did not take long before a communication intelligence section, manned with navy, army and air force personnel, became fully operational. Comint groups were sent to various locations on the most important islands, Java and Sumatra. Soon it was possible to intercept the entire communication network of the Indonesian sea-, air- and land forces and used codes were successfully broken.

Started in 1945, Nefis 8 became a large, efficient Comint section by mid-1947. In the military field, Nefis 8 was very successful in producing intelligence on working frequencies and transmission schedules, including names of Indonesian radio operators. Nefis also managed to produce a reliable picture of the military organisation, the order of battle of the military units and the state of military readiness. On the diplomatic and economic front Nefis was successful in gathering intelligence on intended Indonesian diplomatic movements and the purchase of weapons abroad. Due to Nefis 8 the Dutch military commanders received unique intelligence.[11]

PENETRATION OF THE INDONESIAN HIGH COMMAND RADIO NETWORK

In June 1947 the Netherlands planned the first major military operation to overthrow the Sukarno government. In support of this an electronic intelligence operation was set up to penetrate the Indonesian communication network. Thorough electronic intelligence analysis of the keying-characteristics of radio transmitters used in the Indonesian armed forces and the analysis of manual keying characteristics of Indonesian radio operators, rendered the possibility of a successful infiltration into the radio communication net of the Indonesian Army General Staff.

Nefis 8 managed to set up a short-wave transmitter with modulated continuous wave characteristics, identical to the short-wave transmitter used by the Indonesian army headquarters in Yogja. A Nefis radio operator, trained to imitate manual Morse-keying characteristics of Indonesian operators, was tasked to operate the fake transmitter.[12]

A continous wave transmitter has its own unique 'fingerprint'. This is also the case with each individual manual Morse-code operator. Personal deviations from the ideal time-unit proportions will identify the particular operator.[13]

On the day the Dutch troops launched the attack, the fake Nefis transmitter called all radio stations with the message to listen in to a flash signal from the Indonesian supreme army commander to be sent immediately after the call. This was done 10 minutes before the Indonesian HQ transmitter in Yogja was due to become active. After acknowledgement of the call, the Nefis operator sent a fake message, enciphered in the Indonesian army high command code, to all Indonesian regional radio stations. In the message all regional commanders of the Indonesian army were ordered to abstain from 'scorched earth' tactics when attacked by Dutch troops. After receiving acknowledgement of receipt, the Nefis operator warned the Indonesian operators on duty to be aware of fake transmitters to be expected to come on the air after this contact. When the real Yogja transmitter came on air at 0800, all his calls to the regional radio stations remained unanswered. An important beneficial circumstance of this deception operation was the total disruption of the Indonesian Army Staff communications for days.[14]

After the second major military operation in 1948, Nefis was renamed: Central Military Intelligence Service (CMI). The interception section was now called the CMI Comint section. Not only the Indonesian military communications were intercepted by this section but also the Indonesian national and international public radio communication network. Exploitation of this source provided important information on political, economic and military subjects such as the purchase of foreign military hardware by Indonesia.

NEW GUINEA

When in December 1949 the transfer of sovereignty to the Republic of Indonesia, with the exception of the western part of New Guinea, was a fact, a part of the CMI Comint section was transferred to the Intelligence Section of the Territorial Command New Guinea in the capital Hollandia. Some time before this, a small radio interception detachment of Nefis 8 was already operating in this area. It was arranged that Marid 6 in Amsterdam and section G2 of the General Staff of the Royal Netherlands Army in the Hague should support the detachment in case intercepted codes could not be broken locally. Due to a persistent lack of support from G2 this arrangement never worked satisfactory.[15] The CMI Comint section became Office 5 of the MID, the Army Intelligence service.

The remaining part of the CMI personnel was sent to Holland. There, some of these experts were employed in the new communication-intelligence section in Amsterdam, Marid 6, later called the WKC (Wiskundig Centrum). Marid is the Dutch abbreviation of Navy Intelligence Service. Shortly before the transfer of sovereignty archives and equipment of Nefis 8 were shipped to the Netherlands. Useful material, such as 22 'Boehme' receivers, was disposed to Marid 6. The Nefis archives were shipped to the Ministry of Overseas Territories.[16]

By the end of 1949 the Netherlands faced a completely changed situation in the Far East. An 'empire' was lost and Indonesia was very clear about its intentions towards Netherlands New Guinea. The need for Sigint remained. In a time when the Indonesian armed forces improved the communication security by introducing new code systems, Office 5 in Hollandia lacked experienced cryptanalysts. The Indonesian Air Force codes were especially difficult to decipher.[17] The Army Sigint section was equipped with 15 CR88B receivers, 3 Marconi diversities and 3 Hell recorders. In those days the Royal Netherlands Navy in New Guinea had only one small intelligence unit: the Staff Officer Intelligence with a limited number of personnel. The Army provided the Navy with Sigint information of Indonesian maritime movements when available.

THE ROYAL NETHERLANDS NAVY

By a decision of the Netherlands Government taken on 15 April 1954 the territorial defence of Netherlands New Guinea, including intelligence, became a responsibility of the Royal Netherlands Navy.[18] Experts from the WKC in Amsterdam were deployed to Hollandia and a new Sigint section called Marid 6 Netherlands New Guinea (NNG) became operational in April 1955.

A total of 15 military (3 officers, 6 non-commissioned officers, 6 clerks) and 21 civilians (18 radio operators, 1 assistant cryptologist and 2 translators) took over from their army colleagues and a 'hot line' for crypto- and analysis support between Amsterdam and Hollandia was set up. The first head of the navy Sigint section was Lieutenant-Commander Ruud Gout, a former secret agent who had participated in Nefis operations inside Jappanese occupied New Guinea. Soon the difference in culture and mentality between the military and civilian employees of Marid 6 NNG

FIGURE 9.2

Madrid 6 at Hollandia, New Guinea, 1955.

Van Heutzkump

caused a conflict about the quality of lodging and food. Better financial conditions solved the problems. In the first year Marid 6 NNG intercepted an average of 185 plaintext and 21 ciphered messages daily. While the workload increased gradually, problems were caused by the lack of sufficient manpower.[19]

From 1955 onwards until the end in 1962, it was always very difficult to man the necessary functions. The solution was found in augmentation by civilians, as well as with army and air force personnel. In 1958 radio operators of the former Nefis Section 8 and four additional operators of the Royal Netherlands Air Force were sent to New Guinea.

In September 1957 the radio interception unit was moved to Biak. For the territorial defence of New Guinea, this island, with its international airport, naval airbase, logistic support and repair facilities was of great strategic importance. As a result of the far better facilities at Biak the production of Marid 6 NNG soon increased significantly. The Sigint section was organised in five departments: Traffic analysis (tasking), Cryptoanalysis (codebreaking), Deciphering (technical processing), Translation and a support department for administrative processing and distribution.

The section performed very well. The commander of the Netherlands Forces in New Guinea received four editions with intercepts weekly.

DUTCH SIGINT AND THE CONFLICT WITH INDONESIA 297

FIGURE 9.3

Madrid 6 at Biak, New Guinea, 1957.

Intercepts with an urgent character were sent as an immediate Top Secret message to Hollandia. The intelligence section was now capable of locating the positions of the most important elements in the Indonesian armed forces, the transfer of key personnel, the construction of facilities such as airbases, fuel depots and naval support facilities.[20]

POLITICAL DEVELOPMENTS

One year after the establishment of the independence in Indonesia in 1949, the status of the western part of New Guinea began to influence the diplomatic relationship between the Netherlands and the former Dutch colony in a negative way. Both the Netherlands government and a significant majority in the parliament agreed on the fact that New Guinea should remain part of the Kingdom until the time the Papuan people would be capable of deciding their own fate. Indonesia disagreed with this policy. Their point of departure was: 'Irian Barat belonged to the Indonesian territory and was illegally possessed by the Netherlands.'

When in the 1950s Indonesian attempts to force the Netherlands via the United Nations to participate in negotiations about the transfer of authority failed, it became clear that President Sukarno would use military force to reach his goal. Dutch policy was then aimed at reducing this threat by international diplomacy. At the same time much effort was put in trying to get promises from the United States, the United Kingdom and Australia for military support in case of a large-scale armed conflict. Some vague statements of possible support were given but President Eisenhower stated very clearly that the United States would firmly support the principle of non-violent territorial changes. American Secretary of State, John Foster Dulles, warned his Indonesian colleague Subandrio that the United States would not tolerate the use of military means to solve this political conflict.

The government of the United Kingdom declared that an official statement about military support in case of a military conflict was out of the question now the United States refused to make such a statement. London warned Jakarta that, in case of aggression, it would help the Netherlands in an *appropriate* way. This statement did not contain a promise for military support. In the New Guinea conflict Australia had always stated that Indonesian authority over the western part of the island was unacceptable. Australia changed this policy when it became clear that a military confrontation between the Netherlands and Indonesia had become a serious possibility.

In the late 1950s it became clear that, despite the fact that the United States, the United Kingdom and Australia principally recognised the Netherlands sovereignty over Western New Guinea and strongly opposed Indonesian military actions, military support was practically out of the question. If the Netherlands wanted to stick to its policy of maintaining sovereignty until the self-determination of the Papuans, it had to take its own military responsibilities and measures. From 1959 onwards the emancipation of the Papuan people was speeded up in order to bring it to self-determination as soon as possible. A well-equipped hospital in Hollandia, a shipyard in Manokwari, agricultural research sites and plantations as well as a military unit, the Papua Volunteer Corps, were set up to provide the necessary facilities.

In 1960 the New Guinea Council was established and the next year members of this council were voted in during general elections. An invitation for the official installation of the council was first accepted by the United States but a few days later was countermanded. Jakarta realised that the Netherlands was determined to continue the emancipation policy which would lead to a national consciousness of the Papuan people and a drive towards an independent status. Time was no longer in favour of Jakarta and President Sukarno came to the conclusion that he could not wait until the completion of the democratic process in western New Guinea. Action should be taken soon to incorporate Irian Barat, including the military option. On the diplomatic front Australia was still a Dutch ally but Australia was willing to accept any compromise, except military support, to solve the problem.

In March 1961, the young and dynamic John F. Kennedy became the new president of the United States. The Dutch assessment was that the policy of the new administration would be influenced by the following factors: an ambition for a new and fast approach to solve old problems, the traditional anti-colonial attitude in the United States, the fear of an Indonesian drift towards Communism and the Dutch efforts to create the right of self-determination for the Papuans. History learned that in The Hague this last factor was overestimated. The Kennedy administration was expected to take steps to eliminate the possibility of a potential conflict quickly. Such a solution would most probably not be in favour of maintaining the status quo, which The Hague was aiming at.

Diplomacy failed. The Netherlands government was not able to find international support for its policy and President Sukarno manoeuvred himself smartly into a favourable position. With the political and material support of not only the Soviet Union and its allies but also of some prominent Western countries, Sukarno could constantly increase the military pressure knowing that nobody in the world would like to burn its fingers in supporting the Netherlands. The Netherlands forces in New Guinea had to face military action in a political conflict that was impossible to win.

INTERNATIONAL ISOLATION

The fact that the Netherlands policy of bringing New Guinea to an independent status under Dutch guidance was not generally supported, caused a certain international isolation of the Netherlands. In that time the United States for instance tried to appease President Sukarno in an effort to block Soviet influence in the Asian theatre. This isolation had its effects on the routine intelligence exchange between the Netherlands and its NATO allies. 'We have to do this on our own' was the general thought and 'our national intelligence assets and capabilities should be protected at any price'. When a message was intercepted from an Indonesian warship, on its way to visit the US naval base at Guam, in which the commanding officer informed the Indonesian Naval Command about a cholera outbreak onboard, this intercept was not copied to the US. Instead a message was sent to the US Naval Attaché in The Hague containing the report from a 'so-called' human source in Indonesia about health problems on board that particular ship.[21]

In 1962 the Royal Netherlands Navy deployed a task group, consisting of the aircraft carrier Hr.Ms. *Karel Doorman* escorted by two anti-submarine destroyers, to the Caribbean. The secret part of this operation was the plan to send the two destroyers to the waters around New Guinea when deemed necessary. After passing the Panama Canal the task group visited the US naval base at San Diego. The received order to sail to the conflict area was classified Top Secret and caused the Netherlands commander some painful moments when the hosting US Navy admiral insisted on finding out the real destination of the ships. What the Dutch commanding officer did not know was that The Hague already had notified the US Navy about the real destination. He had hardly enough time to apologise when he received a copy of this message shortly before departure.[22]

NETHERLANDS INTELLIGENCE LIAISON IN SINGAPORE

The intensifying conflict between Indonesia and the Netherlands over the western part of New Guinea required additional reliable intelligence sources in the area. In 1957 the Royal Netherlands Navy posted an intelligence officer in Singapore. Singapore was important because of the presence of the headquarters of the British Far Eastern Forces and the intelligence service MI6. In these years British intelligence had to cover the entire Far East including Indonesia. As a result of this huge area of interest, the British services were initially not very well informed. Fortunately this changed rapidly as a result a significant increase in technical intelligence support facilities.

The Netherlands saw the United Kingdom, as a NATO member and colonial power, as a *natural ally* in a future military conflict with Indonesia. Besides this it was known that the interest of the British intelligence services in Indonesia was growing due to the increasing political influence of the Indonesian Communist Party. After all, British forces had just finished a ten year campaign to overcome the Chinese guerrillas in Malaya. London and The Hague agreed that the exchange of intelligence should take place in Singapore and due to the common interests the mutual intelligence flow started quickly. The Dutch liaison officer used an exclusive code system for radio communication with The Hague, while a diplomatic courier was used for shipment of other material.

The British services produced superior intelligence on the western part of the republic and on various anti-Sukarno movements in Indonesia. For Java and the islands in the eastern Indonesian archipelago, the Netherlands Naval Intelligence Service was in a favourite position to produce intelligence. The anti-Sukarno movements proved to be very important intelligence sources. In 1957 they became involved in a struggle for power in Indonesia. The most important movements were the Darul Islam, the PRRI on Sumatra and the Permesta on Celebes. Representatives of these political groups operated in Singapore trying to acquire political and military support from Western countries. Indeed Singapore became an ideal fishing ground for a professional intelligence officer. A continuous flow of valuable intelligence reports on the situation in Indonesia, the build-up and operational readiness of the Indonesian armed forces and the purchase of foreign military equipment, found its way to London and The Hague.

Intelligence reports in 1958 unveiled that Poland was about to deliver Soviet-built destroyers and frigates as well as MiG-17 fighters to Indonesia and that Czechoslovakia was willing to sell over 100 fighters, bombers and helicopters. In mid-1958 the Indonesian armed forces began to operate against the PRRI rebels on the island of Sumatra and in June 1959 the city of Menado on Celebes was recaptured by the republican troops. In February 1959 a large offensive started against the Permesta in the North Celebes. In this campaign the recently-delivered Soviet-type aircraft, among others flown by pilots trained in Egypt, were used for the first time. During the various actions, the Indonesian forces surprised the world in showing a very remarkable progress in executing air and amphibious operations.

Soon, intelligence reports indicated that Indonesia was capable of carrying out such operations successfully. For the Dutch troops in New Guinea infiltration actions by Indonesian military units, up to the size of a battalion, became a real threat. Another lesson was learned from the operations against the anti-Sukarno rebels: the preparation of these actions was very difficult to detect. By the end of 1959 the situation of the rebels worsened as a result of the republican military operations on the islands of Sumatra and Celebes. The rebellion did not succeed due to the lack of foreign support. Gradually the policy of the United States, the United Kingdom and the government of Malaya changed. The visas of the representatives of the rebellion movements were cancelled upon which most of them moved to Bangkok. The Dutch intelligence officer was now forced to maintain *under cover* contact with his sources.

In 1959 it was assessed that the pro-Western and popular General Nasution would be successful in his attempts to change the Indonesian policy into a more anti-Communist direction. As a result of this the Western allies began to appease Indonesia and offered him not only political support but also large amounts of military material which was not particularly helpful for the Netherlands. The United Kingdom delivered army trucks and Gannet anti-submarine aircraft and also American, French and Japanese military salesmen became frequent guests in Jakarta.[23]

The successful Dutch intelligence liaison officer in Singapore (1957–60) was Vice-Admiral (retired) F. E. (Frits) Kruimink (born 1917), a very remarkable man in the history of Netherlands intelligence. He finished his study at the Royal Netherlands Naval Institute in April 1940, one month before the German occupation of the Netherlands. In German internment he refused to sign a loyalty declaration and was taken prisoner of war. In 1943 he escaped from the POW camp at Stanislau in Poland and joined the counter-intelligence group of the Polish Resistance.

As a member of this organisation he *travelled* in 1944, via Berlin and Brussels, to Paris where he joined the FFI (Forces Françaises de l' Interieur) the French underground resistance. Until the end of the war he was involved in intelligence operations against German East–West road and rail troop movements. In August 1944 he returned to the Netherlands. From 1952 until 1955 he was assigned to the intelligence organisation of the Supreme Allied Commander Atlantic in Norfolk, Virginia. In 1962, he returned to the New Guinea conflict area as commanding officer of the anti-submarine destroyer Hr.Ms. *Groningen*. From 1964 until 1965 Admiral Kruimink was head of the Netherlands Naval Intelligence Service. In 1973 he retired as Vice-Admiral and became, until 1983, Coordinator of the Netherlands Intelligence and Security Services, a function with a direct responsibility to the Prime Minister.

THE THREAT AGAINST NEW GUINEA INTENSIFIES

In 1958, when the Indonesian threat of a military confrontation against Netherlands New Guinea intensified, the Commander of the Netherlands Royal Navy tasked Marid 6 NNG with the responsibility of 'early warning' by means of intelligence collection and electronic warfare in case of an Indonesian attack.[24] Because of the limited military capabilities, extremely long supply lines and the geographical extensiveness of the island, Netherlands plans for a successful emergency deployment of military units in the defence of New Guinea were completely dependent on timely intelligence of enemy plans, state of readiness, forward deployment and logistic support. Again in the history of the Dutch military presence in South East Asia, Holland had to rely on Sigint.

A clear sign of Indonesian intentions was seen in November 1957, when Dutch companies in Indonesia were boycotted, while the government initiated mass rallies and confiscation of Dutch properties. The ending of diplomatic relations was demanded. A message intercepted on 19 December 1957, unveiled that Indonesia was about to confiscate all ships and other facilities of the Dutch-owned merchant shipping company KPM, a company with a long history of services in the Indonesian archipelago.

The Dutch authorities had to move quickly in order to evacuate all 83 merchant ships out of Indonesian waters. Despite this effort, Indonesia confiscated 38 merchant ships. For three months Dutch navy ships were involved in successful operations to recapture the merchant ships. The intelligence support of the naval units involved in these operations proved to be very adequate and useful.[25]

During these operations the working periods of Marid 6 NNG were extended resulting in a daily average of 267 intercepted Indonesian messages, half of these were ciphered.[26] The cooperation between the WKC in Amsterdam and Marid 6 NNG at Biak was very intensive. Amsterdam provided crypto-analysis support and was extensively involved in intercepting and analysing Indonesian diplomatic and economic radio traffic. At Biak all available manpower and equipment had to be applied to focus on military traffic.[27]

In 1958 it became clear that Indonesia was preparing for military operations against Netherlands New Guinea. Intelligence assessments, among others based on Comint, indicated the Indonesian capability to conduct small-scale amphibious and air assault operations. In the same year a 'Front for the liberation of West Irian' was formed. By 1959 Indonesia was considered to be able to deploy armed forces up to the strength of a battalion. Not only the Soviet Union and other Warsaw Pact countries supplied large amounts of military equipment, including, submarines, destroyers, fast missile-carrying patrol boats and modern offensive aircraft, but also the United States, the United Kingdom, West Germany and Japan delivered modern military equipment on a large scale. In the meantime attempts to break the Indonesian Air Force codes became successful. The number of intercepted messages increased gradually reaching an overall daily average of 149 plaintext and 155 ciphered messages in July 1962 when the conflict reached the culminating point.[28]

In 1962 the crew list of MARID 6 NNG included 4 officers, 11 non-commissioned officers, 11 clerks, 2 translators and 22 civilian radio operators. As mentioned earlier, for the commander of the Dutch armed forces in New Guinea 'early warning' of Indonesian attempts to infiltrate

FIGURE 9.4

NUMBERS OF INTERCEPTED INDONESIAN MESSAGES DURING THE NEW GUINEA CONFLICT
TOTALS AND NUMBERS OF ENCIPHERED MESSAGES (DARK BAR)

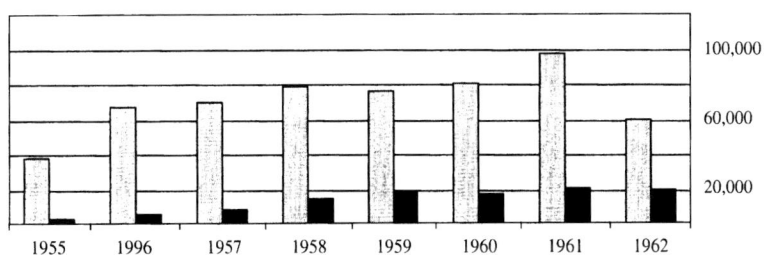

the territory of Netherlands New Guinea was of crucial importance. In 1962 a two-day warning time for a large-scale amphibious assault was considered to be the minimum requirement for a successful deployment of military units. In the meantime President Sukarno had given an order to his military commanders to be ready for action against Netherlands New Guinea by April that year.

NAVAL ENGAGEMENT NEAR 'VLAKKE HOEK'

An interesting example of timely and accurate intelligence support is the operation against Indonesian motor torpedo-boats on 15 January 1962. Intercepted messages indicated that in the evening hours, from the nearby Aroe islands, four fast MTBs would attempt to land a unit of 150 marines on the south coast near Vlakke Hoek.

Precise data on the deployment of the Indonesian motor torpedo-boats, the underway refuelling points, the rendezvous embarkation point and the position of the Indonesian units deployed for diversion purposes was derived from analysis of intercepted Indonesian messages. Three Netherlands destroyers were sent to the area in time.

At 2030 three Indonesian boats were spotted heading north by a Neptune maritime patrol aircraft. The fourth MTB was forced to withdraw due to mechanical problems. It was also known that as part of the operation an Indonesian tank landing ship would be deployed in a more westerly area in an attempt to deceive the Dutch on-scene-commander. One of the destroyers on a 'so-called' routine patrol spotted the landing ship and sent an 'open' flash enemy report. Most probably the Indonesian task group commander interpreted the message as a confirmation of a successful attempt to mislead the Dutch units and that he could continue his northerly course towards the coast of New Guinea undetected and safely. For the Indonesian Navy, the operation ended in a disaster. Two pre-positioned Dutch destroyers, soon accompanied by the third one, opened fire at 2100.

One boat sank immediately, two were badly damaged and reversed course. Many crew members and embarked marines were killed, among others the deputy chief of the Indonesian Navy Staff, Commodore Sudarso. Some 55 survivors were taken prisoner. This time the tactical use of Sigint prevented Indonesia hoisting a flag on Dutch territory and thus weakening The Hague's position at the ongoing United Nations New Guinea negotiations in New York.[29] After the conflict, the Indonesian General Hidajat declared that it was very remarkable that the Netherlands Navy knew the position of the Indonesian units. He was convinced then that treason caused the dramatic failure of the operation.

WAR WITH TIED HANDS

Available intelligence could not always be used for military action. In mid-June 1962 Comint analysis indicated that recently delivered C-130 Hercules aircraft were being prepared for action on the Indonesian air base of Amahai on the island of Ceram. Additional intelligence unveiled that the air force staff at Makassar had a requirement for detailed maps of the area around Merauke, the most southern town in New Guinea. On 24 June, Netherlands naval units tracked four C-130s on radar until they dropped 213 paratroopers in the vicinity of Merauke. There were no means available to intercept or to attack the aircraft before they reached their target. A pre-emptive strike against Amahai air base, from a military point of view the most appropriate operation to counter the threat of airborne assaults, was politically not discussible.[30] In 1962 a total of 1,200 Indonesian paratroopers landed in New Guinea and 340 infiltrators landed by boat.

THE FINALE

In the summer of 1962 it became clear that Indonesian armed forces prepared a large-scale amphibious and air assault against Netherlands New Guinea, most probably to be launched on 1 August. The operation was called 'Djajawidjaja', which means in English: 'Victory over colonialism'. The Netherlands intelligence organisation was able to produce a reliable and accurate order of battle of designated units, detailed information on logistic readiness and the timing of the attack. For some time there has been doubt about the main target of the assault: Sorong (in Vogelkop) or Biak. The best intelligence assessment was that Biak, for without its facilities the Netherlands forces would be crippled, would have the highest Indonesian priority. After the conflict, Indonesian officers confirmed this assessment.[31]

The Indonesian landing force, ATA-17 was organised according to the US Navy amphibious Task Organisation. The task force commander, Captain Sudomo, was embarked in the command and control ship *Multatuli*. The force consisted of 60 ships, organised as follows:

Task unit 17.1 Battle group	–	2 ex-Soviet 'Skoryy'-class destroyers
	–	4 ex-West German *Jaguar*-class MTBs
Task unit 17.2 Escort group	–	2 Italian-built frigates
	–	2 Italian-built corvettes
	–	4 ex-US coastal escort vessels
	–	5 Soviet-built coastal escort vessels
	–	4 ex-Yugoslavian coastal escort vessels
Task unit 17.3 Transport group	–	2 attack troop transport ships
	–	6 tank landing ships

	– 5 cargo ships with landing craft
	– 10 cargo ships
Task unit 17.4 Minesweeper group	– 4 ex-West German minesweepers
Task unit 17.5 Support group	– 4 oil tankers
	– 3 ocean going tugs
	– 3 hospital ships.[32]

In mid-1962, 5 of the 12 delivered ex-Soviet 'Whiskey'-class submarines were considered to have an operational status. At the same time, the Indonesian Air Force was assessed to have the capability to deploy 10 of the 14 delivered Tupolev-16 bombers and 35 of its arsenal of 50 MiG-17 fighters for offensive operations.[33]

Operation 'Djajawidjaja' was scheduled to be carried out in four phases:

Phase 1: Actions to gain sea- and air superiority including bombing raids against airfields at Biak, Noemfoer, Sorong, Manokwari and Kaimana, coastal commando raids and offensive submarine patrols.

Phase 2: Assault by a paratrooper brigade at Biak, followed by an amphibious assault by marines and army units.

Phase 3: Landings of paratroopers near Sentani as a preparation for the attack on Hollandia.

Phase 4: The attack on Hollandia by army units directly after the completion of phase 3.[34]

In 1962, the Netherlands naval forces consisted of 5 modern anti-submarine destroyers, 2 frigates, 3 submarines, 1 survey vessel, 1 supply ship and 2 oil tankers. For maritime patrol duties, nine Lockheed Neptune aircraft were based at Biak. Some 15 Hawker Hunter jets and army anti-aircraft gunnery elements were responsible for air defence. Five companies of the Royal Netherlands Marine Corps and three army infantry battalions were available for ground operations.

Early in 1962 the Netherlands Commander in New Guinea considered executing Elint operations, such as done by Nefis in 1947 and 1948, with the purpose of disrupting the Indonesian communication system in case of a large-scale attack. Due to the enormous effort necessary to execute this operation the plan was cancelled.[35] The large assault did not take place. It is difficult to speculate about the tragic military events that would have taken place if Indonesia had launched the attack. In his memoirs the Indonesian Admiral Sudomo, in 1962 commanding officer of the main amphibious task group, speaks about the planned operation as a 'one way ticket'.[36]

DUTCH SIGINT AND THE CONFLICT WITH INDONESIA

FIGURE 9.5

Hollandia: Staff buildings Commander Netherlands Forces in New Guinea. The intelligence section was located in the centre building.

The Netherlands historian P. B. R. de Geus concludes that it is out of the question that the Indonesian armed forces would have been able to carry out a surprise attack and that, without any doubt, they would have suffered heavy losses, before reaching their target. Marid 6 at Biak would certainly have given a timely early warning, enabling the commanding officer of the Netherlands Forces to deploy his units in the most favourable position to counter the assault. Most probably Indonesia would have failed to gain a decisive victory as was demanded by President Sukarno. But the Netherlands would not have been able to sustain the military confrontation for a long period.[37]

The fact is that on 15 August 1962 at 0930 local time a ceasefire agreement was signed in Washington. Two days later Marid 6 NNG intercepted a signal from Jakarta to the Indonesian armed forces containing the message that Operation 'Djajawidjaja' was cancelled. Indonesia paid a high military price for this political success, about 220 infiltrators were killed in action and 500 were taken prisoner of war.

Marid 6 at Biak was officially closed at 2400, local time, on 27 September.[38] Shortly after this, the crew of 50 men were repatriated leaving the tropical heat of New Guinea. During the intensifying Cold War in Europe the need for Sigint remained undiminished and many of them resumed their employment at the WKC in Amsterdam. In the history of the Royal Netherlands Navy, the intelligence support of the operations during the New Guinea conflict has officially been recorded as outstanding and decisive for the successful deployment of the limited available assets.

EPILOGUE

At 2300 on 15 August 1962 the agreement between Indonesia and the Netherlands about the future of the western part of New Guinea was signed in the presence of the Secretary-General of the United Nations, U-Thant and the American negotiator Ellsworth Bunker. Former ambassador Bunker drafted the plan for the agreement.

Elements of the plan were:

- the Netherlands will turn over the authority to a Temporary Executive Authority under the supervision of Secretary-General of the United Nations
- the Secretary-General nominates a mutual acceptable administrator who will be in charge for a period not shorter than one year and no longer than two years
- Netherlands officials will be replaced by UN personnel and after one year by Indonesian officials
- after an undefined period Indonesia will, in cooperation with the United Nations, create the possibility for the people of Western New Guinea to express their will for self-determination
- after signing of the agreement diplomatic relations between the Netherlands and Indonesia will be restored.[39]

Part of the agreement was that the hostilities should end on 18 August at 0001 GMT.

The New Guinea drama was over. In the last part the Netherlands played only a minor role. Leading actors were Indonesia and the United States. Sukarno had gambled and won. Obviously President Kennedy granted Sukarno the political victory, hoping that by doing so Indonesia would not be pushed into the arms of Moscow. This hope turned out to be vain. One year later, under the increasing influence of the Soviet Union, President Sukarno mobilised public opinion in Indonesia for a confrontation with Malaysia. In 1965 the political situation changed dramatically. An unsuccessful Communist coup took place in which six Indonesian generals were killed. The army took over and President Sukarno was relieved from all his functions.[40]

DUTCH SIGINT AND THE CONFLICT WITH INDONESIA 309

MOSCOW AND THE NEW GUINEA CONFLICT

Situating the conflict in the framework of the East–West confrontation in the early 1960s it is obvious that both the Soviet Union and the United States had their own specific interests in Indonesia. In the heat of the political confrontation, both powers were desperately trying to safeguard and to extend their influence in the Asian theatre. President Sukarno offered both parties an ideal and justifiable scenario to support his ambition: the elimination of the remains of an ancient colonial regime in the Third World.

The Soviet Union and other Warsaw Pact countries delivered large quantities of modern war equipment and placed the conflict in the spotlight of the international warfare of the working class against imperialism. President Kennedy had to balance carefully between the interests of a small, opinionated European ally involved in a far from home and unpopular 'post colonial' conflict and the reality of politics on a global scale. In February 1999 the Moscow correspondent of one of the leading Dutch newspapers *De Volkskrant* interviewed three former officers of the Red Fleet. His attention was drawn by a programme of the Russian national TV station ORT called '*Kak eto byla*' (How it was) about Indonesia and the New Guinea conflict.

In this programme it was stated that in 1962 a Soviet military force of 6 'Whiskey-class' submarines with auxiliary units, 30 Tupolev-16 bombers (NATO codename 'Badger') and a total of 3,000 military personnel was deployed to Indonesia to operate together with the Indonesian units in the final assault against New Guinea. Commanding officer of the Soviet force was Rear-Admiral Tsjernobajs. The orders were to attack Dutch units and facilities from midnight 1 August. Direct operational control of the Soviet units was carried out by the Pacific Fleet high command in Vladivostok.

The three former officers are Aleksey Droegov, who functioned as Indonesian-Russian translator. Presently he is a member of the Oriental Institute in Moscow. The second is Rudolf Ryzjikov who was the executive officer onboard the *S-236*, one of the six submarines involved. The third officer is Gennadi Melkov, commanding officer of the submarine *S-235* and presently professor in the faculty of international law at a Moscow university. The former Soviet Naval Attaché in Washington (1962–66), Y. Tsoebasjev, confirmed the story. In 1998 he put his findings on paper and tried to sell it to the Netherlands Embassy in Moscow.[41] After the publication in *De Volkskrant*, one of the three former Navy officers visited the Netherlands Defence Attaché in Moscow. He had written down his personal experience and asked for help to get the story published in the Netherlands. So far this attempt has remained unsuccessful.

In 1962 the Netherlands naval intelligence service was aware of the presence of a large number of Soviet military instructors in the Indonesian

Air Force and Navy. In fact the fast influx of large amounts of modern military equipment, and the rapid introduction of this material into the operational elements designated for the assault, could only be carried out successfully with an extensive support by foreign expertise. However there is no intelligence available to support the assessment of a direct and large-scale Soviet participation in the planned operations in 1962. The Director of the Dutch Institute of Naval History has strong doubts about it.[42] Nevertheless the Russian story could be true. After all, First Secretary Nikita Khrushchev proved to have the courage to provoke the Western world to the limit in sending missiles to Cuba the same year.

ROYAL NETHERLANDS NAVY SIGINT ACTIVITIES IN THE WEST INDIES

With the transfer of authority to the republic of Indonesia, the Netherlands naval presence in South East Asia came to an end. For centuries the Dutch tricolour had ruled the waves in the Indonesian archipelago. This was all over now and the withdrawal of the Dutch forces also ended the Netherlands intelligence activities in this part of the world. With the closure of Marid 6 at Biak, the communication intelligence section of the Royal Netherlands Navy (WKC) in Amsterdam got disposal of some well-experienced Sigint experts. Finding a job for most of these men was not difficult. At the end of 1962, the need for communication intelligence operations against the Soviet Union and its allies was growing, not only in Europe but also in the remaining Dutch national overseas area of interest, the Netherlands Antilles in the Caribbean.

The military tension in this region caused a growing concern about the safety of this part of the Kingdom. After the successful revolution in Cuba, Fidel Castro was very clear about his ambition to extend his influence in the area and to export the Marxist victory. In 1962 the world almost witnessed the prelude of World War III when the Soviet Union deployed offensive weapons as a direct threat to the United States on Cuban soil. The territorial defence of the Netherlands Antilles has always been a responsibility of the Royal Netherlands Navy. The main facility is the naval base at Willemstad on the island of Curaçao. The military presence consists mainly of a frigate-type station ship, maritime patrol aircraft and units of the Netherlands Marine Corps.

In 1963 a decision was taken by the Commander-in-Chief of the Royal Netherlands Navy to improve the intelligence position by establishing the 'Radio Control Service Netherlands Antilles' at Curaçao. Research had proved the good possibilities for Comint collection. Especially the communication network of the Venezuelan armed forces offered attractive opportunities for intelligence collection. Venezuela never posed a direct military threat to the Netherlands Antilles. But it is obvious that, due to an unstable internal political situation and the fact that the Dutch Leeward Islands Curaçao, Bonaire and Aruba are situated within the Venezuelan economic sphere of influence, the Netherlands Navy Intelligence Service had its interests in Caracas. Beside this, in 1963 the Venezuelan armed forces had the potential capability for offensive, overseas operations. Civilian and military communication personnel from the WKC in Amsterdam manned the new communication intelligence site. Technical problems were caused by the fact that in the beginning the site had to use the antennas of the Navy world-network which were mainly aimed at Europe.

Despite this it was possible to obtain a reliable order of battle of target countries in the area by monitoring military communication networks. Not only the communication network of the Venezuelan armed forces was successfully intercepted but also the Cuban diplomatic network. Another task of the Sigint site was to monitor the high frequency world net of the Soviet Navy, mainly with the purpose of completing the material collected in the Netherlands. Targets were Soviet Navy communications facilities in Vladivostok and Tashkent and the Soviet Navy support facilities in Angola and at Port Mariel in Cuba. The geographical dispersion offered good interception results and provided reliable positional data of inter-fleet ship movements. During the Falklands War the station was capable of providing a good insight into its dramatic events.

Despite the successes, the radio monitor station at Curaçao was closed, mainly for budgetary reasons. On 1 July 1990 the overseas Sigint activities of the Royal Netherlands Navy came to an end. What started in 1933 in Batavia when Japan prepared for war ended 57 years later when Fidel Castro lost the support of the Soviet Union and consequently his appetite to export his glorious revolution.[43]

NOTES

1. Karl de Leeuw, *'Henri Koot'*, Biographical Dictionnaire (Institute for Netherlands History, The Hague) version 2 Sept. 1998.
2. J.F.W. Nuboer, *Marineblad 1981*, No. 6, p.268.
3. Ibid. p.269.
4. Ibid. p.271.
5. Robert D. Haslach, *Nishi no kaze hare* (Weesp: Van Kampen & Zn 1990) p.77.
6. Ibid.
7. F.T.S. van der Laan, *Marineblad 1992*, No. 12, p.459.
8. K.W.L. Bezemer, *Verdreven doch niet verslagen* (Hilversum 1967) p.1.
9. Ibid. p.184.
10. M.W. Jensen and G. Platje, *De Marid* (Den Haag: SDU 1997) p.73.
11. Ibid. Chapter 6.

12. Ibid. p.75.
13. Interview with J.F. Bastiaans, 23 July 1998.
14. Jensen and Platje, *De Marid* (note 10) p.76.
15. The Netherlands Military Intelligence Agency, Marid VI/Wiskundig Centrum 1947–1965, Top Secret, 31 Oct. 1965. S.Marid 6110/000780/65/ZG (hereafter History *WKC/TIVC*) p.38.
16. *History WKC/TIVC* (note 15) p.7.
17. Interview with J.F. Bastiaans, 23 July 1998.
18. *History WKC/TIVC* (note 15) p.21.
19. Ibid. p.39.
20. Ibid. p.43.
21. Jensen and Platje, *De Marid* (note 10) p.94.
22. Ibid. p.137.
23. Ibid. pp.135–7.
24. Ibid. p.111.
25. Jensen and Platje, *De Marid* (note 10) p.117.
26. *History WKC/TIVC* (note 15) p.44.
27. Ibid.
28. *History WKC/TIVC* (note 15) pp.45 and 174.
29. R.E. van Holst Pellekaan, I.C. de Regt, and J.F. Bastiaans, *Patrouilleren voor de Papoea's: de Koninklijke Marine in Nederlands Nieuw-Guinea*, 2 vols. (Amsterdam 1989–90) pp.112–18.
30. Jensen and Platje, *De Marid* (note 10) p.141.
31. Ibid. p.143.
32. Pellekaan, Regt and Bastiaans, *Patrouilleren voor de Papoea's* (note 29) p.81.
33. Interview with J.F. Bastiaans, 7 Oct. 1999.
34. Pellekaan, Regt and Bastiaans, *Patrouilleren voor de Papoea's* (note 29) p.80.
35. Interview with J.F. Bastiaans, 7 Oct. 1999.
36. Pellekaan, Regt and Bastiaans, *Patrouilleren voor de Papoea's* (note 29) p.156.
37. Ibid.
38. *History WKC/TIVC* (note 15) p.47.
39. Pellekaan, Regt and Bastiaans, *Patrouilleren voor de Papoea's* (note 29) p.16.
40. Ibid. p.21.
41. Bart Rijs, 'Moskou beraamde in '62 aanval op Nieuw-Guinea', *De Volkskrant*, 10 Feb. 1999.
42. 'Historici aarzelen over Sovjet-onderzeeërs', *De Volkskrant*, 11 Feb. 1999.
43. Jensen and Platje, *De Marid* (note 10) pp.238–41 and 301–2.

10

Conclusions

MATTHEW AID and CEES WIEBES

The central question raised in this work is this: what were the important contributions made to the security of the United States, Canada and various nations of Western Europe by Sigint? The writing of this study was motivated by the desire of the authors to fill the void that currently exists in the public literature regarding the role and importance of Sigint in the years after the end of World War II. By the same token, the purpose of this study was also to broaden in a profound way the public's knowledge about the role of Sigint in countries outside of the United States. In order to realise this goal, we have chosen to follow a multinational and more comparative approach.

In our view, there has been too much 'intelligence history snobbery'. To date, the focus of most intelligence studies was the role and activities of the world's largest intelligence services, such as those of the United States and the former Soviet Union. The writers of this work have tried to construct a new approach regarding intelligence history. We think that in the future, serious academic historical studies dealing with all aspects of intelligence should be far more 'internationalist' in approach, taking into account the accomplishments of the world's smaller services, from whom the world's larger intelligence organisations derived much of their information during the Cold War. In this regard, one can even postulate that it is impossible to write a comprehensive history of American or Soviet intelligence activities during the Cold War without talking at the same time about the liaison relationships with their partners around the world. For instance, in much that has been written in the past the Canadian and European Sigint organisations have usually been portrayed as subordinate entities acting in lockstep under the leadership and tutelage of the National Security Agency. In fact, this publication and other recently declassified documents clearly shows for the first time that the Sigint sharing relationships between the US,

Canada and the European allies during the Cold War were far more complicated than previously known.

Each of the authors has given in this work a sketch of the establishment and growth of the national Sigint service that they were asked to write about. Important matters like successes, failures, national acceptance, international cooperation, intra-European liaison, budgetary problems, national rivalry and targeting have been dealt with extensively. The purpose of this final contribution is to draw these national experiences together and try to present a coherent picture of what we deem to be the primary revelations stemming from this study.

COMMON GROUND

There were certain common threads that bound the United States, Britain, Canada and their Western European allies together in their collective Sigint effort against the Soviet Union and its Eastern European allies.

The *first thread* was the jointly held belief with the intelligence communities of the US and European partners that the Soviet Union and its allies posed a real and substantive military threat to the security of Western Europe. As such, all of the nations that are discussed in this work, even those who were nominally neutral, devoted the vast majority of their Sigint collection resources to monitoring the military and political activities of the Soviet Union and its Eastern European allies. For example, throughout the Cold War the percentage of NSA's collection resources dedicated to monitoring the Soviet Union never fell under the 50 per cent mark. The figures of America's European Sigint allies, although still a secret, were almost certainly higher. For example, it is rumoured that the Norwegian collection resources still dedicates about 80 per cent of its resources to monitoring the former Soviet Union and its allies. The same probably holds true for the Danish and Swedish Sigint organisations.[1]

The *second thread* that bound the UKUSA Comint organisations together throughout the Cold War was mutual dependence on each other. The period of fiscal austerity following World War II forced the US and Britain to pool their limited Comint resources in order to cover the Soviet Union and its allies in Europe and Asia. This led to the 1946 revision of the wartime British-US Comint Agreement (or BRUSA). This was followed by a more inclusive Comint sharing and collaboration agreement signed in June 1948 called the UK-USA Comint Agreement (or UKUSA), which added the nascent Comint organisations of Canada, Australia and New Zealand but retained Britain as the dominant partner among the Commonwealth nations.

Throughout the Cold War the complex and interlocking relationships between the UKUSA nations were so close that it was (and remains today) difficult to separate them into individual and distinct national intelligence collection efforts. Referring to the intimate Sigint relationship between the US and Great Britain, the former British ambassador to the United States, Sir Peter Ramsbotham, said in 1976 that 'We have all grown so accustomed to the closeness of the relationship between our two countries, that those of us who live and work in the United States have come to take it for granted.'[2]

The relationship between NSA and GCHQ became so closely interwoven that it was oftentimes difficult for American and British intelligence officials to determine whether the SIGINT reports they were reading every morning came from NSA or GCHQ.[3] An 28 April 1948 memorandum from the chief the US Air Force Air Intelligence Requirements Division reported that 'At the present time, there is complete interchange of communications intelligence between the cognisant United States and British agencies. It is not believed that the present arrangements on the interchange of this information could be improved.'[4]

There was also a parallel Elint sharing relationship between the three US military services and their British military counterparts. In the summer of 1951, GCHQ entered the Elint collection field. Specifically, GCHQ was given control over the Elint intercept units in the field, and was responsible for transmitting the collected intercepts to a unit in London called the Technical Radio Investigation Committee (TRIC) for analysis and reporting. TRIC was also responsible for providing finished Elint intelligence reports to the US Navy and Air Force 'Special Intelligence Exchange Officers' assigned to the Ministry of Defence in London.[5] It was not until 1958 that NSA was given operational control of the American national Elint effort by virtue of a National Security Council directive.[6]

As Richard Aldrich wrote in his contribution on British signals intelligence, the intimate intelligence collaboration between the US and Britain extended to sensitive submarine reconnaissance missions performed off the coast of the Soviet Union and its allies. There was also close collaboration between the CIA and FBI on one hand, and the foreign intelligence and internal security services of Britain on the other in the field of clandestine intelligence collection of foreign cryptographic systems.

The Sigint sharing and collaborative relationships between the UKUSA 'First Party' nations, namely the US and Britain, and the Sigint units of the 'Second Party' states of Canada, Australia and New Zealand were somewhat more limited. For example, the terms of the 1948 Canadian – US Communications Intelligence Agreement (CANUSA) was nowhere near as generous as the comparable 1946 BRUSA COMINT agreement between the US and Britain. The CANUSA Agreement limited the exchange of

Comint translations and gists among the two countries to a 'need to know' basis because, as one American official at the time argued: 'The Canadians have no information to exchange.' The Canadians reluctantly agreed to the American terms (in truth they had no choice but to accept), and CANUSA was born in June 1948.[7]

The relationship between the US and the UK on one hand and the so-called UKUSA 'Third Party' nations was even more limited. During the 1950s, the US and Britain signed a series of bilateral Sigint sharing agreements with European and Asian countries. Declassified documents show that by the mid-1950s, the US and Britain had established formal bilateral Sigint collaboration and sharing agreements with Norway, Denmark, West Germany, Italy, and Turkey in Europe; and Pakistan, Burma, Thailand, the Republic of China (Taiwan) and South Korea in Asia.[8] In the mid-1960s, additional foreign Sigint organisations became UKUSA Third Parties, including those of Austria, Greece, South Vietnam, the Philippines and Japan.[9]

The strategic value of these countries to the UKUSA Sigint community was largely due to their geographic location adjacent to the Soviet Union, Eastern Europe and China. For example, the Sigint sites in northern Norway provided unique and invaluable access to Soviet military communications emanating from the Kola Peninsula. Danish Sigint sites were perfectly situated to provide in-depth coverage of Soviet and Polish military activities in and around the Baltic Sea. The multitude of Sigint stations in West Germany and West Berlin permitted unparalleled access to radio traffic emanating from East Germany, Poland and Czechoslovakia. The best Sigint coverage of communications traffic coming from eastern Czechoslovakia, Hungary and northern Yugoslavia was obtained from Austria, which complemented the Sigint coverage obtained from Italy and Greece. Turkey was perfectly positioned geographically to allow unfettered Sigint coverage of military radio traffic coming from the much of the Balkans as well as much of the southern Soviet Union.

Some nominally neutral European states that were not UKUSA Third Parties informally engaged in Sigint collaboration and/or information sharing with the UKUSA nations. For example, there was some Sigint co-operation between the French and the US during the first war in Indochina in the 1950s, despite the fact that France was not a signatory to the UKUSA agreement. In return for direction-finding bearings from the Army Security Agency listening post at Clark Air Base in the Philippines, the French provided the American embassy in Saigon with copies of their Sigint reports on Viet Minh military activities.[10]

Despite being a member of NATO and possessing a small but accomplished Sigint service, the Dutch never joined UKUSA despite repeated requests to participate. As detailed in Dr Wiebes' study on the Netherlands, the Sigint service of Dutch Naval Intelligence, the TIVC, did accept relatively small amounts of financial and technical assistance from their American counterparts, but succeeded in maintaining its full independence throughout the Cold War. This also pertained to the Sigint services of the Dutch Army and Air Force.

Nominally neutral Sweden (who some European officials referred to during the Cold War as 'the secret member of NATO') maintained a close working relationship with the US and Britain in the Sigint field throughout much of the Cold War, but managed to retain its independence despite the somewhat intimate nature of the relationship.[11] During the 1940s and 1950s, US military attachés based in Stockholm were routinely given the results of Sweden's Sigint efforts against the Soviet Union and its Warsaw Pact allies by senior officials in the Swedish Defence Staff. For instance, as early as January 1949 the Swedish Defence Staff was passing to American and British military attachés in Stockholm the results of their ELINT collection efforts against Soviet radars in the Baltic Sea region.[12] In April 1951 Swedish military intelligence officials gave the US military attaché the results of Sweden's Sigint monitoring of a large Soviet air defence exercise conducted in Soviet Karelia north of Leningrad (now St Petersburg).[13]

Informal liaison between the Swedish Comint organisation, the FRA, and GCHQ continued well into the 1960s, including a joint operation to spy on the Russian cruiser *Ordzhonikidze* during her visit to Sweden in 1959.[14] In 1959, the FRA agreed to provide GCHQ with copies of Soviet military intelligence (GRU) messages transmitted by Russian spymasters from Britain during World War II that had been intercepted by the Swedes.[15] In 1969 a covert listening post operated by the CIA's Office of Elint on the Swedish island of Gotland, called 'Briar Patch', intercepted the first emissions from the Soviet 'Hen House' missile early warning radar at Skrunda in Latvia.[16] Vice Admiral Bobby Inman, who had served as naval attaché in Stockholm, worked hard to improve relations with Sweden when he was the head of the Office of Naval Intelligence from 1974 to 1976. This assistance included helping to get secret US funding for a new Swedish spy ship called the *Orion*, which was commissioned in 1984. The quid pro quo was that the intelligence information generated by the *Orion* was to be shared with the US.[17]

But as time went by, NSA grew increasingly large and powerful, reaching almost 100,000 men and women by the late 1960s. But the other UKUSA member organisations were not able to match the size and scope of NSA's worldwide Sigint collection efforts. Starting in the 1950s, fiscal austerity and operational considerations forced all of the Second Party UKUSA Comint organisations to scale back their Sigint intercept and cryptanalytic activities. Budget cuts forced GCHQ to cut back on the scale of its Sigint collection operations during the 1950s and 1960s, forcing the financially-strapped British Sigint Agency to become increasingly dependent on NSA for money and high-tech equipment.[18] In a harsh assessment of the changing dynamics of the US-British Sigint relationship, two British authors wrote: 'From a post-Second World War partnership of equals it [the UKUSA relationship between NSA and GCHQ] has evolved into a master-servant arrangement of convenience.'[19]

In 1957 Canada's CBNRC ceased conducting cryptanalytic work, reducing Canada to the position of being a mere supplier of intercepts for NSA and GCHQ. Because it had no cryptanalytic capability, the Canadian Sigint organisation had to rely on NSA to process and analyse much of the materials collected by the Canadians. As the Scandinavian contribution by Alf Jacobsen showed, the Norwegian Defence Intelligence Staff (FO/E) also ceased its cryptanalytic efforts in the late 1950s because of the poor results coming out of the effort, making the FO/E almost completely dependent on NSA for high-level intelligence derived from Comint. In some cases, the degree of dependence on the US took on alarming proportions. For example, by 1970 about 80 per cent of the budget for the Norwegian Sigint effort was being secretly paid for by the US. This contrasts with other European countries like Germany, France and the Netherlands, where cryptanalytic work was continued and is flourishing today.

But some members of UKUSA, however, have over time become accustomed to being dependent on the US for leadership and resources. Indeed, some have arguably even benefited from it. The seductive power of access to American fiscal power and technological wherewithal is indeed addictive. Former British intelligence official Michael Herman has written that 'For Britain and others, access to the United States' weight of resources, technology and expertise is an overwhelming attraction.'[20] The relative decline in strength and capabilities of GCHQ since World War II has meant that over time the British have increasingly obtained more out of the special relationship with NSA than they have contributed. Naturally, this means that over time Britain has become even more 'strongly wedded to the special relationship' with NSA.[21] The 2000 annual report of the British Parliamentary Intelligence and Security Committee revealed that 'The quality of the intelligence gathered [by GCHQ] clearly reflects the value of the close coordination under the UKUSA agreement.'[22]

But fear of being 'marginalised' out of UKUSA because of diminished capacity and resources scared the British intelligence community during the 1970s and 1980s. An in-house GCHQ document, entitled the 'Strategic Direction Summary', argued that GCHQ's contribution to UKUSA had to be necessarily 'of sufficient scale and of the right kind to make a continuation of the Sigint alliance worthwhile to our partners'.[23]

Subsidisation of the British Sigint effort, direct and indirect, appears to have been more pervasive than previously believed. In the early 1980s, all of the UKUSA Sigint organisations adopted the R-2174 HF intercept receiver made by the British defence contractor Racal for use at all HF intercept stations around the world. This was done despite the fact that American defence contractors had offered NSA better and less expensive equipment.[24]

In the case of Canada, as shown by Martin Rudner, NSA provided the Canadian Sigint organisation with the latest Sigint equipment, trained Canadian personnel, and gave the Canadian Sigint organisation entrée to some of NSA's most sophisticated and sensitive operations.

There were tangible benefits for UKUSA's 'Third Party' members as well. The US secretly provided tens of millions of dollars of assistance in the form of training and equipment for the Third Party Sigint units. For example, during the late 1950s and early 1960s, the German BND and its Sigint service took advantage of American largesse to get the US taxpayer to foot the bill for building and equipping a huge Sigint intercept site in Augsburg. In the Netherlands, the Dutch Sigint service was more than happy to accept the American gift in the form of a Granger HF intercept antenna at the Eemnes listening post, which is still intercepting diplomatic and military HF traffic today. In the late 1960s, the Italian Navy manned what superficially looked like fishing boats equipped with Sigint and other technical intelligence collection gear supplied by the US Navy's Office of Naval Intelligence (ONI). These boats were used to covertly monitor Soviet naval activities in the Mediterranean.[25]

In some cases, Third Party Sigint personnel worked side by side with their American counterparts at NSA listening posts, such as in West Germany, Turkey, Thailand, and South Korea. In the 1960s, Turkish Sigint personnel were integrated into the Army listening post at Sinop, Turkey, where they performed many of the same Comint intercept and processing tasks as their American counterparts.[26] Also NSA eavesdroppers were sometimes welcome on board Dutch ships, or could operate on Dutch territory, such as from the Dutch-controlled islands in the Caribbean. The same happened in a not so neutral Sweden, where from the 1950s American personnel worked hand in hand with the Swedish Sigint service. NSA officials even reportedly participated in Swedish Sigint flights over the Baltic Sea.[27]

But the pricetag for taking NSA's money and equipment was the inevitable loss of independence for the UKUSA member organisations. NSA typically demanded and got virtually all of the intercepts collected by its partner agencies. By the mid-1980s, some British authors suggest that GCHQ had become nothing more than an extension of NSA, with an internal GCHQ document stating 'This may entail on occasion the applying of UK [Sigint] resources to the meeting of US requirements.'[28] A former USAF Sigint analyst wrote that the Third Party nations received 'absolutely no material from us, while we get anything they have, although generally it's of pretty low quality'.[29]

NSA also could largely dictate which targets they wanted their allies to cover, which inevitably meant a focus on the Soviet Union at the expense of targets of interest to the nations in question. An example of this loss of independence can be found in the section of this study dealing with Scandinavia. Another good example is the case of the West German Sigint services, which strove mightily to preserve their independence from NSA and GCHQ. It was only when the BND and the West German military began to dramatically expand their Sigint infrastructure during the 1970s that the Germans reluctantly sought NSA's financial and technical assistance. A step taken solely because they discovered that the cost for a sophisticated Sigint collection infrastructure was too high to try to 'go it alone'. The cost of accepting NSA's help was the loss of a certain degree of their independence, as well as having to agree to NSA requests, such as providing NSA with more raw Sigint materials.

A *third thread* was the fact that by the early 1950s, NSA, GCHQ and the other UKUSA Sigint organisations were producing intelligence on what Richard Aldrich has described in his contribution as 'an industrial scale'. Unlike the CIA, MI6 and other Western Humint organisations, the UKUSA Sigint organisations pioneered the use of the still nascent computer and electronic technologies. They continued the process begun during World War II of turning out intelligence on an assembly line basis, using as their model Henry Ford's automobile production plants in Detroit, Michigan. As a result, the UKUSA Sigint organisations quickly became more efficient and efficacious producers of intelligence than their better-known counterparts in the Western intelligence communities. This made Sigint production organisations an attractive showcase for other European nations who wished to participate in the Western intelligence community.

While many Western European nations dedicated the majority of their intelligence resources to Sigint collection, such as the Scandinavian countries of Norway, Sweden and Denmark, other European countries invested most of their resources in Human Intelligence (Humint). For example, this study has shown that the West Germany intelligence service, the Bundesnachrichtendienst (BND), backed by the CIA, dedicated the vast majority of its resources to Humint collection at the expense of Sigint. It was not until the retirement in April 1968 of the BND's founder, General Reinhard Gehlen, that the BND dramatically shifted its emphasis from Humint to Sigint collection.

For its part, the West German military, the Bundeswehr, began to build up its Sigint collection capabilities after the 1962 Cuban Missile Crisis. We have also seen in the contribution on France by Roger Faligot that the French intelligence service, the SDECE, did not begin investing heavily in Sigint collection until the 1970s. This collection also showed that the situation in the Netherlands differed somewhat compared to other nations in that much more money was poured into its Sigint endeavours than into Humint.

As said earlier, in much that has been written in the past about the UKUSA Sigint agreement, the member organisations are often portrayed as a single monolithic entity acting in lockstep under the leadership of NSA. In fact, this collection and other recently declassified documents for the first time prove that the Sigint sharing relationship between the US, Canada and the European UKUSA allies during the Cold War was far more complicated than previously known.

Often, the US imposed severe limits on what, if any Sigint it was willing to share with its European allies. There were even limits on American Sigint collaboration with Britain. A series of 1949 American decrypts of Czech internal security messages detailing the arrest of a number of US Army Counterintelligence Corps agents in Czechoslovakia were all marked 'THIS MATERIAL TO BE SEEN BY US PERSONNEL ONLY' in order to keep news of these Humint failures away from the British.[30] In March 1954, the US Far East Command in Tokyo was authorised for the first time to share Comint information with British intelligence officers serving on the command's intelligence staff in Tokyo. However, American intelligence officials were ordered to excise (with scissors) any technical information from the materials that were to be given to British, which presumably would include such technical data as frequencies being monitored, date and time of intercept, case notation information, etc.[31]

Moreover, it was not until the summer of 1954 that Far East Command was permitted to give Elint reports to the British naval attaché in Tokyo, but Washington only permitted the release of finished intelligence reports classified no higher than Secret and only for the Far East theatre. Furthermore, the material was to be released only on a quid pro quo basis, meaning that the British in return had to give Far East Command the results of their Elint collection efforts in the region, and allow American military intelligence officials access to their Elint facilities in the region. The Top Secret Elint materials generated by Far East Command, which included a great deal more technical detail than was contained in the Secret reports, were specifically barred to the British intelligence representatives in the Far East. The Pacific was, after all, an 'American Lake'.[32] British personnel who worked at the American listening post at RAF Chicksands, England, were barred from entering the part of the listening post called Joint Operations

Center, Chicksands (JOCC), which was manned by NSA civilians and USAF intercept personnel.[33]

NSA did not have a monopoly on this sort of behaviour. Britain's GCHQ also withheld information from NSA. For example, after the end of World War II GCHQ refused to give to the US sensitive decrypts of Soviet clandestine radio traffic between Moscow and the Soviet mission with Mao Tse-tung's Chinese Communist forces at Yenan. It was not until March 1946 that GCHQ began sending a single copy of these translated decrypts to Washington, but on the condition that they were to be read only by the deputy chief of US Army Intelligence, General Carter Clarke.[34]

Not surprisingly, NSA's treatment of its 'Third Party' Comint allies was often even worse. The sharing of Sigint with the UKUSA 'Third Party' Comint organisations was far more restricted because of security considerations. For example, West German intelligence officials complained about being treated as second-class citizens by NSA in their own country. German Sigint personnel stationed at NSA's Augsburg listening post in West Germany found that their requests for direction-finding missions were almost always placed behind American tasks, and they were denied access to large parts of the station by American security officials.

Sigint sharing between the US and France was terminated in the 1960s after it was discovered that the French foreign intelligence organisation, the SDECE, had been penetrated by the KGB. Sigint sharing with France was not resumed until a decade later.[35]

Moreover, the Dutch intelligence community was not pleased with the unwillingness of the US to fully share Sigint with it. But in all fairness, the Dutch were also not pleased by the reluctance of GCHQ to share the results of their Sigint operations, such as during the Dutch colonial struggle with Indonesia over New Guinea in the 1950s and 1960s.

There have also been numerous complaints from America's allies that NSA got far more from its European partners than it gave in return. One specific complaint that is still often heard from European intelligence officials is that NSA was not particularly forthcoming when it came to passing to its allies high-level intelligence derived from Comint.[36] A British analyst recently wrote that: 'America's allies have long complained that it is particularly mean with its intelligence.'[37] A former senior Norwegian intelligence official, Knut Willy Kval, was quoted as saying that '[W]here it was not in the interest of NSA that we should possess cryptographic insight, they did not have to share such matters with us.'[38] In 1951, the Dutch Sigint Agency stopped the weekly shipment of raw intercepts of Soviet radio and telegraph to the CIA station in The Hague because NSA's predecessor organisation, the Armed Forces Security Agency (AFSA), refused to provide the Dutch with the results of its analysis of these messages. These complaints continue to this day. Intelligence officials working for the United Nations verification mission in Iraq (UNSCOM) complained often that they were forwarding to the US all of the raw Sigint they were collecting inside Iraq, but getting little back.[39]

What America's European allies did not realise was that the order to not provide the NATO allies with Sigint came from the very top. A declassified USAF intelligence document revealed that: 'The National Security Council does *not* want the end product of USAFSS [US Air Force Security Service] operations to be given to foreign nations in NATO commands. It is recognized that most of the NATO member nations will benefit from USAFSS operations. Officially, the NATO member nations, *with the exception of Great Britain* [editor's emphasis], will not be given the details of USAFSS operations although they must realise, unofficially, how and where the end product was secured.'[40]

Another problem area, and thereby a *fourth thread*, was that NSA has in the past, and continues to this day to conduct its Sigint sharing relationships on a bilateral country-by-country basis, with some of NSA's partners getting more information and resources from NSA than others depending on geopolitical and geographic considerations. For example, the Norwegian intelligence service was treated particularly well by the US because the telemetry intelligence that was collected by Norwegian listening posts from Soviet missile and space launches at Plesetsk and Soviet naval missile test firings from Nenoksa in the White Sea were deemed to be invaluable by the US intelligence community.[41] Other nations, such as Italy and Greece, were not treated as well. These bilateral relationships led to considerable resentment on the part of those countries who were the 'have-nots' against those who were particularly favoured by NSA, with some European intelligence chiefs believing that NSA was playing one partner organisation off against the others.[42]

This problem remains to this day. During the war in Bosnia between 1992 and 1995, and during the 1999 war in Kosovo, NSA experienced serious problems sharing intelligence information with its NATO allies during the conflict, in large part because many of these partners were not members of UKUSA. Bill Nolte, formerly the head of NSA's Legislative Affairs office, stated that 'compartmentalization of intelligence doesn't really work anymore in modern coalition operations and complained about the current problems of getting NSA to modernize both its practices and mentality'.[43]

One avenue that the European nations pursued in order to maintain at least a modicum of independence from NSA and UKUSA was to establish separate Sigint-sharing arrangements with other European countries outside the UKUSA framework. For example, Britain and France have shared

Sigint information with each other, albeit not consistently, despite the fact that France left NATO in 1966 and has never been a member of UKUSA. French Sigint reportedly proved useful to London during the 1982 Falklands War, and there has also been increased sharing of Sigint between the two countries regarding international terrorism since the 1970s.[44]

The essay on French Sigint by Roger Faligot in this collection showed that France also maintained a separate bilateral Sigint sharing relationship with the German foreign intelligence service, the BND. From the late 1960s, the Dutch Air Force Sigint organisation established an unofficial Sigint sharing relationship with its counterparts in the West German Air Force, while the Dutch Army Sigint cooperated with the BND and the West German Army to locate clandestine radio transmitters operating inside East Germany. Today, the level of cooperation between the Dutch Sigint community and its counterparts in Germany and France is growing at a rapid pace. Up until they joined NATO in 1949, Norway and Denmark maintained a particularly close Sigint-sharing relationship with neutral Sweden.

Another means used by UKUSA 'Third Party' nations to maintain a semblance of independence was to keep NSA and GCHQ out of their countries. The most successful practitioners of this were the Scandinavian nations of Norway and Denmark, who despite becoming members of UKUSA in the 1950s, insisted on running their own intelligence operations and did not permit foreign Sigint personnel to operate on their soil. Also the Netherlands and France remained 'free' from any American intrusion on their territory. However, this did not preclude these nations from taking American equipment and money in return for providing NSA with raw intercepts.

It may come as a shock for many reading this book to learn that the intelligence chiefs of many of the European Sigint organisations sometimes kept many of the details of their intelligence collaboration with the US and Great Britain a secret from senior civilian officials in their own governments. For example, the first American Sigint sites in Turkey were established pursuant to a secret agreement between American military officials in Ankara and the Chief of the Turkish General Staff, General Nuri Yamut. No Turkish civilian government officials were informed of the secret agreement. This meant that senior American military officials in Turkey had to remind their colleagues back in Washington '... not to mention [this agreement] in any discussions with Turkish political or diplomatic authorities'.[45]

In another poignant example, in 1954 the US Air Force attaché in Copenhagen entered into an oral agreement with the head of the Danish Defence Intelligence Service (FET), which allowed USAF RB-50G 'Ferret' aircraft to secretly overfly Denmark without the knowledge or consent of the

Danish government. In May 1954, the head of the FET, Colonel Hans Mathiesen Lunding, informed the US Air Attaché in Copenhagen that he was willing to assist the US in facilitating overflights of Denmark by American Ferret aircraft, which could not be cleared through normal political channels because of 'security considerations'. This meant that Danish political and diplomatic officials were not to be trusted. These 'Ferret' flights, codenamed 'Fluorescent', flew over the Danish island of Zealand on their way to and from the Baltic Sea, all the time covered by a Comint collection aircraft of the Royal Danish Air Force (RDAF), which monitored communications between East German radar stations. The standard procedure for the 'Fluorescent' missions called for the US air attaché in Copenhagen to submit a request for the overflight to Colonel Lunding, who in turn would inform the commander of the Danish Air Force of the impending USAF flight. Colonel Lunding, his deputy, the commander of the RDAF and his deputy would then monitor Danish military communications circuits to prevent reports of the American overflights of Denmark from being disseminated within the Danish military or government.[46]

A *fifth thread* was the independent use of Sigint by various European powers in their colonial wars. The contribution on Dutch Sigint in Indonesia by Wies Platje demonstrates how Sigint did yield excellent results of great use to Dutch military and political decisionmakers. It also showed that Sigint and Humint must work hand in hand in order to obtain excellent results. The successful French Comint collection effort against the Viet Minh in Indochina by the *Services Techniques de Recherches* (STR) is also worth mentioning. Also the French eavesdroppers were quite successful during the early stages of the war in Algeria.

And in recent years, as was demonstrated in the study on France, French special operations in Africa continued unabated under President Mitterrand. In 1987, the operation to combat Colonel Gaddafi's Libyan troops in Chad, Operation 'Epervier', was supervised by elite intelligence personnel from the French Army. Also involved in these operations were French Navy Breguet Atlantic Sigint collection aircraft. A Breguet Atlantic aircraft was used by the French as a mobile headquarters. The intercepted Libyan transmissions monitored by the aircraft were immediately decoded and translated by Arabic linguists, which permitted more accurate and efficient attacks on the Libyan forces.

Less is know about the British Sigint successes and failures during the various uprisings in their colonial possessions in Asia and the Middle East. This issue is a topic for further future research by intelligence historians.

A *sixth thread* was the issue of the extreme political sensitivity in many of the West European nations concerning their intelligence collaboration with the US and Britain, especially in Scandinavia. For example, there was a close bilateral Sigint-sharing relationship between the US and Norway, as memorialised in two agreements signed in 1952 and 1954, which made Norway an integral part of the UKUSA network. But while the working relationship between NSA and the CIA on one hand and the Norwegian military intelligence organisation, the FO/E, on the other hand was quite close, the political leaders of Norway were constantly fearful that this relationship would become a matter of public record. For instance, following the Soviet downing of a US Air Force RB-47 Elint aircraft off the Kola Peninsula on 1 July 1960, Norwegian Foreign Minister Dr Halvard Lange met with CIA Director Allen W. Dulles on 6 October 1960, and demanded that 'information concerning the RB-47 shootdown which had been obtained from Norwegian sources not be used by the US in the United Nations debate'. Minister Lange was referring to Norwegian Sigint intercepts of Soviet air defence communications, which showed that the American reconnaissance aircraft was flying well outside Soviet airspace at the time it was attacked by the Russian fighters. Dr Lange's demand meant that the US could not publicly use the Norwegian intercepts to prove its case before the United Nations. Declassified documents in American archives make clear that Dr Lange knew very little about the nature and extent of American intelligence gathering activities in and around Norway.[47]

We have also seen in the essay on the Netherlands in this study that the Dutch Cabinet was very reluctant in 1970s in agreeing to intelligence sharing arrangements with the NSA and BND. Apparently this was considered as 'too hot to handle' despite the fact that both nations were trusted and long-time NATO allies.

The presence of sensitive American Sigint collection facilities in countries bordering the Soviet Union was a constant source of diplomatic tension between the US government and the host nations. For example, as part of a negotiating ploy to obtain more military aid from the US, in March 1961 the Turkish General Staff imposed severe restrictions on reconnaissance overflights of Turkey, banning any flight from coming closer than 100 kilometres to the Russian border, or flying higher than 40,0000 feet. These restrictions, for all intents and purposes, killed all US and British Sigint reconnaissance flights over Turkish airspace, and the Turks knew it. As detailed in the contribution on GCHQ, when American and British defence officials came to Ankara to discuss these onerous restrictions with General Sunay, the Chief of the Turkish General Staff, they discovered that it was all a ploy designed to extract more money and equipment from the US for the Turkish military and intelligence service.

During the 1950s, relations between the US and the Chinese Nationalist government on Taiwan were severely strained over the question of how much access the Chinese would have to the materials generated by the

American Sigint sites on their territory. The Chinese demanded that the US provide them with 'all information obtained through the operation of the [American Sigint] units on movements and disposition of the Communist forces and on other related matters of mutual interest to the two governments'. This provision was unacceptable to the US government, and the Chinese refused to modify their demand for complete and total access to the materials being gathered by the American Sigint sites in their country. For three years the two sides refused to budge, leading to suspicions within the Chinese government that perhaps the American sites were being used to spy on them.[48] Finally, in June 1959, the US gave the Taiwanese what amounted to a face-saving solution to the imbroglio: Washington agreed to pass to the Chinese government 'information vital to the security of the GRC obtained through the operation of these units'. But the US would not agree to complete access to the raw intercepts being gathered by the American sites.[49]

The American response to this sort of pressure in some instances was to immediately lessen dependence on that country. In the case of Turkey, beginning in the mid-1960s, the US began closing several NSA sites in that country and transferring their mission elsewhere as a means of lessening US dependence on Turkey. In this way the US began pressuring the Turkish government to soften its negotiating demands on access to the American bases in that country.[50]

A *seventh thread* in this collection is the fact that many European nations, as well as Canada, are deeply involved in economic espionage against other nations, including their allies within NATO or the European Union. This study shows that Sigint is one of the most important methods used to collect this information.

On the basis of recently declassified documents it was shown in the contribution dealing with the Netherlands that Sigint was used to target financial and economic goals. For example, there was a secret arrangement with the main Postal and Telegraph office in Amsterdam where all open but also coded and ciphered telegrams were brought. Here copies were made of most if not all of the important telegrams and then delivered to the cryptanalysts of the TIVC. In this manner extremely valuable economic intelligence was gathered because commercial secrets and other trade information flowed into the hands of intelligence officials. Very often the economic traffic (cipher or encoded) by telegram and later telex posed few problems for the codebreakers. Via intercepts the TIVC collected much intelligence for Dutch industry. In particular intelligence about foreign tenders and contracts was of importance to companies and firms. Also developments in the international tourist industry were monitored via Sigint. In addition, Sigint on large construction projects like airports, harbours, irrigation, etc. was also looked for. Also various Dutch multinationals like Philips and Shell received this economic intelligence. Intelligence was also gathered on behalf of the Ministry of Agriculture. The TIVC forwarded information about the agricultural output of various European countries.[51] The TIVC also intercepted German contract information dealing with the construction of frigates for foreign governments. This enabled Dutch shipyards for example to present bids, which were more attractive than the German ones.

However, the Dutch service was not alone in collecting economic and financial intelligence with the utilisation of Sigint. For example, the German services like the BND sometimes also outsmarted the Netherlands by intercepting Dutch tenders enabling the Germans finally to get the contract. Thus Germany was manifestly involved in economic espionage. Also Canadian Sigint was involved in targeting non-security related economic targets of opportunity as part of Operation 'Aquarian' aimed at foreign embassies and consulates even those of friendly or indeed allied countries. As was ascertained in this collection, CSE intercepts were said to have been instrumental in enabling Canada to out-compete the United States in a US$5 billion wheat sale to China in 1981.

As was shown in this work, France is also heavily involved in economic espionage. In 1996, the former head of the DGSE, Claude Silberzahn, admitted to German reporters that the DGSE not only collected economic intelligence, but also that this intelligence was passed to French corporations.[52] Exposés in the French press have revealed that the DGSE and its predecessor organisations have spied on the US since the early 1960s, including conducting economic intelligence collection against American multinational corporations such as Boeing, IBM, Corning Glass, and Texas Instruments. Some of these operations were conducted by French agents in New York City and Washington. But the majority of the intelligence collected was derived from Sigint intercepts of American commercial communications. The US, however, was not the only commercial target of French intelligence. In 1967, the GCR Sigint station at Mont Valerien outside Paris began intercepting the commercial communications of major West German corporations. In 1974, the listening post at Boullay-ces-Trois was tasked with intercepting West German and British commercial communications.[53]

Also GCHQ has been deeply involved in economic espionage since the end of World War II. It monitored for example from 1945 French diplomatic traffic, which offered insights into subjects as diverse as French economic negotiations with the United States and French plans for exploiting the Saar coal mines in Germany. About later years not much is known yet and more research has to be done. However, the simple fact that GCHQ is openly advertising on its website that it is looking for linguists who speak the languages of the EU is an indication that this Sigint service is still involved in targeting its allies. At last, as the study on Scandinavia already implied, countries like Denmark and Sweden are probably also involved in intercepting economic targets.

Finally, what about the United States? Former CIA director James Woolsey vigorously denied that the United States was engaged in industrial espionage in the sense of collecting or even sorting intelligence that it collects overseas for the benefit of or to be given to American corporations. He took this remarkable step in view of the 'tumult' in Europe as regards the revelations about 'Echelon' and economic espionage. Despite this reassurance the suspicions remain regarding the main targets of NSA in Europe. This scepticism prompted the German Under-Secretary for Intelligence Affairs, Ernst Uhrlau, and BND President, August Hanning, to visit Bad Aibling Station in November 1999. After the visit, Uhrlau stated that NSA Director General Michael V. Hayden had given him a guarantee that no German interests were being violated by the station's intelligence work. Uhrlau told the German weekly magazine *Der Spiegel* that Bad Aibling Station conducted no economic espionage against German firms. However, according to *Der Spiegel*, while Uhrlau and Hayden's statements in November 1999 were correct, in fact NSA's economic espionage against Germany had been transferred to Menwith Hill Station in Britain. The main economic focus of the Bad Aibling listening post was now Switzerland and Liechtenstein and in particular banking transactions and money laundering.

Finally, an *eighth thread* in this work is probably the most revealing one: there are no friendly intelligence services. In the world of Sigint all members of NATO and the EU spy on each other. The Americans and British took the lead during World War II and never terminated spying on their European allies. With the establishment of UKUSA a third partner Canada stepped in this business too. The major services are probably reading the ciphers and codes of the other and often smaller members on NATO and the EU. For example, the diplomatic code of the Netherlands was broken in 1943 but during the 1950s the language schools of NSA and GCHQ were still teaching Dutch. That GCHQ was openly advertising in the year 2000 on its website that it is looking for linguists who speak Dutch is an indication that this Sigint service is still targeting the Netherlands. Also CSE in Canada is able to read the Dutch traffic.[54] It is known at the official level in the Netherlands that this country ranks very high on the list of targets of the NSA site at Menwith Hill.

However, spying via Sigint on friends and allies is also done by the smaller services. As has been shown in this work, Germany has spied on other EU members, like France, Italy and the Netherlands. The Netherlands was eavesdropping on the diplomatic traffic of NATO allies like France, Belgium, Italy and Turkey. France and the Scandinavian Sigint organisations were listening to the traffic of their European allies as well. A book recently published in France revealed that in 1993 the French intercepted the radio conversation between an American ambassador aboard a US Air Force plane and an American military official in Brussels, Belgium.[55]

This work also shatters the myth long cherished in Paris that 'France was solely targeted by the Americans'. France is targeted not only by NSA, but also by its main European allies. This work at the same time also demolishes the long held myth that only the NSA is 'the principal evil' in the world of Sigint. It turns out that no targets are safe from the Sigint services whether they originate from small or large countries.

NOTES

1. Confidential interview.
2. *Cryptolog*, Spring 1996, p.5.
3. Michael Herman, *Intelligence: Power in Peace and War* (Cambridge: Cambridge UP 1996) p.203.
4. Memo for the Chief, Air Intelligence Policy Division, Subject: Half Moon, 28 April 1948, RG-341, Entry 214, File 2-1400-2-1499, NA, CP.
5. Memo for Record, 11 May 1951, RG-341, Entry 214, Box 57, File 2-19600 – 2-19699, NA, CP; Letter to Chief, Supplemental Research Branch, USAF Directorate of Intelligence from Major William S. Trites, *Exchange of Information Between US Services and Technical Radio Investigation Committee (TRIC)*, 8 Aug. 1951; Memo for Record, 6 Sept. 1951; Letter to Lt. Col. William S. Trites, US Air Attaché Office, London, *Exchange of Information Between US Services and TRIC*, 7 Sept. 1951, all in RG-341, Entry 214, Box 59, File 2-20700 – 2-20799, NA, CP.
6. CIA Historical Staff, *Allen W. Dulles as DCI*, Vol. II, pp.89–90, RG-263, NA, CP; Philip K. Edwards, 'The President's Board: 1956–60', *Studies in Intelligence* (Summer 1969) p.121, RG-263, Entry 27, Box 16, NA, CP; O.D. Dickey, *The Development of the US ELINT Effort*, p.4, AIA FOIA; NSA/CSS Manual 22-1, 21 Oct. 1986, p.2, NSA FOIA; Memorandum for the Special Assistant, Office of the Secretary of Defense, 9 Dec. 1980, p.1; and Assistant to the Secretary of Defense (Special Operations) to Secretary of Defense, 30 July 1958, both provided to authors by Dr Jeffrey T. Richelson.
7. Memo, Agee to Coordinator of Joint Operations, Proposed US-Canadian Agreement, 7 June 1948, RG-341, Entry 214, Box 40, File 2-1200 – 2-1299, NA, CP; John Bryden, *Best-Kept Secret: Canadian Secret Intelligence in the Secold World War* (Toronto: Lester 1993) p.296.
8. William F. Friedman Catalog, 5 Sept. 1957, RG-457, Yardley Collection, Box 2, Document 21, NA, CP.
9. For South Vietnam, see Army Security Agency, *When The Tiger Stalks No More: The Vietnamization of SIGINT: May 1961 – June 1970*, p.1, INSCOM FOIA. For Japan, see 'US Electronic Espionage: A Memoir', *Ramparts*, Aug. 1972, p.45; FBIS-EAS-96-043, 'Japan: Defence Agency Setting up Intelligence Agency', 4 March 1996.
10. John D. Bergen, *Military Communications: A Test for Technology* (Washington DC: US Army Center for Military History 1986) p.20.
11. Mats R. Berdal, *The United States, Norway and the Cold War, 1954–60* (London: Macmillan Press 1997) p.39.
12. Memo, Everest to Chief of Staff, Daily Activity Report, dated 26 Jan.1949, RG-341, Entry

214, File 2-6100 – 2-6199, NA, CP; 'Lost Plane was Spying for US, Former Official Says', *The Associated Press*, 26 Sept. 1990.
13. Rad, AFC 444, USAIRA STOCKHOLM SWEDEN SGN WERTHENBAKER to CSAF WASH DC, 26 April 1951, RG-319, Entry 58 US Army G-2 Top Secret Incoming/Outgoing Cables 1942–1952, Box 172, Sweden, NA, CP.
14. Peter Wright, *Spycatcher: The Candid Autobiography of a Senior Intelligence Officer* (NY: Viking 1987) pp.113–14.
15. Ibid. pp.185–6.
16. Gene Poteat, 'Stealth, Countermeasures, and ELINT, 1960–1975', *Studies in Intelligence*, pp.53–4, 57–8, via Dr Jeffrey T. Richelson.
17. Ola Tunander, 'The Uneasy Imbrication of Nation-State and NATO: The Case of Sweden', *Co-operation and Conflict*, June 1999, pp.183, 186.
18. Richard Norton-Taylor, 'GCHQ's Service to US Crucial', *The Guardian*, 17 May 1994, p.8.
19. Hugh Lanning and Richard Norton-Taylor, *A Conflict of Interest: GCHQ 1984–1991* (Cheltenham: New Clarion Press 1991) p.33.
20. Herman, *Intelligence* (note 3) p.204.
21. Charles Grant, *Intimate Relations: Can Britain Play a Leading Role in European Defence – and Keep its Special Links to US Intelligence?* (London: Centre for European Reform 2000) p.3.
22. CM 4897, Intelligence and Security Committee, *Annual Report 1999–2000*, 2 Nov. 2000, located at www.official-documents.co.uk/document/cm48/4897/4897-02.htm.
23. Norton-Taylor, 'GCHQ's Service to US "Crucial"' (note 18).
24. Confidential interview. For the use of the Racal intercept receiver at NSA intercept sites, see Department of the Army, FM 34-40-12, Morse Code Intercept Operations, 26 Aug. 1991, pp.4–7, INSCOM FOIA.
25. *Naval Intelligence Professionals Quarterly*, Spring 1992, p.15.
26. *Annual Historical Report, TUSLOG Detachment Four, Fiscal Year 1970*, Vol. I, p.19, INSCOM FOIA.
27. Confidential interview.
28. Norton-Taylor, 'GCHQ's Service to US "Crucial"' (note 18).
29. 'US Electronic Espionage: A Memoir', *Ramparts*, Aug. 1972, p.45.
30. XA-1406, Bratislava to Unit 402-A, intercepted 25 Feb. 1949, solved 29 March 1949; XA-1407, Bratislava to Unk, intercepted 17 March 1949, solved 4 April 1949; XA-1409, Bratislava to Nitra, 17 March 1949; XA-1414, Bratislava to All Stations, 29 March 1949, all in RG-38, Box 2744, NA, CP.
31. Disposition Form, CofS to J3, [subject classified], 6 March 1954; Disposition Form, Chief of Staff to J3, [subject classified], 16 March 1954; and Disposition Form, CofS to Chiefs of Divisions and Separate Branches, [subject classified], 25 March 1954, all in RG-349 Records of Far East Command, Entry 71 J-2 Decimal File 1953–1954, Box 2, File: 311, NA, CP.
32. G-2, GHQ Inter-Office Memorandum, J2 (Foreign Military Liaison) to JSigint (Communications Branch), Release of JEC Reports to UK, 4 June 1954, RG-349 Records of Far East Command, Entry 71 J-2 Decimal File 1953–1954, Box 2, File: 311, NA, CP.
33. Duncan Campbell, 'Over Here and Under Cover', *The Independent*, 6 Oct. 1993, p.24; *House of Commons Hansard Debates*, 25 March 1994, Column 612, located at www.parliament.the-stationery-office.co.uk/pa/cm199394/cmhansard/1994.../Debate-7.htm.
34. 'Comint and the PRC Intervention in the Korean War', *Cryptologic Quarterly*, date unknown, p.6, NSA FOIA.
35. Tom Mangold, *Cold Warrior* (NY: Simon & Schuster 1991) p.134.
36. Confidential interviews.
37. Grant, *Intimate Relations* (note 21) p.4.
38. Olav Riste, *The Norwegian Intelligence Service, 1945–1970* (London and Portland, OR: Frank Cass 1999) p.95.
39. Marian Wilkinson, 'Revealed: Our Spies in Iraq', *Sydney Morning Herald*, 28 Jan. 1999, p.1.
40. Letter for USAF Director of Intelligence, Director of Operations and Director of Plans from Assistant for Air Bases, DCS/Operations, USAF Requirements Versus NATO Requirements, 3 Oct. 1951, RG-341, Entry 214, File 2-21100 – 2-21199, NA, CP.

41. Berdal, *The United States, Norway and the Cold War* (note 11) pp.30–1.
42. Confidential interviews.
43. 'How Co-operation in Balkans Works', *Intelligence Newsletter*, Sigint 385, 29 June 2000.
44. Grant, *Intimate Relations* (note 21) p.6.
45. Rad, TAP 8945, CHIEF JAMMAT ANKARA TURKEY (SGD ARNOLD) to DEPTAR WASH DC FOR G3, 7 Dec. 1951, RG-319, Entry 58 Army G-2 Top Secret Incoming/Outgoing Cables 1942–1952, Box 172, File: Turkey, NA, CP.
46. Air Attaché, Copenhagen, Denmark, Air Intelligence Information Report IR-149-54, Danish Attitude and Co-operation in Regard to USAFE Special Mission, 'Fluorescent', 19 Oct. 1954, RG-341, Entry 267, File 4-4222, NA, CP.
47. Memo, Kohler to The Secretary, Conversation with Norwegian Foreign Minister Lange: U-2 and RB-47, 6 Oct. 1960, RG-59, Entry 5182 Records Relating to the RB-47/U-2 Incidents, Box 2, File: Top Secret RB-47, NA, CP. Memo, Allen W. Dulles to the Secretary of State and Secretary of Defence, Memorandum of Conversation, 7 Oct. 1960; Memo, Foy D. Kohler to The Secretary, Conversation with Foreign Minister Lange: U-2 and RB-47, 10 Oct. 1960, all in RG-59, Entry 5182 Records Relating to the RB-47/U-2 Incidents, Box 1, File: Memo, NA, CP.
48. Msg No. 984, Taipei to Secretary of State, 22 March 1957, RG-59, Entry 5221 Top Secret Files Regarding the Republic of China, Box 2, File: Air Force Communications Project – Formosa, NA, CP.
49. Memo, Kelly to Jefferson *et al.*, Annex J of Taiwan SOF Agreement, 11 June 1959, with attachments, RG-59, Entry 5221 Top Secret Files Regarding the Republic of China, Box 3, File: Air Force Communications Project – Formosa, NA, CP.
50. Letter, Howison to Bronez, 8 Dec. 1966, RG-59, Entry 5258 Records of the Country Director for Turkey, Box 11, File: DEF 15 Bases, NA, CP.
51. Confidential interview.
52. Imre Karacs, 'France Spied on Commercial Rivals', *The Independent*, 11 Jan. 1996, p.11.
53. Peter Schweizer, *Friendly Spies* (NY: Atlantic Monthly Press 1993) p.103.
54. Confidential interview.
55. 'Eavesdropping Between Worst of Friends', *Intelligence Newsletter*, No. 352, 2 Nov. 1999.

Abstracts

Introduction:
The Importance of Signals Intelligence in the Cold War
MATTEW M. AID and CEES WIEBES

This contribution presents a general overview of Signals Intelligence (Sigint), as well as a balanced assessment of the historical strengths and weaknesses of Sigint as an intelligence source, with a focus post the Cold War era. One of the key findings is that Sigint became an essential source of intelligence information on both sides of the Iron Curtain because of the failings of other intelligence sources, especially Human Intelligence (Humint). It is also apparent that Sigint, together with the reconnaissance satellites operated by the US and the USSR, consistently produced the most reliable intelligence available to consumers on both sides of the Atlantic. After weighing Sigint's successes and failings during the Cold War, the authors also conclude that Sigint's true value as an intelligence source can only be achieved when it is effectively combined with information produced by other intelligence sources into an 'all-source' product.

The National Security Agency and the Cold War
MATTHEW M. AID

This study focuses on NSA's 39-year intelligence collection effort against the former Soviet Union during the Cold War. All available evidence indicates that NSA's Sigint product was an essential means by which the US intelligence community was kept apprised of what was going on inside the Soviet Union, despite the fact that NSA was oftentimes unable to solve Russia's most important encryption systems The ability to adapt and apply its superior technological wherewithal to overcome obstacles was a hallmark of NSA's Cold War efforts. When it could not crack the Soviet Union's most important encryption systems, NSA appears to have gone around the problem and exploited less important systems that proved to be important sources of intelligence information. NSA also worked closely with other branches of the US intelligence community, such as its work with the CIA and the FBI to obtain foreign cryptologic materials by clandestine means, its joint effort with the National Reconnaissance Office to field Sigint satellites, and its close collaboration with the US Navy to tap undersea communications cables.

GCHQ and Sigint in the Early Cold War 1945–70
RICHARD J. ALDRICH

GCHQ is without question the least discussed of Britain's Cold War intelligence services. Yet by most indicators – including budget, numbers of staff, or intelligence output – it was the most important. This essay argues that this reflects not only a greater anxiety on the part of GCHQ to maintain secrecy, but also a disinterest among historians in more technical matters. It reviews some of the limited material available in recently released files relating to GCHQ for the period up to 1970, noting its particular importance in the areas of military intelligence and nuclear targeting. The particular security problems presented by large-scale cryptographic operations are also discussed.

Canada's Communications Security Establishment from Cold War to Globalization
MARTIN RUDNER

The Communications Security Establishment (CSE) is Canada's largest and costliest intelligence organization and the main provider of foreign intelligence to the Canadian government. CSE collects, analyses and reports on signals intelligence (Sigint), and participates in the UKUSA alliance with the United States, United Kingdom, Australia and New Zealand and other third parties. During the Cold War Canada's Sigint effort was directed primarily at the Soviet Union and its Warsaw Pact allies, but also served other foreign policy objectives as well. After the Cold War ended, and in response to the changing threat environment, the Canadian Cabinet issued a policy directive that promulgated new priorities for foreign intelligence, notably international terrorism, ethnic and religious conflict, proliferation of weapons of mass destruction, illegal migration, transnational organized crime, economic (counter-)espionage, and trade intelligence. This essay traces the historical evolution of CSE in performing its signals intelligence functions from the Cold War through to today's more diversified and globalized security agenda, utilizing the international liaison arrangements and technological capabilities at its disposal.

The Bundesnachrichtendienst, the Bundeswehr and Sigint in the Cold War and After
ERICH SCHMIDT-EENBOOM

The contribution shows the development of German Sigint after World War II beginning in 1946, when the Organisation Gehlen under the patronage of the CIA performed the first operations, and ending in the late 1990s when the system of Cold War listening stations was restructured. Based on a lot of top secret documents all stations, their tasks and many hidden operations are revealed. Another main focus lies on German foreign relations in the field of Sigint, i.e. joint ventures with NATO partners in and outside Germany and cooperation with countries reaching from South Africa to both Chinas. The radio transmission system for agents, the BND centre for coding and decoding, and the legal foundations are also described in detail.

France, Sigint and the Cold War
ROGER FALIGOT

When the Cold War began in 1947, France used a widespread signal intelligence network set up in the 1930s, mostly in Africa and Asia. The postwar reorganization of the French intelligence services (SDECE, DST, GCR and others) saw former resistance fighters and ex-military and policemen who had served under the Vichy government joined in the common cause to defeat Communism. They were involved in the decolonization wars, in Indochina and Algeria as well as the communication war in Eastern Europe. The diversity of battlefields involved a smaller French Sigint community that linked up with the predominant Anglo-Saxon system. At the same time, they opposed some of the US operations. Following the Gulf War, France devoted more financial means and human power to developing an independent satellite intercept system. It covers new local conflicts in Central Europe and the economic intelligence war. French strategists believe this is a first step towards a united European intelligence system.

Scandinavia, Sigint and the Cold War
ALF R. JACOBSEN

All four Scandinavian countries (Norway, Denmark, Sweden and Finland) were deeply engaged in signals intelligence collection against the Soviet Union throughout the Cold War, albeit in great secrecy. Norway and Denmark, both of whom were members of NATO, worked closely with their American and British Sigint counterparts in monitoring military and civilian radio traffic coming from the Soviet Union and Eastern Europe. But the Norwegian and Danish Sigint services retained a high degree of independence because the American and British Sigint organizations were barred from operating on their soil. But both nations were heavily dependent on the US for financial and technical assistance throughout the Cold War. Neutral Sweden and Finland also devoted the vast majority of their Sigint resources to monitoring the military activities of the Soviet Union, although very little is publicly known about the activities of the Finnish Sigint organization. Available evidence suggests that Sweden, despite its neutrality, maintained a substantial clandestine Sigint sharing relationship with the US and Great Britain, particularly during the early stages of the Cold War.

Dutch Sigint during the Cold War, 1945–94
CEES WIEBES

Not much is known about the Netherlands Signals Intelligence (Sigint) activities, which were executed by the Technisch Informatie Verwerkings Centrum (TIVC). During the Cold War the Sigint effort was not so much directed at the Soviet Union and its allies but primarily against European nations. It also served other foreign policy objectives. Sigint was used in the Dutch colonial struggle with Indonesia over the control of the island of New Guinea but also in Latin America. After the demise of the Soviet Union new priorities were formulated like international terrorism, ethnic and religious conflict, proliferation of weapons of mass destruction, illegal migration, transnational organized crime, etc. This essay traces the historical evolution of the TIVC in performing its Sigint functions during the Cold War.

Dutch Sigint and the Conflict with Indonesia 1950–62
WIES PLATJE

The outbreak of the First World War speeded up the development of cryptology and signals intelligence in the Netherlands. Amid belligerent European neighbors not only neutrality had to be safeguarded but the Netherlands also had to maintain its position in the overseas colonies. Later, when Japan prepared for war a remarkable capability was set up to intercept Japanese communications and to break Japanese codes. This capability became very fruitful during the military campaigns against Indonesian nationalist troops after the Second Word War. During the New Guinea conflict in 1962, Netherlands Sigint operations against Indonesia played a very important and crucial role.

Conclusions
MATTHEW M. AID and CEES WIEBES

In this final contribution a synopsis of the various essays is given and an overview is provided about what important contributions were made to the security of the United States, Canada and several nations of Western Europe by Sigint. In view of the purpose of this study, to broaden in a profound way the public's knowledge about the role of Sigint in countries outside the United States, what lessons can be learned from this collection? The question will also be answered if it is worthwhile to follow an enhanced multinational and more comparative approach regarding intelligence history in the future. Has there been too much 'intelligence history snobbery' in the past? Is it possible to construct a new approach regarding intelligence history? Should serious academic historical studies dealing with all aspects of intelligence be far more 'internationalist' in approach? They should take into account the accomplishments of the world's smaller services, from whom the world's larger intelligence organisations derived much of their information during the Cold War.

About the Contributors

Matthew M. Aid, a native of New York City, served as a Russian linguist and intelligence analyst with the US Department of Defense, then as a senior manager with two international financial research and investigative companies for 13 years. He is presently an Associate Managing Director in the Washington DC office of Kroll Associates. He is currently completing a history of the National Security Agency and its predecessor organisations. He is also the author of a chapter about the National Security Agency in a book published by the University of Kansas Press in 1998 entitled *A Culture of Secrecy: The Government Versus the People's Right to Know*.

Richard J. Aldrich is currently Professor of Politics at the University of Nottingham, United Kingdom. In addition he is Director of the Institute of Asia-Pacific Studies, University of Nottingham and Co-editor of the journal *Intelligence and National Security*. His recent publications include *Intelligence and the War Against Japan: Britain America and the Politics of Secret Service* which was published in 2000 by Cambridge University Press. His current study *The Hidden Hand: Britain, America and Cold War Secret Intelligence* will be published by John Murray in July 2001.

Roger Faligot is a French journalist specialising in intelligence and security matters. He is a former correspondent for *The European* and he is the co-author of the book *La Piscine: The French Secret Service Since 1944* (1985) and *DST, Police Secrète* (1999). He is also the author of several books on the Far East: *The Chinese Secret Service* (1989) and *Naisho: Investigating the Japanese Secret Service* (1998). His most recent publication deals with *The Chinese Mafia in Europe* (2000).

ABOUT THE CONTRIBUTORS

Alf R. Jacobsen is the editor-in-chief of current affairs documentaries at NRK, the Norwegian Public Broadcasting Corporation. Before he entered television he was an investigative journalist with the biggest daily newspaper of Norway. He has published more than 20 books and written screenplays for several feature films. Among his books are *The Prize of Suspicion, Ice Kiss* and *The Moles* about the history of the Norwegian Intelligence Services. Jacobsen has won various prizes for his work. He was elected Investigative Journalist for the year 1991, and he co-produced with Anglia TV the documentary *Secrets of the Gaul*, which won the Royal TV Society Award for best current affairs documentary in Britain in 1998.

Wies Platje, born in 1940 in Eindhoven, is Lieutenant Commander retired Royal Netherlands Navy. From 1961 until 1994, he held various functions within the Netherlands Navy Intelligence Service. He was a junior intelligence analyst in the former Netherlands New Guinea; intelligence analyst Warsaw Pact naval forces and intelligence officer to the group of maritime patrol aircraft at Valkenburg naval airbase. He also served as intelligence officer in the staff of the Commander-in-Chief Eastern Atlantic Area at Northwood, UK and Head of the acquisition and distribution section of the Military Intelligence Service/Navy. He is the author of a book about the Netherlands Navy Intelligence Service, published in 1997. He is since 1998 also a member of the NISA board.

Martin Rudner is Professor of International Affairs and Director of the Centre for Security and Defence Studies at the Norman Paterson School of International Affairs, Carleton University, Ottawa, Canada. He teaches a graduate course on Intelligence, Statecraft and International Affairs in the school. In September 2000 he was elected President of the Canadian Association for Security and Intelligence Studies (CASIS).

Erich Schmidt-Eenboom was born in 1953. In 1973 he joined the Bundeswehr and in the officers' training course studied Sociology and History at the Bundeswehr University in Hamburg. After 12 years of service he left the German Army in 1985 and began to work for the *Forschungsinstitut für Friedenspolitik*. He published various articles and books dealing with the strategy, infrastructure and troop force composition in NATO. Since May 1990 he has been Director of the Institute and concentrated his work on intelligence and security services in Europe, Japan and North America. He has published two books dealing with the Bundesnachrichtendienst (BND).

Cees Wiebes (born in 1950) began to study Political Science at the University of Amsterdam in 1973. In 1983 he received his Doctoral degree (*cum laude*) with as specialisation: international diplomatic relations, contemporary history and international public law. In 1993 he defended his PhD at the University of Leiden. Thesis title: 'Belgium, The Netherlands and Alliances, 1940–1949'. Since 1983 he has been a Senior Lecturer at the Department of International Relations and International Public Law at the University of Amsterdam. Since 1 January 1999 he has been on loan to the Srebrenica research project of the Netherlands Institute for War Documentation. His latest publication was with Bob de Graaff entitled: *Villa Maarheeze: The History of the Dutch Foreign Intelligence Service, 1946–1994*, published in 1998.

Index

Note: Page numbers in *italic* indicate illustrations.

'Able Archer 83' exercise 50
ABM 44, 47
Achille Lauro crisis 50
Adventure border intercept site *8*
Afghanistan 157, 158
   invasion 47, 195
Afghanistan War 9, 51
Ahmed, Hocine Aït 191
Aid, Matthew 1, 27, 77, 311
air force, Soviet 209
air forces, Elint 80–3
aircraft
   Elint 234–5
   Sigint collection 198
airfields, radio traffic 3
Albania 155
Aldrich, Richard 67, 315, 320
Algeria 181–2, 198–9
Algerian Islamic Salvation Front 155
Algerian War 189–92
Allemand, Jean 191
Allen, Richard V. 53
'Alliance' network 179
Ames, Aldrich H. 52
Amethyste submarine 203
Amin, Idi 155
ANBw 140–1, 143–4, 148, 168–9, 171
Anderson, HMS 75, 77–8
Andrew, Christopher 1
Angleton, James 248
antenna array, AN/FRD–10 *48*
antenna systems
   German 144–6
   Granger 256
antenna towers, Wobeck, West Germany *49*
Apel, Hans 140
Arab–Israeli War (1967) 43
Argentina 161, 275–6
Armed Forces Security Agency (AFSA) 19–20, 32, 35, 36
army, Soviet Union 209
Army Security Agency 16, 18, 32, 34, 35, 76, 77
TAREX 44
Arnaud, Col. Paul 179, 183, 190
ASTRAB 194–5
atomic bomb, Soviet 82–3
Australia 103, 108, 109, 118, 298
   Sigint 13–14, 76
   UKUSA 314, 315
Austria 316
AVI 281
AWACS aircraft 198
Ayios Nikolaos 74

Baader–Meinhof Group 164
ballistic missile submarines 45, 47
ballistic missiles, testing 4
Barbieri, Maj. 72–3
Bastiaans, J.F. 292
Baun, Herman 166
Bazeries, Etienne 178
Belgium 277
Belleux, Col. Maurice 188
Ben Barka, Mehdi 192
Ben Bella, Ahmed 189, 191
Benoît, Col. André, 193
Berliat, Paul 186
Berlin Tunnel, cable tapping 41
Berlin Wall, fall 199
Berry spy ship 198–9, 201, 202–3
Bertrand, Cmndt Gustave 179, 180, 182, 183–4
Best, Werner 216–17
Beurling, Arne 214
BfV 132
Biesheuvel, Barend 265
Bistos, François 190
Black Chamber 212, 213, 215, 221
'Black Friday' 18, 35, 37
Black, Lt. Col. 184
Blake, George 19, 92
Blanc, Georges 194
Bletchley Park 67, 69, 71–2

342 SECRETS OF SIGNALS INTELLIGENCE DURING THE COLD WAR

Blötz, Dieter 136–8, 139, 140, 144–6, 147, 151, 159, 160, 163
BND 20, 183, 204, 260, 275, 279, 320
    anti-criminal activities 164–5
    cooperation with Netherlands 265–6, 324, 326
    cooperation with other agencies 138–40, 154–9
    cryptanalysis 159–62
    early years 130–3
    economic intelligence 328
    modernization 135–40
    relations with Bundeswehr 135, 137–40, 141–7, 166–71
    satellite reconnaissance 150–3
Bodin, Col. 184
Boetzel, Friedrich 131–2
Boouterse, Desi 280
Bosnia 323
Boudiaf, Muhammed 191
Bougainville spy ship 203
Bourgogne, Col. Paul 198
Brandt, Willy 135, 146
Bratt, Peter 224
Bridges, Sir Edward 75, 83
British Guyana 262–3
Brook, Sir Norman 83
Brothers to the Rescue 7
Brouwer, Lt Cdr L. 291
Brownell, George 36–7, 79
BRUSA 101, 314, 315
Brynildsen, Lt. Jon Krag 221
bugging 87–8, 158, 163–4
Bundeswehr
    relations with BND 135, 137–40, 141–7, 166–71
    Sigint 133–5, 137–8, 321, 324
Bunker, Ellsworth 308
Burchardt, Robert 135–6
Burrough, Dr John 78
Burrough, John 147
Bush, President George W. 53
BVD 245, 253, 261, 263, 265, 266, 278
    liaison with Americans 251–2

C-130 7
Cabell, Gen. Charles 77, 80
cable tapping, Berlin Tunnel 41
Caillaud, Col. 192
Canada
    economic intelligence 328
    Sigint 11, 20–1, 76–7, 99–103, 103–8, 114–18, 318
    UKUSA 103, 314, 315
Canadian Forces Information Operations Group 107–8

Canaris, Adm. Wilhelm 217
CANUKAS 109
CANUSA 108, 315–16
Canyon satellite 46
Castagne, Col. Maurice 201
Castro, Fidel 310
Castro, Raul 10
Cavendish, Anthony 31
CBNRC 76, 99, 102, 318
Celebes 300, 301
Central Signals Establishment 75
CER 186
CERM 198–9, 201
Ceylon 77–8
CGT 194
Chad 197, 325
Chalet, Marcel 193
Chalet satellite 46
Chechnya 157–8
Chernobyl disaster 51
Chiles, Griffin 225–6
China 5, 39, 115, 157, 158
Chinese Civil War 14
Chinese Communist Party 178
Chopin de Janvry, Col. 194
Christmas–Moller, Wilhelm 217–18, 229
Churchill, Winston S. 70
CIA 5, 6, 7, 9, 17, 19–20, 27, 78, 89, 145, 148, 157, 158, 182, 193, 204, 223, 225, 229, 248, 251–2, 262, 274–5, 315, 317, 329
Division D 41
operations against Soviet Union 30, 38, 40, 42, 52
reconnaisance satellites 42, 46
Ciliax, Adm. Otto 222
cipher machines
    Scandinavian 212–13
    stealing 41
cipher systems
    Cuba 261
    Japanese 288–90
    Soviet 33–5, 37, 41, 43–4, 99–100
ciphers, high-grade 87
civilian personnel, security organizations 89–90
Clarke, William F. 70–1
'Classic Bullseye' surveillance system 45
'Cod Hook' 231, 237
Cold War
    Canadian operations 105–8
    early 31–5
Collin, Bernard 194
Combined Cipher Machine 79
Comint 2–3, 12, 14, 19–20, 27
    BND 161–2

INDEX

decline in importance 46
Finland 230
France 178
Indochina 187–8, 189
Norway 226
Scandinavia 233
Soviet 88
Sweden 223
US agencies 32, 35–6, 77, 78–9
US/Norwegian cooperation 226–7
Comintern, intelligence service 178
Commonwealth Sigint Organization 108
communications cables, tapping 52
communications technology, export 154–5
communications traffic, access to 18
communism, influence 89
computer systems
    DGSE 203–4
    Sigint 111–12, 138, 255–6
Comsec 32, 79, 87–92, 121
Concordia programme 137, 138
Cook, Robin 11
Coote, Cdr John 84
countermeasures, electronic 3–4
Cousseran, Jean–Claude 200
Crabb, Cdr Lionel 'Buster' 83
Cram, Cleveland 261
Crankshaw, Edward 69
Cray supercomputers 111, 112, 203
Cremet, Jean 178
CRIM 114
crime, transnational 116–17, 123, 164–5
CRITICOMM 7
cruise missiles 209, 231, 232
cryptanalysis 68, 107, 122
    BND 159–62
    CSE 104–5, 111
    France 177–8, 187
    Netherlands 288–90
    Norway 226
    TIVC 257–8
cryptographic systems, Soviet 41, 43–4, 46
Cryptological Bureau, Netherlands 287
CSA 32, 34
CSE 11, 97–9, 103–8, 111, 113, 114
    economic intelligence 118–21
    future prospects 121–4
    HQ buildings *98*
    performance 120–1
    post-Cold War priorities 114–18
CSIS 116, 117, 118–19
CST 178–9
Cuba 310
Cuban Missile Crisis 7–9, 10, 42, 43, 134, 260–1
Cunningham, Adm. of the Fleet Sir Andrew 76
Curaçao 261–2, 310–11
Cyprus 78
Czechoslovakia 137, 140, 301
    crisis 1968 17
    Soviet invasion 43, 265

Damm, Arvid and Ivar 212–13
Darul Islam 300
data protection 164
de Gaulle, Gen. Charles 180, 181, 183, 191
de Silva, Pierre 31
Dean, Patrick 83
Defence Intelligence Staff, Norway 211, 221–2
Defence Radio Institute 213
Defence Signals Directorate, Australia 13–14, 76, 103, 118
Defense Intelligence Agency 7, 9, 17, 27
den Uyl, Joop 243, 265, 271, 273
Denmark 139, 161, 209, 225
    Naval Radio Service 217–19
    occupation 216–17
    Sigint 228–9, 233, 237, 316, 320, 324, 324–5
'Desert Storm', Operation 201
Dessille, Edmond 194
Deutch, John 204
'Dew Worm', Operation 105
Dewavrin, André, 180, 182, 183
DGA 203
DGER 182
DGSE 148, 169, 196–7, 198
    economic intelligence 199–200, 328
    spying on USA 199–200
Dictionary computers 112
Dien Bien Phu 188
DND, Canada 97
Douglas–Home, Sir Alec 88
Drake, Lt. Col. Ed 76
DRM 169, 201–2, 203
'Drouot' 180
drug trafficking 117
DSD 76
DST 179, 186, 190, 191, 192–3, 195, 197
Dulles, Allen W. 10, 326
Dulles, John Foster 298
Dur, Adm. Philip 201
Dutch Air Force 254
Dutch Army 254, 259–60
Dutch Navy 247, 259–60, 295–7, 299–300

early warning radar 44, 47
East Germany 15, 136, 137, 140, 143, 260
    communications traffic 44, 194–5
    Sigint 10–11

Soviet bases 3
East Timor 13–14
eavesdropping, electronic 163–4, 165
EC–121 7
'Echelon' system 111–14, 204, 262, 278, 280, 329
economic intelligence 118–21, 122–3, 150
  DGSE 199–200, 328
  TIVC 278–9, 327–8
Eden, Anthony 83–4
Egypt 190, 191, 271
Ehmke, Horst 137, 144
Einthoven, Louis 251–2
Eisenhower, President Dwight D. 31, 36, 298
Electrical Trade Union (ETU) 89–90
Elie, Gen. Bruno 202
Elint 2, 3–4, 27, 37–8, 70, 321
  aircraft 234–5
  Anglo–American cooperation 80–3, 315
  Scandinavia 220–1
  Sweden 222, 223
Elkins, Adm. Robert 84
Enigma cipher machine *100*
'Epervier', Operation 197
Escobar, Pablo 19, 20
Etienne, Capt. Gérard 203
Evang, Col. Vilhelm 220, 221, 226–7

Faligot, Roger 177, 321, 324
Falklands War 161, 275–6, 324
'Farrah' satellite 46–7
Farwick, Gen. Dieter 143
FBI 39, 41, 47, 186, 315
FCR 219
'Ferret' aircraft 80, 81, 85–6, 223–4, 324–5
'Ferret' satellites 44
Ferris, John 53
FET 324–5
Finland 209
  Sigint 213–16, 229–30
Finnish Air Force 230
Fisint 27
Fleury, Col. Jean 180, 182
FLN 190–1
Fluorescent flights 228, 324–5
FOA 222, 223
foreign policy, impact of Sigint 7
FRA 213, 216, 222–5, 226, 233, 277, 317
France 106, 161, 169
  diplomatic traffic 259
  electronic warfare assets 197–9
  post-Cold War intelligence 199–205
  satellite reconnaisance 151–2
  Sigint 148, 177–9, 183–96, 316, 321, 322, 323–4
French Air force 187–8

Frontal Aviation 38

G-2, Germany 72
'Gabriel' spyplane 198, 201
Gates, Robert M. 50
Gaudefroy 180
GCHQ 6, 11, 13, 33, 34, 43, 44, 45, 46, 67–8, 103, 109, 147, 161, 194, 195, 224, 225, 257, 258–9, 261–2, 317, 322
  Comsec 88–9
  economic intelligence 328–9
  Elint 80–3
  Falklands War 275–6
  post-World War II planning 68–71
  security 90–2
  Sigint cooperation 75–80, 108–9, 315, 317–20
  Soviet targets 73–4
GCR 179–80, 184–5, 188, 191, 328
  merged into SDECE 193–5
GCSB, New Zealand 100–1, 113
Geheimschreiber 222
Gehlen, Maj. Gen. Reinhard 130, 132, 135, 161, 166, 183, 320
German Air Force, Sigint 134, 143, 147
German Navy, Sigint 134, 139, 147
Germany, GCHQ exploitation of Sigint materials 71–2
GIC 192
Giraud, Gen. Henri 181
Giscard d'Estaing, President Valéry 195, 196, 197
Glazebrook, George 76
Globke, Hans 166
Golitsin, Anatoly 193
Gout, Lt Cdr Ruud 295
Gouzenko, Igor 99–100
Government Code and Cipher School 67–8, 100–1, 180
Graham, Lt. Gen. Daniel O. 7
'Granger' antenna 256
Greece 323
'Green Door' syndrome 16
Greenland 218
Greenpeace 196
'Grosbeak', Operation 5
Grossler, Gen. Hubertus 136
GRU 13, 20, 178, 225, 252
Gudgeon, USS 39
Guillou, Jan 224
Gulf War (1991) 200–1, 231, 279–80
Güllich, Adm. Gerhard 138
Gyldén, Otto 212
Gyldén, Yves 212

Hagelin, Boris 213, 220, 248

INDEX

Hagelin cipher system 220, 248, 250–1, 257
Halibut, USS 52
Hallamaa, Reino 213–14, 216, 229
Hanning, August 150, 158, 329
Hauptverwaltung III 10–11
Hayden, Gen. Michael V. 150, 329
Heinrich, Gen. Jean 169, 201, 202
Helfrich, Vice Adm. C.E.L. 290, 291
Helios satellite 202, 204
'Hen House' radar system 44, 317
Heusinger, Lt. Gen. Adolf 130–1
HF radio networks 101, 105–6, 187
HFDF systems 45
Hidajat, Gen. 304
Hinsley, Sir Harry 69
'Hippodrome' space collection facility *8*
Hizbul Mujahideen 18
Ho Chi Minh 14, 186, 189
Hong Kong 75, 78
Horton, Philip Clark 182
Humint 2, 5–6, 9, 10, 12, 17, 18, 46, 260, 325
  downgrading 16
  Indonesia 257–8
  operations against Soviet Union 29–31, 42
  risks 5
  West Germany 130, 320–1
Hungary 155
  uprising 1956 40, 255
Hüttenhain, Erich 72, 160

IB 224–5, 249
ICBM 50, 51
IDB 247, 253, 258, 263, 266, 279, 280
Imbot, Gen. René 196
India 15–16, 17, 115
Indochina 14, 316, 325
Indonesia 13–14, 155, 264, 323
  arms supplies 301, 303, 309–10
  claims on New Guinea 297–302
  conflict 254, 256–8, 292–5, 302–8, 325
  independence 250
  planned amphibious assault on New Guinea 305–8
industrial espionage 150
information technology security 120
Inglis, Rear-Adm. John 84
Inman, Adm. Bobby 27–8, 235, 317
intelligence history, new approach 313–14
Intelstat 110–11
international terrorism, Canadian security concern 115–16, 117
Iran 115, 155, 276
Iran–Contra affair 19
Iraq 115, 198, 200–1, 273, 279–80

Isard spy ship 203
Israel 159, 170, 265, 271
Italy 161, 278, 319, 323
  GCHQ exploitation of Sigint materials 72–3
'Ivy Bells', Operation 52

Jacobsen, Alf R. 209, 318
jamming, radio signals 3–4
Japan 170
  Dutch monitoring of naval traffic 287–91
  satellite reconnaisance 151
  Sigint 69–70
JNICO 218
Joint Intelligence Committee 67, 71, 73–4, 86
Joint Intelligence Committee, Canada 76
Jones, Eric 75
Joubert des Ouches, Col. Jean 181
Joxe, Pierre 201
'Jumpseat' satellite 46

KAL 007 7, 51
Kalmus, Andres 214, 216
Karel Doorman Dutch aircraft carrier 299
Kashmir 18
Keller, Bayard 147
Kennan, George 88
Kennedy, President John F. 299, 308
Kepsu, Col. Keijo 230
KGB 6, 9, 13, 18, 20, 50, 192, 252, 323
  security and counter–intelligence 30, 100, 105
Khrushchev, Nikita 86
Kimche, David 159
Kissinger, Henry 19
Kistiakowsky, George 31
Kluiters, Frans 252
Kohl, Helmut 138, 152, 204
Kola Peninsula 209–10, 316
'Kolchose' programme 137
Komsomolets Soviet submarine 232
Koot, Maj. Gen. Henri 286–7, *289*
Korean War 10, 12, 14, 35, 36, 81
Kornblum, John C. 153
Kosovo 134, 153, 169, 323
Kruimink, Vice Adm. F.E. 301–2
Kullback, Solomon 28
Kuntze, Heinz 160
Kuwait, invasion 200–1, 279

Labadie, Col. Maurice 188
Labat, Col. Paul 179, 180
Labrosse, Col. 192
Laccariere, Michel 200
Lacoste, Adm. Pierre 196

Lange, Halvard 326
Latin America, Sigint operations 260–3
LAUS operation 137
Le Pivain, Capt. Jean 198
Lecoeur, Auguste 186
Lecot, Capt. 180
Lehtonen, Major Lasse 230
Leiberich, Otto 162
LeMay, Gen. Curtis 80
Liberty, USS 5
Libya 115, 155, 159, 197
London Communications Security Agency 89
London Signals Intelligence Board 67
Lossberg, Gen. Bernhard von 221–2
Lubbers, Ruud 280
Lunding, Col. Hans 217, 219, 228–9
Lundquist, Åke 214
Luns, Joseph 258, 259
Lyalin, Oleg 42

MacArthur, Gen. Douglas 291
Macmillan, Harold 85, 86
'Magic' 17, 100
Maihofer, Werner 164
Maistre, Albéric de 180
Malinovsky, Marshal Rodion 86
Männchen, Maj. Gen. Horst 11
Mao Tse–tung 14
Marcy, Oliver M. 248–9
Marenches, Count Alexandre de 193–5
Marid 6 291, 294–5, 310
    Biak New Guinea *297*
    Hollandia New Guinea *296*
Marid 6 NNG 295–7, 302–4, 307
Marjata spy ship 236
Marr-Johnston, Col. 81
Marsch, Günther 163–4
Martin, Command Sgt. Maj. John 13
Martin, William H. 18, 92
McConnell, Vice Adm. J.M. 'Mike' 28, 53
McNamara, Barbara 275
Medellin Cartel 19
Meir, Golda 271
Melnik, Constantin 191, 192
Menzies, Maj.-Gen. Sir Stewart 70
merchant marine, Soviet, communications traffic 43
Mermet, Gen. François 197, 200
MI5 41, 192
MI6 5, 19, 30–1, 41, 179, 225, 229, 275, 276, 300
    operations against Soviet Union 42
MID 245, 280–1
    HQ *246*
Middle East 264–5, 271–2

Middle East War (1973) 17
military bases, radio traffic 3
military intelligence, Netherlands 286–7
Miller, Gen. Yevgeniy 178
Millis, John 7
missile testing, Soviet Union 231
Mitchell, Bernon F. 18, 92
Mitterrand, President François 196–7, 325
MNA 190
Monge missile ship 203
'MORE' project 274
Mork, Paul Adam 217–19, 229
Morocco 191
Mossad 159, 225, 271, 275
Mozambique 155
Murmansk 84–5
Muslim nationalists 181

NACSI 276
NAFTA 119
Nasution, Gen. 301
National Elint Program 38
National Reconnaisance Office 42
National Security Agency 2, 5, 6, 7, 27–8, 78–9, 92, 109, 158, 194, 225, 226, 236, 257
    1950s 35–41
    1960s 41–5
    1970s 45–7
    1980s 47–53
    cooperation with Netherlands 265–6, 275, 326
    cooperation with West Germany 144–50
    computer systems 112–14
    relationship with CSE 103, 106–7
    relationship with GCHQ 33–5, 315, 317–18
    Scandinavia Sigint 225, 226, 236
    Soviet target 29–31
    speedy delivery of Sigint 7–9
National Technical Processing Center 38
    restrictions on information 321–3
NATO 79, 109, 147, 252, 260, 276, 323
Naval Communications Intelligence Organization 32
Naval Intelligence Division, UK 38
Naval Radio Service, Denmark 217–19
Navy
    joint UK-US Sigint operations 84–5
    Soviet Union 38–9, 46, 210
Nefis 291, 291–3
Nefis 8 *293*, 294
Netherlands
    Sigint 20, 244–51, 254–6, 316–17, 321, 322–3
    Sigint cooperation 258–60, 319, 326

INDEX 347

Sigint materials 6–7
World War I 286–7
Netherlands Antilles 261, 264
    Sigint activities 310–11
Netherlands-Indies
    Japanese occupation 291
    Sigint 287–90
New Guinea 254, 257, 262, 264, 292, 295–7, 323
    diplomatic conflict 297–9
    Indonesian operations against 302–8
    planned amphibious assault 305–8
New Zealand 103, 109
    UKUSA 314, 315
Ngo Dinh Diem 189
Nixon, President Richard M. 18–19
NKVD 178
Noemia computer system 203
Nolte, Bill 323
non-proliferation 117–18
Nordentoft, Lt. Col. Einar 216, 217
North Korea 5, 39
North Vietnam 12–13
NORUSA II 231
Norway 77, 209
    Defence Intelligence Service 30
    Defence Intelligence Staff 221–2, 225–7, 238
    Elint 230–1
    German occupation 215
    intelligence data 211
    Sigint 212, 219–22, 225–7, 237, 316, 320, 322, 323, 324, 326
    SOSUS 231–2
Norwegian Security Service 227–8
Nuboer, Lt Cdr J.F.W. 288–9
nuclear testing 15–16, 231
nuclear weapons, proliferation 117–18

objectivity, Sigint 5–6
O'Donnell, Justin 248, 251
Office of Naval Intelligence, USA 38, 274–5, 319
oil companies, monitoring communicatoions 272
oil crisis (1973) 271–3
Okinawa 78
Olshausen, Gen. 143–4
O'Neill, Lt. Stasia 182
OP-20-G 32
Ordzhonikidze, Soviet cruiser 83, 225
Organisation Gehlen 30, 72, 130–3, 155, 159
Orion spy ship 235, 236, 317
Osipovich, Lt. Col. Gennadi 51
OVIC 281

Paasonen, Col. Aladar 216, 229
Page, Gen. Karl Heinz 145
Pakistan 15–16
Pale, Erkki 214, 216
Palestinian Liberation Organisation 271
Palme, Olof 235–6
Pan Hannian 178
Panier 180
Paschke, Adolf 160
'Passy' 180, 182, 183
PCR 197
Pearl Harbor 290–1
Pelton, Ronald W. 52
Pendred, Air Vice-Marshal Lawrence 80
Penkovskiy, Col. Oleg 6, 14, 42
People's Liberation Army, China 14
Permesta 300, 301
Perry, William J. 53
personal messages, clandestine
    communications 180–1, 195
Petersen, Joseph 247–9, 250
Philby, Kim 68
phone-tapping 163–5, 192
Photint 12
'Pilgrim', Operation 106–7
Platje, Wies 250, 258, 285, 325
Poland 50, 276, 301
    Danish intelligence target 229
Polisario guerrillas 196
Politburo, Soviet 13
political intelligence 137
Pompidou, President Georges 192, 196
Porzner, Konrad 143, 167
positive vetting, GCHQ 90–2
Poulden, Cdr Teddy 76
Powers, Gary 85
'Prawn', Operation 292
Prime, Geoffrey 46, 91, 92
Proforma 4
Profumo Scandal 68
PRRI 300, 301
PTT, Netherlands 253, 265, 266, 269
Pueblo, USS 7
Purnell, Capt. William R. 290
Putin, President Vladimir 157–8
PVO 37–8, 51–2
Quebec, separatism 106, 115, 238

radar, early warning 44
radar systems, monitoring 3–4, 38, 44, 47, 230–1
radio communications 153–4
Radio Intelligence Committee 266
radio networks, HF 101, 105–6
radio traffic
    decryption 160–1, 177–8

Indonesian 293–4
monitoring 3, 38–9, *48*, 132, 190–1, 253–4
radio transmission, propaganda 190
RAF intelligence 80–3
Rainbow Warrior 196
Ramsbotham, Sir Peter 315
Reagan, President Ronald 53
reconnaisance
 aerial and maritime 5
 satellites 42
REG 179
Reilly, Patrick 75
Renamo 155
Resistance, Danish 217
'Rhyolite' satellite 46
Ridgway, Gen. Matthew B. 35
Rocard, Michel 200
Romon, Lt. Col. Gabriel 179, 180
Rondot, Gen. Philippe 201
Room 14, cryptological bureau 288–90
Roscher Lund, Alfred 212, 213, 215, 219–20, 221, 226
Rossby, Åke 223
Royal Air Force 81
Royal Navy 81, 84
 Sigint 75, 89–90
RUARA-1 34
Rudner, Martin 97, 319
Rühe, Volker 152
RUMUA-1 34
RUNLB-1 34
RUNLC-1 34
RUNMD-1 34
Russell, Brian Ashford 72
Russell, Sheridan 72
Russia 211
 foreign intelligence service 157–8
Russo–Japanese War 178
Rust, Mathias 51–2

SACEUR 252
SACLANT 252
Saïd, Mohammedi 181
Salan, Gen. Raoul 188
Salm, Rear Adm. G.B. 291
SALT 1 47
'Sambo' direction finding system 45
SAMs 233
Sarigue 198, 201
satellite interception dishes *110, 270*
satellites
 interception, Netherlands 269–70
 reconnaissance 42, 44, 204
 Sigint 17, 46–7, 109–12, 201
 West Germany 150–3

'Satyr', Operation 105
Saudi Arabia 155
Sauterne Mark I cipher machine 34
Scandinavia, Sigint 209–10, 236–8
Scharping, Rudolf 1
Schauffler, Rudolf 160
Schlesinger, James R. 146
Schlichter, Franz 159
Schmidt, Helmut 135, 144, 146
Schmidt-Eenboom, Erich 129
Schnell, Karl 145
Schröder, Gerhard 167
SDECE 183–6, 321, 323
 Algeria 190–2
 control of GCR 193–5
 Indochina 186–9
 reorganisation under Mitterrand 196–7
Second World War, Sigint 1
Secord, Richard 19
security
 GCHQ 90–2
 Sigint operations 12–14
Selchow, Kurt 160
Serre, Henri 196
Service de Renseignement 178–9
SHAPE 79–80
Shell 272
Si Nacer 181
Sigint
 1950s 35–41
 1960s 41–5
 1970s 45–7
 1980s 47–53
 Algerian War 190–2
 based at diplomatic posts 106–7
 Bundeswehr 133–5, 137–8, 139–40, 141–3, 143–4
 Canada 11, 20–1, 76–7, 99–108
 cooperation 75–80, 258–60, 313–14
 Commonwealth network 108
 cost-effectiveness 10–11
 early Cold War 31–5
 equipment 154–5
 flaps and shoot-downs 83–7
 fragmentary nature 16–17
 France 177–9
 future trends 121–4
 importance 4–11
 Indochina 186–9
 Italy 72–3
 joint operations, West Germany 144–50
 lack of co-ordination 19–20
 Latin America 260–3
 local operations 105
 long–distance operations 105–6
 Netherlands 20, 244–51

over-reliance on 14–16
post-Cold War 114–18
Scandinavia 209–10, 236–8
Scandinavian/US relations 210–11
secrecy 12–14
snobbery 16
speed of 7–9
US-German cooperation 144–50, 323
US/Danish cooperation 228–9
US/Sweden cooperation 223–5
vulnerability 18–19
West Germany 131–2, 135–40
Signal Security Agency 31–2
Silberzahn, Claude 200
Silver cipher system 43
Singapore, Netherlands intelligence 300–1
SIS 68–9, 70
SLBMs 209, 231, 232
Smith, Gen. Walter Bedell 20, 35
SOE 180
Sokolov, Marshal Sergei L. 52
Sokolowski, Claudine 194
Sorge, Richard 178
SOSUS 47, 231–2, 237
Sound Surveillance System 47
Soustelle, Jacques 182
South Africa, Sigint equipment 161–2
South Korea, invasion 39–40
Soviet Union
    air activity 80–3
    atomic bomb 73–4
    Canadian Sigint operations 99–103, 105–8
    cipher systems 18, 33–5, 37, 213–15
    closed society 6
    end of 201
    infiltration 5
    intelligence target 28–31, 314
    proximity to Scandinavia 209–10
    Sigint 10–11, 13, 32–5
    support for Indonesia 309–10
Soviet–Chinese fighting (1969) 43
Spain 152, 161
    secret service 156
Spanjaard, J. 251, 252, 255, 258, 259
SPOT satellite imagery 201
spy planes 233–4
spy rings, Soviet 40
spy ships 236
    French 198–9
    Swedish 235
SR 169, 180
    Denmark 217–19
SS-24 50
STASI 19, 195
Steichen, Commander Michel 192

Stimson, Arthur 144
Stoph, Willy 135
Stordahl, Nils 221
STR 182, 187, 188
Strategic Air Command, USA 37–8
strategic bomber force, Soviet 37, 43
Strauss, Franz Josef 135
Stripp, Alan 74
Studeman, Vice Adm. William O. 28, 275
Stützle, Walther 153
Subandrio 298
submarines
    ballistic missile 45, 47
    French 203
    reconnaissance 39
    Sigint operations 84–5
    Soviet, tracking 231–2
Sudomo, Adm. 305
Suez Canal 78
Suharto, Gen. 264
Sukarno, President 155, 264, 292, 293, 297, 298, 304, 307, 308
    political movements against 300–1
Sumatra 300, 301
Superb, HMS 81
Surinam 262–3, 280
Sutela, Gen. Lauri 230
Sverdrup, Erling 221
SVIC 281
SVR 157–8
Sweden 77, 209, 277
    cooperation with USA 223–5
    codebreaking 214, 216
    Sigint 212–13, 222–5, 234–6, 237, 317, 320
    spy planes 233–4
Swedish Air Force 233–4
SYGDASIS 276–7
Syria 272

tactical air force, Soviet 38
Taiga computer system 203–4
Taiwan 156–7, 326–7
'Taper' traffic, Soviet 73
Teicher, Howard 122
telecommunications, covert interception 116–17
telex traffic, interception 265
Telint, Scandinavia 220–1
terrorism
    Canadian security concern 115–16
    international 117
Thant, U- 308
Thorén, Cdr Torgil 222–3
Thorneycroft, Peter 88
Tibet 5

TICOMs 71–2
Tirante, USS 39
TIVC 20, 243, 244, 247–51, 280–1, 317
  cooperation with Britain 275–6
  diplomatic traffic 277–8
  economic intelligence 278–9, 327–8
  financial problems 253–6, 263–4, 273–4
  Gulf War (1991) 279–80
  Indonesian traffic 256–8
  Latin America 260–3
  liaison with Americans 251–2, 274–5
  Middle East operations 264–5, 271–2
  oil crisis 271–3
  targets 266–9
Tonkin Gulf incidents 7
Tordella, Louis 225–6, 228
Tovey, Brian 195
trade unions
  CGT 194
  GCHQ 89–90
transnational crime 116–17, 123
Traube, Klaus 164
Travis, Edward 73, 76
TRIC 315
Truman, President Harry S 36, 76, 79
Tunander, Ola 235
Tunisia 181, 191
Turkey 87, 161, 278, 316, 319, 324, 326, 327
  secret service 156
Turner, Vice Adm. Stansfield 5

U–2 spyplane 40, 83–4, 85–6
Uganda 155
Uhrlau, Ernst 150, 329
UKUSA 33–5, 41, 45, 50, 76–7, 91, 109, 118, 192, 195, 204, 229, 261–2, 280
  Canadian participation 103
  collaborative relationships 314–20
  future of 122–3
  satellite network 109–12
Ulbricht, Walter 135
Ultra 17, 100
United Kingdom
  and conflict over New Guinea 298, 300–1
  cryptanalysis 33–5
  Sigint 11, 13
United States
  and conflict over New Guinea 198–9
  economic intelligence 329
US Air Force, intelligence 80–3, 230
US Army, origins of BND 131–2
USAFFS 32

van der Stoel, Max 243, 273
Venezuela 261–2, 310–11

Venona decrypts 40, 91, 100, 103, 192–3, 215, 225
Ventre, Jérôme 200
Verkuyl, Col. J.A. 247, 250, 251, 291
Vichy 179–80
Viet Minh 186–9, 316, 325
Vietnam 5
Vietnam War 10, 12–13, 41, 43, 45, 195, 235–6
Villa, Danish Sigint 218–19, 225, 228–9, 233
VKL 230
Vlakke Hoek, naval engagement 304
Vo Nguyen Giap, Gen. 186, 188
'Vortex' satellite 51

Wagner, G.A. 272
Wahoo, USS 39
Waibel, Col. Max 183
Walker, John 91
Wang Zheng, Gen. 188
weapons, mass destruction 117–18
Weinberger, Caspar 19, 235
Weisband, William W. 18, 35, 91–2
Weizäcker, Richard von 167
Welchman, Gordon 69
Wenger, Gen. Günther 143
Wenger, Vice Adm. Joseph N. 4
Wennerström, Col. Stig 224
Wessel, Gerhard 135, 138–9, 164
West Germany 15
  French zone 185
  Sigint 20
'White Cloud' satellite 46
Wiebes, Cees 1, 243, 313, 317
Wigg, George 90–1
Wilson, Harold 90
WKC 243, 244, 291, 303, 307, 310
  New Guinea 295
Woolsey, James 329
Wright, Peter 225, 259
Wust, Gen. Harald 140, 151
Wybot, Roger 186

Yamut, Gen. Nuri 324
Yardley, Herbert 212
Yellow Stroke Reports 160–1
Yom Kippur War (1973) 195
Young, Senator Milton 7, 53
Yugoslavia 272

'Zebra' computer 255–6
ZfCH 159–60, 162
Ziare 196
Zimmermann, Friedrich 140–1

Books of Related Interest

British Military Intelligence in the Crimean War, 1854–1856

Stephen M Harris

'Stephen Harris's book is the most formidable and scholarly account yet written about British intelligence operations in the Crimean War. The basic conception is original and interesting; the issues are real and clearly conceived; the sources used unrivalled in extent; and the entire treatment rests on an exceptional combination of perspicacity and sensitivity, both of thought about, and feeling for, intelligence operations.'
Journal of Military History

208 pages illus 1999
0 7146 4671 7 cloth
Studies in Intelligence Series

British Military Intelligence in the Palestine Campaign 1914–1918

Yigal Sheffy, *Tel Aviv University*

'Yigal Sheffy's study of military intelligence joins a very short shelf of large-scale studies. ... It is to be welcomed, and not just for its rarity value. It is an admirably thorough, straightforward, professional piece of research, which may well claim to be the last word on its subject.'
Journal of Imperial and Commonwealth History

408 pages maps 1998
0 7146 4677 6 cloth
0 7146 4208 8 paper
Studies in Intelligence Series

FRANK CASS PUBLISHERS
Crown House, 47 Chase Side, Southgate, London N14 5BP
Tel: +44 (0)20 8920 2100 Fax: +44 (0)20 8447 8548 E-mail: info@frankcass.com
NORTH AMERICA
5824 NE Hassalo Street, Portland, OR 97213 3644, USA
Tel: 800 944 6190 Fax: 503 280 8832 E-mail: cass@isbs.com
Website: www.frankcass.com

Intelligence and the Cuban Missile Crisis

David A. Welch, *University of Toronto* and **James G Blight**, *Brown University* (Eds)

'The book brings together contributions from well-known scholars and practitioners and is the best study of the relationship between the intelligence community and the decision-makers in the missile crisis and should be recommended to all students and scholars of international relations.'

International Affairs

248 pages   1998
0 7146 4883 3 cloth
0 7146 4435 8 paper
Studies in Intelligence Series
A special issue of the journal Intelligence and National Security

The Norwegian Intelligence Service, 1945–1970

Olav Riste, *Norwegian Institute for Defence Studies, University of Oslo*

'This book is unique because it is the first time that a Western intelligence service has opened its most secret archives from the Cold War period to an independent historian.'

Cryptologia

336 pages illus maps   1999
0 7146 4900 7 cloth
0 7146 4455 2 paper
Studies in Intelligence Series

FRANK CASS PUBLISHERS
Crown House, 47 Chase Side, Southgate, London N14 5BP
Tel: +44 (0)20 8920 2100 Fax: +44 (0)20 8447 8548 E-mail: info@frankcass.com
NORTH AMERICA
5824 NE Hassalo Street, Portland, OR 97213 3644, USA
Tel: 800 944 6190 Fax: 503 280 8832 E-mail: cass@isbs.com
Website: www.frankcass.com

Knowing Your Friends
Intelligence Inside Alliances and Coalitions from 1914 to the Cold War

Martin S Alexander, University of Salford (Ed)

'A pathbreaking work ... This is an extremely interesting collection of essays, well worth reading by students of intelligence. It is certain to live up to its editor's hopes of stimulating further detailed research into the ways in which allies, rather than adversaries, have used intelligence in their evaluation of each other.'
Journal of Military History

320 pages illus 1998
0 7146 4879 5 cloth
0 7146 4433 1 paper
Studies in Intelligence Series
A special issue of the journal Intelligence and National Security

The Clandestine Cold War in Asia, 1945–65
Western Intelligence, Propaganda and Special Operations

Richard J Aldrich, Gary Rawnsley and **Ming-Yeh Rawnsley**, all at the Institute of Asia-Pacific Studies, University of Nottingham (Eds)

'A useful, even exhilarating account of many subplots and sideshows to the histrionics at the apex of the international security system that largely defined post-1945 history.'
RUSI Journal

312 pages 7 figs 2000
0 7146 5045 5 cloth
0 7146 8096 6 paper

FRANK CASS PUBLISHERS
Crown House, 47 Chase Side, Southgate, London N14 5BP
Tel: +44 (0)20 8920 2100 Fax: +44 (0)20 8447 8548 E-mail: info@frankcass.com
NORTH AMERICA
5824 NE Hassalo Street, Portland, OR 97213 3644, USA
Tel: 800 944 6190 Fax: 503 280 8832 E-mail: cass@isbs.com
Website: www.frankcass.com

American-British-Canadian Intelligence Relations 1939–2000

David Staffor and **Rhodri Jeffreys-Jones**, *both at the University of Edinburgh* (Eds)

Penetrating the myths, this volume provides a clear-sighted view of the complexities of allied exchanges in this most sensitive area of inter-state relations.

288 pages   2000
0 7146 5103 6 cloth
0 7146 8142 3 paper
Studies in Intelligence Series
A special issue of the journal Intelligence and National Security

Intelligence for Peace
The Role of Intelligence in Times of Peace

Hesi Carmel, *Former Assistant Chief of Mossad* (Ed)

'An important collection of articles by distinguished experts in intelligence and security ... This unique collection provides insights into the role of intelligence in times of conciliation and political process and offers an insider's view of how intelligence and secret diplomacy serve in times of peace.'

Cryptologia

288 pages   1999
0 7146 4950 3 cloth
0 7146 8009 5 paper
Studies in Intelligence Series

FRANK CASS PUBLISHERS
Crown House, 47 Chase Side, Southgate, London N14 5BP
Tel: +44 (0)20 8920 2100 Fax: +44 (0)20 8447 8548 E-mail: info@frankcass.com
NORTH AMERICA
5824 NE Hassalo Street, Portland, OR 97213 3644, USA
Tel: 800 944 6190 Fax: 503 280 8832 E-mail: cass@isbs.com
Website: www.frankcass.com